THE McGRAW-HILL
Construction
Management
Form Book

THE McGRAW-HILL
CONSTRUCTION MANAGEMENT FORM BOOK

Robert F. Cushman, Esq.
Pepper, Hamilton, & Scheetz
Philadelphia, Pennsylvania

Alan B. Stover, Esq.
General Counsel, The American Institute of Architects
Washington, D.C.

William R. Sneed, III, Esq.
Pepper, Hamilton, & Scheetz
Philadelphia, Pennsylvania

William J. Palmer, C.P.A.
Arthur Young & Company
San Francisco, California

McGraw-Hill Book Company
New York St. Louis San Francisco Auckland Bogotá
Hamburg Johannesburg London Madrid Mexico
Montreal New Delhi Panama Paris São Paulo
Singapore Sydney Tokyo Toronto

Library of Congress Cataloging in Publication Data

Main entry under title:
The McGraw-Hill construction management form book.
  Bibliography: p.
  Includes index.
  1. Building—Contracts and specifications.
2. Construction industry—Management—Forms.
I. Cushman, Robert Frank, date.
II. McGraw-Hill Book Company.
TH425.M38 1983   624'.068'5   82-22882
ISBN 0-07-014995-X

  4567890 KGP/KGP 89876

ISBN 0-07-014995-X

The editors for this book were Joan Zseleczky and Claire Trazenfeld,
the designer was Naomi Auerbach, and the production
supervisor was Sally Fliess. It was set in Baskerville
by Bi-Comp, Inc.
Printed and bound by The Kingsport Press.

In recent years, the role of women in the construction industry has seen tremendous growth, and this is indeed desirable. However, our industry terminology commonly includes such words as "foreman," "manpower," "journeyman," and "workmanlike." In addition, the English language has no neuter personal pronoun. Therefore, where words such as the above or the pronouns "he" and "his" are used, we hope they will be understood to refer to persons of either gender.

DISCLAIMER

In this book, statements are made and forms are presented with the understanding that the publisher, the editors and the co-authors are supplying practical guidelines for educational purposes. Neither the statements nor the forms are intended as substitutes for legal advice from experienced attorneys in specific fact situations.

The editors acknowledge with appreciation the cooperation of the American Institute of Architects, the National Society of Professional Engineers, the Associated General Contractors and Arthur Young & Company, CPA in providing permission to reproduce their forms and documents herein.

FOREWORD

In today's world of complex contracting methods, procedures, and options that compound the decision-making process, owners small and large alike face the question, "How do I successfully complete my project on time, within budget, and in a quality manner?"

The authors of *The McGraw-Hill Construction Management Form Book* have come to the owner's rescue with a complete, concise account of what construction management is, what it is not, its benefits, its pitfalls, and how it should and should not be used.

The authors make their point with a liberal measure of practical examples and applications. References to actual case histories allow readers to increase their understanding of the contractual forms presented and evaluated.

To complete their effort, the *Form Book* authors provide an extensive list of record-keeping forms essential to ensure the maintenance of good, sound project documentation. This aspect is particularly beneficial when addressing the activities associated with disputes, claims, and court actions that are prevalent today.

I heartily recommend the addition of *The McGraw-Hill Construction Management Form Book* to the libraries of those responsible for planning and managing construction projects. I am sure they will find it informative reading and a ready reference to the numerous questions that arise in today's expanding construction management industry.

JAMES P. ROWAN

JAMES P. ROWAN *is senior vice president, Construction Services, Day & Zimmermann, Inc., Philadelphia, Pennsylvania. Day & Zimmermann is responsible for the professional construction management of multimillion-dollar projects worldwide. Mr. Rowan has over twenty years' experience in the construction industry. He holds a Bachelor of Engineering from Villanova University and is a registered professional engineer.*

PREFACE

This is a form book designed to be a practical working tool for owners, developers, design professionals, construction managers, contractors, and attorneys to construction management approaches contemporarily in use and to the selection of a specific agreement most appropriate for a contemplated project.

In today's economy the proper use of construction management concepts and techniques can be, from the owner's viewpoint, the force that makes the project fly. From the construction manager's viewpoint, it can be the difference between continuing success and a memory of a once venerable name in the archives of a trade association. Ask ten owners, construction managers, or lawyers to define construction management and one will wind up with ten complicated and diverse explanations. The reason—there is no single standard in construction management.

Construction management is indeed a lot of things, but its raison d'être is threefold.

First, it enables the use of phased (fast-track) construction procedures to dovetail a construction project's design and construction phases. It simply reduces the time between design and occupancy, which reduces the project's cost and allows the owner the beneficial use of the project at the earliest date possible.

Second, it allows the use of multiple prime contracts for various portions of the construction work (and in doing so necessitates the engagement of a construction manager to coordinate and schedule).

Third, it integrates more sophisticated estimating and "constructability" analysis during the decision and design phase.

This book will be devoted to an analysis of construction management with particular emphasis on the current, prevalent construction management contractual arrangements. Chapter 1 will trace the evolution of the construction management concept, analyze fundamental definitions of construction management, and review an early classic example. Chapter 2 will identify basic contractual arrangements utilizing construction management and their relationship with current preferred forms of delivery and will identify basic services an owner may obtain from a construction manager. Chapter 3 will be devoted to a critical analysis of the owner–construction manager relationship, the fiduciary duties which arise therefrom, and potential conflicts of interest of which all parties should be aware. Chapter 4 will review the current standard construction management contracts. Chapter 5 will review special construction management contracts. Chapter 6 will review construction management administrative forms. A bibliography follows. No other book of this type presently exists.

We believe that this guidebook will clarify the construction management concept, direct the decision maker, and provide peace of mind to the scrivener.

Robert F. Cushman, Esquire
Alan B. Stover, Esquire
William R. Sneed, III, Esquire
William J. Palmer, CPA

CONTENTS

THE McGRAW-HILL
CONSTRUCTION
MANAGEMENT
FORM BOOK

CHAPTER ONE

INTRODUCTION TO CONSTRUCTION MANAGEMENT

During the past decade public and private owners have increasingly utilized construction management concepts. In 1980, the leading construction and program manager, Brown and Root, won approximately $4.3 billion in construction management contracts.[1] During 1979, nearly one-third of the construction companies on *Engineering News Record's* top 400 list obtained significant revenues from construction management contracts, compared with less than one-fourth in 1978.[2] In 1979, the top fifty United States general contractors acting as construction managers collected construction management fees in excess of $869.7 million;[3] the top fifty consulting architectural and engineering firms acting as construction managers obtained revenues in excess of $432.3 million.[4]

Despite the current prevalent use of construction management and the substantial economic resources committed to it, the concept is relatively new. The private sector commenced utilizing construction management and other innovative construction concepts such as phased construction and multiple prime contracts during the middle and late 1960s. These construction innovations first received official federal government sanction in 1970. On March 17, 1970, a landmark study entitled *Construction Contracting Systems* was submitted to the Honorable Robert L. Kunzig, Commissioner of Public Buildings Service (PBS) of the General Services Administration (GSA).[5] The PBS study was the product of a study group organized and commissioned by PBS for GSA. The PBS study group was directed to examine all reasonable alternative approaches to construction contracting for the government and to identify the kinds of approaches to construction contracting which would be most advantageous for the construction of public projects. The goal of the PBS study was fourfold: first, to obtain higher-quality design to ensure bids within budgets; second, to reduce delivery or construction time; third, to reduce delivery or construction costs; and fourth, to modify a system of construction contracting that seemed to encourage contract disputes, claims for change orders and delay damages, and cost-cutting actions that were incompatible with quality construction.

In a letter accompanying the PBS study, its authors noted that the cost of construction projects had risen steadily during the 1960s. After a thorough review of the private sector of the construction industry, the PBS study observed that private industry and commercial builders had attacked rising costs in construction by adopting new approaches which reduced design and construction time and maximized the use of other cost-saving techniques. The PBS study concluded that PBS/GSA had not followed comparable courses of action, with the result that its construction projects took significantly longer to construct at significantly greater costs. The PBS study recommended that the PBS/GSA construction contracting system be overhauled, that new approaches to construction contracting be utilized on selected projects, and that PBS/GSA's internal management be substantially revised.

In essence, the PBS study found that the government's traditional approach to construction contracting was still a sequential one—namely, the government's decision to construct a project, its retention of an architect and development of a design, its formal advertisement for bids, its execution of a fixed-price contract with a general contractor, and finally the delivery or construction of the project. The PBS study concluded that this rigid, sequential system of construction had numerous disadvantages which had been identified, analyzed, and remedied in the private sector during the late 1960s by the use of construction management, phased construction, and multiple prime contract concepts.

The PBS study concluded, among other things, that PBS/GSA should experiment with the use of phased construction and multiple prime contracts. In order to attempt to ensure designs within budget, shorter delivery or construction time, reduced delivery or construction costs, and coordination of phased construction and multiple prime contracts, the PBS study recommended: "Use construction manager system for multistory office buildings, complex design projects, and other projects over $5,000,000."[6]

Thus, in 1970 construction management received its first official imprimatur. That imprimatur was meaningful since PBS/GSA managed approximately 220 million square feet of space, had an annual budget exceeding $1 billion, employed over 25,000 people, including 900 professional architects and engineers, and had 1200 to 1500 construction projects in process ranging from $25,000 to $102 million per project.[7]

During the 1970s the use of construction management increased. In October of 1977 PBS/GSA revised and reprinted its handbook, *The GSA System for Construction Management.* In the foreword GSA enthusiastically embraced the construction management concept as follows:

> Construction Management is an idea whose time has come. Construction Management not only rationalizes the entire construction process, but it provides us with an interdisciplinary approach for handling large and complex projects. It pulls together a fragmented industry and allows architect-engineer, construction manager, urban planner and a variety of specialists to work together on a "project team."
>
> The construction manager is the "renaissance man" on a project overseeing it from the "cradle to the grave." The construction manager wears several hats working closely with both the architect-engineer and the owner. To the owner, an important difference between Construction Management and more traditional construction procedures is that the owner—the man who pays the bills—has control.
>
> Effective management is the key to the construction industry's success in the coming years. Construction Management gives us the tools we need to modernize our management techniques and insure that they keep pace with our rapidly expanding building technology. Through this publication, we hope to inform industry, the public sector and other interested groups about GSA's Construction Management program and philosophy. I encourage everyone involved in the construction industry to read "The GSA System for Construction Management" and consider the use of Construction Management on their projects to the fullest extent possible.[8]

The very term "construction manager" is difficult to define. As the concept has evolved in the construction industry, it has embraced many roles played by various parties. But regardless of specific definitions of construction management, most of the relevant literature maintains that the construction manager provides professional services ranging from pure agency to the owner/client for a fee with no entrepreneurial profit motive—professional construction management—to a role where the construction manager contracts with the owner/client for a fixed-price contract with the traditional general contractor's risk for the cost and timeliness of construction and the corresponding entrepreneurial profit motive—contractor construction management. In between those roles is one where the architect/engineer which designs the project contracts with the owner to provide construction management services during delivery—architectural construction management. In any of these three roles as a provider of professional services to the owner as opposed to actual construction work, the construction manager's liability to the owner most closely resembles that of the architect, the traditional provider of professional services in the construction context.[9]

EVOLUTION OF THE CONSTRUCTION MANAGER

In the construction industry there is not yet a generally agreed-upon definition of construction management or an agreed-upon set of professional services which constitute construction management. In order intelligently to understand and negotiate construction management contracts, it is first necessary to trace the construction manager's evolution in and affect upon the construction industry.

Architectural literature classically defines three steps in the construction process: first, the owner's or client's *decision* to build a project; second, the *design* of the project; and third, the *delivery* or construction of the project.[10] The traditional construction approach in both the private and public sector to decision, design, and delivery is a sequential one—each step is entirely completed prior to the commencement of the next step.[11] In the traditional, sequential construction approach the owner first identifies its needs, estimates the cost of construction to meet those needs, assesses available financing, and then decides whether to undertake a building project.[12] Next, the owner hires an architect in order to obtain a design for the project. Then, the architect produces a set of construction drawings and specifications that is sufficiently definitive to permit a solicitation of bids for construction of the entire project.[13] Although the architect has generally used its best efforts to design a project that will be bid within the owner's budget, neither the owner nor the architect knows by the end of the design step whether a general contractor will submit a bid with respect to construction of the entire project even near the owner's budget. Thus, in the traditional sequential approach two entire steps often take place without the advantage of professional advice as to the design cost of construction materials, methods, and labor. After bids are solicited and opened, the public owner must accept the lowest bid provided that it is responsive to the solicitation.[14] The private owner has greater flexibility since it is not required to select the lowest bidder. Traditionally a fixed-price contract is then executed with a general contractor, which proceeds to negotiate its final subcontract prices with its subcontractors which will perform a majority of the physical construction work on the project.

During the late 1960s and 1970s two significant factors began affecting the construction industry's sequential approach to construction: first, the economy became less predictable and this country was confronted with surging inflation approaching and entering the double-digit range;[15] second, as with other industries, the technology of construction systems rapidly advanced and became increasingly specialized.[16]

Disadvantages of the Traditional Approach

These two factors resulted in three principal disadvantages to the traditional, sequential approach to construction.[17]

First, the owner's decision to build a project of a certain scope to meet its needs within its budget was made without any input from the contractors which were most knowledgeable with regard to the cost of construction materials, methods, and labor. Similarly, an architect would produce a detailed set of construction drawings and specifications to meet the owner's needs without consideration of market factors affecting contractor's bid prices. Architects frequently failed to anticipate the effect inflation would have on the cost of construction even during the time it took to complete a detailed set of construction drawings and specifications.[18] Consequently, owners and their architects would reach the end of the first two steps of decision and design with a completed design and would receive bids that were frequently far in excess of the owner's budget.[19] Thus, the owner either went ahead at a substantially increased price, or required the architect to redesign the project, resulting in further delay, or abandoned the decision to build altogether. The first disadvantage of the traditional, sequential approach is a failure to incorporate at the decision and design phases expert construction cost analysis of materials, methods, and labor.[20]

The second disadvantage of the traditional, sequential approach is the delay inherent in that approach. The design effort does not start until the owner's decision to build is complete. Delivery or construction does not start until a comprehensive set of construction drawings and specifications is completed and bid upon.[21]

The third principal disadvantage of the traditional, sequential approach is its effect on the construction process and the relationship between the owner and its agent, the architect, on one hand, and the general contractor and its subcontractors on the other hand. In order to compete successfully in the bidding market, a general contractor must submit the lowest bid.[22] Since general contractors frequently subcontract most of the contract work and share the subcontractor pool of bids with other general contractors bidding on the same project, the general contractor only has two variables within which to meet the low bid: the cost of its portion of the work, for example, general cleanup, and the cost of its management.[23] Since the general contractor's cost of doing its own work is relatively fixed with regard to material and labor, it must cut costs with respect to its management.[24] Moreover, the general contractor often bases its bid on an interpretation of the drawings and specifications which permits the lowest-quality construction intended by the architect.[25] Since even the most detailed construction drawings and specifications contain errors and ambiguities susceptible of varying interpretations, general contractors frequently are forced to interpret them as calling for the minimal quality and quantity of work.[26] In order to make a profit, general contractors frequently rely on claims regarding the scope of work, changed conditions, change orders, extra work, and delays.[27] Both owners and general contractors have increasingly been willing to assert their respective legal rights. During contract performance an adversarial relationship frequently develops between the owner and its general contractor, creating an atmosphere of claims, controversy, and mistrust throughout the performance of the work.[28]

As a result of inflation and rapidly advancing technology and the three disadvantages summarized above, each of the four principal parties in the traditional sequential approach to construction—owner, architect, general contractor, and subcontractor—developed new needs with respect to the construction industry.

As the party ultimately paying the cost of an inefficient, uneconomical, sequential approach to construction, the owner needed to reduce the cost of designed delivery. Owners discovered that they needed someone with expertise with respect to the cost of construction materials, methods, and labor to advise them in deciding when to build and how to build. Similarly, owners discovered that their architect needed this same expertise in order to obtain bids for the design near the owner's budget. Owners discovered that they needed to shorten the sequential process of decision, design, and delivery by overlapping those steps utilizing phased or fast-track construction. Phased construction is the process of dividing the construction work into bid packages or individual phases and completing the design of the early bid packages such as site preparation, excavation, and foundation construction so that they can be bid and delivery or construction of those packages can be performed while the remaining project is designed.[29] An owner's need for phased construction requires that it employ multiple prime or trade contractors. Thus, the owner and architect need the advice of one with expertise in determining the most efficient method of dividing the bid packages and awarding the multiple prime contracts. Additionally, the owner requires some party, whether it be the architect or a construction manager, to schedule, coordinate, and supervise the multiple prime contractors. An owner's use of phased construction and multiple prime contracts achieves an additional significant monetary gain: it avoids the general contractor's markup of all its subcontractors' contracts as well as obtaining the subcontractors' real bid prices in the first instance, thus obtaining for its benefit the general contractor's profit as a result of the general contractor's bid shopping.[30] Finally, owners discovered that the kind of expertise they needed to perform these varying functions ought to be from a professional owing an allegiance only to them, in other words, their agent with respect to the design and delivery of the project. In effect, this turns one, for example, the general contractor, acting as construction manager from an adversary into an agent with a fiduciary duty to act in the owner's best interest in obtaining timely and quality construction near budget.

With respect to the traditional, sequential approach, architects discovered that it was increasingly difficult for them to predict the cost of their designs.[31] Architects discovered they needed the same kind of expertise the owner required with respect to experienced construction cost analysis of materials, methods, and labor. Although architects traditionally have played a significant role with respect to scheduling, coordination, and supervision of construction projects as agents for owners, architects found themselves increasingly subject to liability for such responsibilities.[32] In 1957 the leading decision abolished the architect's traditional defense of lack of privity with respect to all suits against them except those of an owner.[33] Since then, architects have been sued frequently by both those in privity of contract and those without privity of contract with them for the negligent preparation of plans and specifications, negligent supervision of the construction process, and negligent certification of payments to contractors.[34] Through the American Institute of Architects and revisions to the AIA standard architect-owner and owner-general contractor contracts, and based on the philosophy that responsibility follows control, architects have attempted to reduce their exposure to liability by carefully defining and limiting their contractual rights and duties to stop the work, to make continuous site inspections, to guard the owner against defects in the construction means and methods of trade contractors, to coordinate, and to certify the legitimacy of progress payments.[35] As architects limited their involvement in these duties, the owner was left with no one to perform these very functions as the owner's agent. Although the general contract traditionally requires the general contractor to perform some of these functions, such as coordination, the general contractor is not an agent of the owner; rather, the general contractor is a party who attempts to profit from its entrepreneurial risk, and therefore the owner cannot fully rely on the general contractor, particularly in the case of inspection and certification of progress payments. Thus, owners incurred a new need for a professional as their agent to perform these tasks.

With regard to the traditional, sequential approach, general contractors also discovered significant problems and new needs. First, the cost of current projects frequently became so great that many general contractors simply did not have the financial ability to take the entrepreneurial risk of submitting a low bid.[36] Additionally, due to technological advances, many general contractors commenced relying on speciality subcontractors to perform most of the work.[37] Thus, general contractors frequently became subcontractor brokers. The cost of actually assembling bids for a project became inordinately expensive when compared with the probability of obtaining the low bid.[38] Thus, many general contractors simply withdrew from the bidding process and attempted to rely upon contracts which were not fixed-price contracts, but were negotiated in the private sector. Finally, general contractors frequently experienced significant cash flow problems due to their entrepreneurial risk and their middle position in the construction industry cash flow. At the conclusion of a project, owners frequently attempted to retain control of the general contractor's retainage for as long as possible. Consequently, the general contractor was caught between the owner on one hand and its subcontractors on the other. The general contractor's efforts to avoid this cash flow dilemma by utilizing contingent payment clauses to shift the risk of nonpayment by the owner to the subcontractors frequently failed to provide adequate protection to the general contractor.[39] Thus, general contractors began looking for a way to reduce their financial risk by offering their expertise as a professional service—the knowledge they possessed with regard to the construction cost of materials, methods, and labor, along with their expertise at scheduling, supervision, and coordination of the delivery of the project.

With respect to the traditional, sequential approach, subcontractors commenced active lobbying efforts to require the use of multiple prime or trade contracts. Their most persuasive argument was that such a system eliminates the profit a general contractor makes on its subcontracts.[40] Accordingly, many states enacted statutes which require state and municipal governmental entities constructing public projects to enter into separate prime contracts for general construction and various trade work.[41] Subcontractors persuasively pointed out that under the traditional, sequential approach, the owner did not necessarily receive the benefit of competitive subcontractor prices on the bid date.[42] In order to calculate the general contractor's bid to the owner, the general contractor takes bids from the subcontractors just prior to bid time and uses those prices in making its bid to the owner.[43] The general contractor bases its price to the owner in large part on the subcontractors' bid to the general. Once the general contractor enters into a contract with the owner, the general contractor proceeds to negotiate the subcontractor's price downward—a process called bid shopping. This has two effects: first, the owner pays the general contractor's markup on the subcontractor's prices; and second, the owner does not receive the general contractor's profit of its ultimately successful bid shopping.[44] Thus the traditional, sequential approach frequently results in a higher cost to the owner.[45] Additionally, under the traditional, sequential approach the owner pays the general contractor's performance and payment bond costs as well as the subcontractors' performance and payment bond costs. Similarly, the owner pays two levels of insurance costs. Thus, subcontractors legitimately argue that removal of the general contractor from the traditional, sequential contracting approach results in a savings of one-half of the owner's expense for bonding and insurance.[46] Subcontractors further argued that the retainage held by the owner on their contracts was used to finance the construction. In other words, in a traditional, sequential approach the excavation and foundation subcontractors complete their work in the first months of the project, but the owner and general contractor respectively hold their retainage of 10 percent of their contract amounts until the entire project is complete—perhaps years later. Since most subcontractors simply cannot finance that delay, their bids are correspondingly higher, with the result being a higher bid to the owner.[47] Trade contractors argue that under multiple prime contracts as soon as their work is adequately completed, they obtain their retention. They point out that this creates a great incentive to complete their punch list items early in the course of the project, since they know they will not have to wait until the end to receive their retention. This can result in a more expeditious completion schedule for the owner and less costly bids by subcontractors.[48]

WHAT IS CONSTRUCTION MANAGEMENT?

Thus, a careful analysis of evolving requirements of the four traditional parties in the construction industry—owner, architect, general contractor, and subcontractor—reveals a central, common need for a professional with expertise and knowledge of the construction cost of materials, methods, and labor combined with expertise in scheduling, coordination, and supervision to act as an agent of the owner and apply its professional expertise during the decision, design, and delivery phases of construction working closely with the owner, the architect, and the trade contractors.[49] This, in its most general sense, is one definition of construction management. In its purest form, the construction manager is *solely* the agent of the owner working for a fee for its professional service as opposed to possessing any entrepreneurial risk and consequent potential profit. We shall refer to this form as *professional construction management*. Theoretically, under this professional form of construction management the construction manager should not be liable for the cost, the timeliness, or the quality of construction. This professional construction managment role and definition was advocated for PBS/GSA by the PBS study, which defined construction management as follows:

> A construction manager is a prime contractor (professional services) who will work with PBS and the architect to formulate the project budget, furnish the architect with information on construction technology and market conditions to insure their building designs stay within the budget, manage the procurement effort, supervise the construction of the building, and provide, if desired, a wide range of other services. . . . The construction manager functions as a member of a team with the PBS project manager and the architect. While use of a construction manager eliminates the need for a general contractor, competitive pricing is retained since all major segments of the work will be procured on a competitive bid basis. A close cooperative working relationship

will exist between the architect and the construction manager. Differences of opinion will be resolved expeditiously by the PBS project manager having total responsibility for performance of the project. Management of construction, including adherence to project schedules, is a primary responsibility of the construction manager, subject to necessary decisions by PBS project manager, on such matters as issuance of changes, settlement of large dollar changes, as recommended by the construction manager, and interpretations of contract documents.[50]

Another early recognized definition of professional construction management came from a graduate student thesis on the topic:

The construction manager, as the name implies, is a professional consultant who offers his services for a predetermined fee as opposed to a general contractor who makes an entrepreneurial profit on the project he handles. The basic difference, therefore, is the shifting of risk from the contractor to the owner. The construction manager guarantees neither project cost nor completion time, and since he assumes no risk in this area, he therefore expects no related profit.[51]

An early *Construction Methods and Equipment* special report defined professional construction management as follows:

The CM is the qualified general contracting organization which performs Construction Management under a professional services contract with the owner. As the construction professional on the construction team, the CM works with the owner and the A-E from the beginning of design through completion of construction, providing leadership to the construction team on all matters relating to construction, keeping the construction team informed, and making recommendations on construction technology and construction economies. The CM proposes construction alternatives to be studied by the construction team during the planning phase, and accurately predicts the effects of these alternatives on the project cost and schedule. Once the project budget and schedule have been established, he monitors subsequent development of the project to ensure that those targets are not exceeded without the knowledge and concurrence of the owner. The CM manages procurement of material and supplies, coordinates work of all trade contractors, assures conformance to design requirements, provides current cost and progress information as the work proceeds, and performs other construction-related services as required by the owner.[52]

A *Harvard Business Review* article on professional construction management defined it as follows:

Construction Management. Although this approach has several different variations, the essence of the concept centers around the introduction of a construction manager as the owner's agent and manager of the entire building process. The position is similar to that of the "master builder" of ancient times, whose responsibilities spanned both the design and construction phases of the building process.

Today, however, "he" is more commonly a group, a company, or a partnership with two paramount characteristics: construction know-how and management ability. The construction manager assists the owner in arranging for the contractors and the architects who will actually do the work, seeing to it that their efforts are coordinated right from the very start of the design process to the final delivery of the completed facility.

Not surprisingly, this approach involving the combined and coordinated efforts of construction manager, architect, and contractor is sometimes referred to as the "team approach." With it, the architect is assured of having resources available which allow him to foresee the cost consequences of his design decisions. At the same time, the owner is made more aware of the aesthetic-cost trade-offs.

Since the construction manager has overall management responsibility, there is no prime or general contractor on the job. Each segment of construction is contracted separately with the owner—not the construction manager—for the advisement of the construction manager. . . .

A principal benefit of the CM approach is that it lends itself to "fast track" or phase construction—which can often result in major time savings.[53]

Partly as a result of the confusion surrounding the role of a construction manager, the American Institute of Architects and *Architectural Record* commissioned a book by William B. Foxhall, senior editor of *Architectural Record*, with regard to professional construction management. The 1976 edition of that book, *Professional Construction Management and Project Administration*, is now recognized as a seminal work on construction management. Mr. Foxhall defines construction management as follows:

What is a construction manager? "He" is a firm that applies knowledge of construction techniques, conditions and costs to the three phases of decision, design and delivery of a project. First, as construction consultant he clarifies the time and cost consequences of decision and design options as they occur. Second, as construction manager he enters, still as a professional, into construction scheduling, prepurchasing of critical materials, advising on the method of obtaining contractors and awarding contracts, and coordination direction of all construction activities, including those of producers of systems and sub-systems.[54]

Thus, the consistent theme of literature defining construction management through the middle of the 1970s focused on professional construction management, its purest form, where the construction manager is an agent of the owner with a fiduciary duty to act in the owner's best interest and apply its construction expertise during all three phases of the construction process for the benefit of the owner for a fee. In its professional form the construction manager takes no entrepreneurial risk for cost, timeliness, or quality of construction, and the trade contractors contract directly with the owner.

During the course of the 1970s, however, as owners paid increasing fees for construction management services, various parties attempted to fulfill that role, resulting in numerous variations on professional construction management. In one principal variation the construction manager after offering its expertise during the decision and design phases submits a bid as general contractor and contracts directly with trade or subcontractors, thus assuming some entrepreneurial risk for the cost, timeliness, and quality of construction. We shall refer to this form as *contractor construction management*.

This form arguably raises serious conflict-of-interest problems and may involve a substantial shift in the potential liability of a construction manager. During the first two phases the construction manager owes a fiduciary duty to the owner to review and provide effective cost analysis of the architect's design. At the same time, however, the construction manager knows that it will be submitting a bid for a fixed-price general contract or a contract with a guaranteed maximum price pursuant to which it will directly contract with subcontractors. This scenario arguably presents an inherent conflict of interest and undercuts the value of the construction manager's advice to the owner during the decision and design phases. Moreover, during the construction phase the construction manager becomes a potential adversary of the owner. With respect to the negotiation of change orders, for example, the construction manager may be asked to assess a change order required by an ambiguous or faulty design which was partly its responsibility which will increase the cost of the contract. Consequently, PBS/GSA does not permit a construction manager to perform work on a project it is managing.[55] These conflict issues will be analyzed more fully in Chapter 3.

In another variation the architect for the project also acts as construction manager, and the owner employs a traditional general contractor or multiple prime or trade contractors to perform the actual delivery or construction of the project. We shall refer to this form as *architectural construction management*. Architectural construction management also arguably poses classic conflict-of-interest problems. The architect has the primary responsibility for design of the project. As construction manager the architect has an increased responsibility and may have a discrete duty for assessing that design and providing construction cost analysis thereof. The Association of General Contractors (AGC) argues that a conflict of interest exists for an architect acting as a construction manager on a project it designed as follows:

> AGC's position on the conflict of interest matter is that a construction manager who is retained to assist with, among other things, design review and evaluation for a fee over and above the design fee paid to the architect or architect/engineer, is akin to retaining a second doctor for an opinion. The first doctor, no matter how good or dedicated, would not serve the patient well by offering to provide the second opinion himself for an additional fee. The object of the second doctor's opinion is to obtain an independent review, free of unintentional bias which can develop from close association to the patient. The object of the CM's design review services is to call the owner's and design professional's attention to areas of the plans that contain possible ambiguities,

errors, omissions, or features that, although correct, may not serve the owner's interest because of construction costs, material costs, or expenses.

AGC believes that the architect/engineer who designs a project not only is prevented from giving the detached review required by a CM, but in fact, as CM, may face a situation in which he may expose himself to liability as an architect if he does deliver an objective review. If he discovers an error or omission as a CM and does not disclose it, however, then he has violated his contractual duty as a CM.[56]

Again, these conflict issues will be analyzed more fully in Chapter 3.

THE WORLD TRADE CENTER—A CLASSIC EXAMPLE OF CONSTRUCTION MANAGEMENT

Prior to turning to a review of the basic professional services construction managers offer under any of these three forms of construction management, in order to understand how a construction manager's services are utilized on a project, it is useful briefly to review one of the classic, early construction projects upon which the construction management concept was utilized—the World Trade Center of the Port of New York Authority (PA). This project is particularly significant since it was one which was extensively studied by the construction industry in order to assess the effectiveness of construction management techniques. In particular, the PBS study carefully examined the PA's use of construction management.[57] Moreover, the World Trade Center is of particular interest because it was created by a quasipublic authority operating with less freedom than that enjoyed by commercial developers, but with more freedom than that available to PBS/GSA.[58]

The World Trade Center is the largest office building ever constructed, consisting of two 110-story office towers and four smaller buildings containing 10 million square feet of space on a 16-acre site in lower Manhattan.[59] The PA utilized two architects in order to achieve its desired balance of capabilities: one architect had a reputation for aesthetic designs, and the other had significant experience in designing office buildings.[60] After the conceptual design phase had been completed, the PA paid four general contractors $25,000 each to submit proposals to be construction managers for the project.[61] After careful analysis of the construction managers' bids, the PA selected the Tishman Company to act as construction manager for the project.[62] The PA selected Tishman not only because of its attractive budget and CPM capabilities, but as a result of its experience operating large building complexes.[63] The PA paid Tishman a professional service fixed fee of $3 million plus reimbursement of special costs.[64]

Tishman's construction management contract required it to continually coordinate with the architect during the design phase, provide assistance with respect to purchasing, supervise construction, monitor the CPM, manage the budget, administrate change orders, and provide inspection.[65] The major element of Tishman's construction management work was preparation of a detailed budget which contained analytical backup material in support of various elements of the construction cost of materials and methods.[66] In order to accomplish this, Tishman assigned some ninety in-house professional, administrative, and field people and supplemented this team with outsiders it knew and retained directly.[67] In addition, budgets were requested from several producers of major subsystems for the building.[68] The most important of these was the structural steel which constituted approximately 14 percent of the total price of the project. This detailed budget information was used to create a firm project budget which was used from the inception of the project.[69]

The construction manager was largely responsible for developing the phases in which the project would be constructed, and it developed approximately 170 bidding packages as soon as the necessary design work had been completed.[70] The early packages, such as excavation and foundations, were awarded providing for immediate commencement of the work.[71] In contrast, several of the later packages were put out to bid far in advance of the time when work would commence, with the result that the PA gained a long period of time in which to evaluate bids and make changes in the products used or the contracting arrangements altogether.[72] In one case, this period of time enabled the PA to handle procurement on a multisource rather than a single-source basis and thereby save over $30 million.[73]

In addition to preparing the bid packages, the construction manager assisted the PA in evaluating bids and sought from each bidding contractor creative ways to improve the project at a lower cost or shorter delivery time. Careful review of steel bidding by the construction manager saved substantial sums for the PA. When the lump sum bid for approximately 200,000 tons of steel was substantially higher than the budget, the construction manager brought the bidding within budget by suggesting that the package be divided into smaller elements, each of which could be competitively bid by several smaller firms. As a result, the PA ended up awarding fifteen separate steel fabrication and erection elements, which brought the price of the steel within budget.[74]

With respect to scheduling and coordination, the construction manager designed and used a CPM technique that covered design, procurement, logistics, and construction and applied it to activities of the PA as well as the trade contractors.[75] In addition to a CPM, the construction manager used a computer program to schedule and cost control the budget. The construction manager's scheduling and programing system produced information in English for field foremen pointing out areas where a foreman had fallen behind in schedule.[76] The construction manager carried out inspection work in two respects: first, checking to see that assignments were performed in the right areas and that trade contractors were working at the required rate at the designated location; and second, inspecting for quality control and deficiencies.[77] The World Trade Center was completed substantially on budget.[78]

FOOTNOTES

[1] *Engineering News-Record,* Apr. 16, 1981, p. 85.

[2] Ibid., Apr. 17, 1980, pp. 77–78.

[3] Ibid.

[4] Ibid., July, 31, 1980, p. 30.

[5] General Services Administration, Public Building Service, *Construction Contracting Systems: A Report on the Systems Used by PBS and Other Organizations,* cited as Mar. 17, 1970 (hereinafter PBS study). The Commissioner of GSA organized the PBS study and authorized its conduct by William J. Gregg, Chairman of PBS, Phillip G. Reed, Deputy Director of Federal Procurement Regulations Staff, Office of the General Counsel of GSA, and Ralph C. Nash, Associate Dean of the George Washington University Law Center, a recognized expert on federal procurement law.

[6] Ibid., "Recapitulation of Conclusions and Recommendations," p. 6.

[7] B. Hart, "Construction Management, 'CM for Short': The New Name for an Old Game," *Forum,* vol. 8, p. 210.

[8] General Services Administration, Public Building Service, *The GSA System for Construction Management,* October 1977. Recently, however, PBS/GSA has apparently revised its opinion of construction management as applied to PBS/GSA projects. On Oct. 20, 1980, PBS/GSA order ADM/P545039B was issued to GSA's regional offices. This order provided that no construction management contract could be initiated or awarded without the prior approval of the commissioner of PBS/GSA. GSA's former assistant commissioner for construction management, Mr. David R. Dibner, explained in an interview that the PBS/GSA directive is designed to preclude the use of construction management except on rare projects. PBS/GSA has not yet issued a formal written explanation of this change in policy, however, Mr. Dibner explained that PBS/GSA's problem with construction management as utilized in accordance with the GSA handbook was that the construction manager did not possess sufficient responsibility to enforce its scheduling and coordination activities. Mr. Dibner maintained that this failing was largely a result of problems inherent in the bureaucracy of PBS/GSA and the use of PBS/GSA's contracting officer and project manager systems. Mr. Dibner expressly stated that this change in policy by PBS/GSA should not be understood as a condemnation in any respect of the construction management system per se by PBS/GSA. Indeed, this change in policy highlights an important point—even though an owner utilizes a construction manager, the owner must provide direction and authority to ensure an efficient, productive contract administration.

[9] R. Cushman, B. Ficken, and W. Sneed, *Construction Litigation,* Practising Law Institute, New York City, 1981, pp. 338–380; Note, "Architectural Malpractice: A Contract-Based Approach," *Harvard Law Review,* vol. 92, 1979, p. 1075 (hereinafter cited as Harvard note). Since their liability in the construction context is dictated by the same legal theories, this book will use the term "architect" to mean both architects and engineers.

[10] R. Hastings, "Proposal: A New and Comprehensive System for Design and Delivery of Buildings," *Architectural Record,* November 1968, pp. 135–138. See also W. Foxhall, *Professional Construction Management and Project Administration,* 2d ed., McGraw-Hill, New York, 1976, pp. 4–5.

[11] PBS study, p. 2–4; Ralph C. Nash, "Innovations in Federal Construction Contracting," *George Washington Law Review,* vol. 45, 1977, p. 310.

[12] Nash, loc. cit.

13 Ibid., p. 311.

14 Ibid., p. 312.

15 Ibid., p. 314; PBS study, 2-6; Harvard note, p. 1079.

16 Foxhall, op. cit., pp. 10–14; Harvard note, p. 1079.

17 Nash, op. cit., p. 314.

18 Ibid.

19 Ibid., pp. 314–315.

20 Ibid., p. 315

21 Ibid., pp. 315–316.

22 Ibid., p. 317.

23 Foxhall, op. cit., p. 14.

24 Ibid.

25 Nash, op. cit., p. 317.

26 Ibid.

27 Ibid.; PBS study, pp 3-9–3-11.

28 Nash, op. cit., p. 317; PBS study.

29 Nash, op. cit., p. 318; PBS study, p. 3-11.

30 McNeill Stokes, "Statement on Construction Management by American Subcontractors Association," American Subcontractors Association, Inc., Landover, Maryland, pp. 5–12.

31 Nash, op. cit., p. 314.

32 Harvard note, pp. 1076–1077.

33 Ibid., p. 1075.

34 Ibid., pp. 1083–1101.

35 Ibid., p. 1086.

36 Foxhall, op. cit., p. 13.

37 Ibid., p. 10.

38 Ibid.

39 See, e.g., *Thos. J. Dyer Co. v. Bishop Intern Eng' Co.*, 303 F. 2d 655 (6th Cir. 1962); *Schuler-Haas Electric Corp. v. Aetna Casualty & Surety Co.*, 49 A.D..2d 60, 371 N.Y.S.2d 207 (1975), *aff'd*, 40 N.Y.2d 883, 389 N.Y.S.2d 348, 357 N.E.2d 1003 (1976).

40 Stokes, op. cit., p. 1.

41 The Commonwealth of Pennsylvania statute is typical of those requiring state and municipal governmental entities constructing public projects to enter into separate prime contracts. The Pennyslvania act provides in pertinent part:

> Hereafter in the preparation of specifications for the erection, construction, and alteration of any public building, when the entire cost of such work shall exceed two thousand five hundred dollars, it shall be the duty of the architect, engineer, or other person preparing such specifications, to prepare separate specifications for the plumbing, heating, ventilating, and electrical work; and it shall be the duty of the person or persons authorized to enter into contracts for the erection, construction, or alteration of such public buildings to receive separate bids upon each of the said branches of work, and to award the contract for the same to the lowest responsible bidder for each of said branches.

Act of May 1, 1913, Pub. L. 155, No. 104, §1, *as amended,* Act of Sept. 1, 1967, Pub. L. 296, No. 124, §1, *as amended,* 1978, Oct. 4, Pub. L. 1051, §1, 71 P.S. §1618.

42 Stokes, op. cit., p. 2

43 Ibid.

44 Ibid., p. 3.

[45] Ibid., p. 4.

[46] Ibid.

[47] Ibid., pp. 8–9.

[48] Ibid.

[49] Foxhall, op. cit., pp. 6–7.

[50] PBS study, pp. 5-1–5-2.

[51] J. Lammers, "Construction Manager: More Than a Hard-Hat Job," *A.I.A. Journal,* May 1971, p. 31.

[52] "Special Report: Construction Management: Putting Professionalism into Contracting," *Construction Methods and Equipment,* March 1972, p. 70.

[53] W. Davis and L. White, "How to Avoid Construction Headaches," *Harvard Business Review,* March–April 1973, pp. 89–90.

[54] Foxhall, op. cit., p. 6.

[55] C. Reed, "CM's Conflict of Interest Question," *Constructor,* June 1980, p. 18.

[56] Ibid., pp. 18–19.

[57] PBS study, p. 3-17; Foxhall, op. cit., p. 47.

[58] Foxhall, loc. cit.

[59] "Special Report: Construction Management: Putting Professionalism into Contracting," *Construction Methods and Equipment,* April 1972, p. 116.

[60] PBS study, p. 3-18; Foxhall, loc. cit.

[61] PBS study, loc. cit.; Foxhall, loc. cit.

[62] Foxhall, loc. cit.

[63] Ibid.

[64] Ibid.

[65] PBS study, p. 3-19.

[66] Ibid.

[67] "Special Report," *Construction Methods and Equipment,* April 1972, p. 116.

[68] PBS study, p. 3-19; Foxhall, loc. cit.

[69] PBS study, p. 3-19; Foxhall, op. cit. p. 48.

[70] Foxhall, loc. cit.

[71] PBS study, p. 3-20; Foxhall, loc. cit.

[72] Foxhall, loc. cit.

[73] Ibid.

[74] "Special Report," *Construction Methods and Equipment,* April 1972, p. 116.

[75] PBS study, p. 3-22; Foxhall, op. cit., p. 49.

[76] "Special Report," *Construction Methods and Equipment,* April 1972, p. 116.

[77] Ibid.

[78] Ibid.

CHAPTER TWO

CONTRACTUAL RELATIONSHIPS WITH THE CONSTRUCTION MANAGER

CHOICE OF DELIVERY METHOD

Once an owner conceptualizes a project, it must organize teams of professionals to assist it in making the critical decision of whether to proceed to the design and delivery phases. The owner's teams will typically consist of three general categories of professionals. The owner/developer team will include the owner/developer, its general and/or limited partners, its accountant, its real estate agent for assemblying the land package, its leasing agent for ultimate marketing of the project, and its attorneys. The owner's financial team will include the owner's general and limited partners, its permanent lender, its gap lender, its construction lender, and if the project has public dimensions, perhaps a government lender. The owner's construction team will include its architect and the architect's consulting engineers, its construction manager if it decides to utilize one, and a general contractor and its subcontractors or prime contractors and their subcontractors, depending upon whether general or prime contractors are utilized.

Traditionally, during the decision phase an owner consulted primarily with its architect and the architect's consulting engineers with respect to the project concept and the construction cost of materials, methods, and labor.

As noted in Chapter 1, however, the need for construction management arose, in part, as a result of owner's needs for more accurate, cost-tested expertise during the decision phase with respect to the construction cost of materials, methods, and labor. On complex projects it is now common for the owner to consult with both its architect and its construction manager during the decision phase. Indeed, a construction manager's input during the decision phase can be most helpful to an owner with respect to the development of the initial concept and design, and the development of budgets for utilization by other professional members of the owner's teams, for example, the owner's accountant, real estate agent, leasing agent, and lenders. Thus, an important early question for an owner to answer is whether to retain its construction manager during the decision phase.

Once an owner decides to retain a construction manager, the owner must begin an analysis with respect to the authority, responsibility, and accountability the owner, its architect, and its construction manager will undertake during the decision, design, and delivery phases. It is critical for the owner, its architect, and its construction manager to resolve together their respective authority, responsibility, and accountability for the project during the negotiation of their contracts. In effect, these contract negotiations serve as the parties' collective allocation of duties and risks relating to their project involvement.

In order to define the roles the architect and construction manager will fulfill during the course of the project, the owner must make a fundamental decision with respect to which type of construction management it intends to utilize: architectural construction management, professional construction management, or contractor construction management. If the owner decides to utilize architectural construction management, where the architect performs the basic design function and then also acts as construction manager for the project, the owner may negotiate with the architect one contract covering both functions or it may decide to negotiate two contracts to separate the functions: one for design and one for construction management. If the owner decides to utilize professional construction management, the owner will negotiate a contract with the architect for the basic design of the project and will negotiate a contract with the construction manager for the services it will perform during the decision, design, and delivery phases of the project. In this situation since the professional construction manager is not guaranteeing the cost, timeliness, or quality of construction, the owner will also negotiate a contract with a general contractor or multiple prime contractors to perform the actual, physical construction work. If the owner decides to use contractor construction management, the owner will negotiate a contract with the architect for basic design services and will negotiate a construction management contract with the contractor which may undertake to guarantee the cost, timeliness, or quality of construction in various permutations. If the contractor construction manager operates as a general contractor, the owner may negotiate one contract covering both construction management and actual, physical construction work or it may decide to negotiate two contracts: one for construction management services and a second for construction work. Regardless of which construction management arrangement the owner selects, it is imperative that the basic contract(s) for the actual, physical construction work and the general conditions thereto are reviewed carefully to ensure that the provisions relating to the duties of the architect and construction manager with respect to the project are consistent with their respective duties in their contracts with the owner.

Once the owner has decided on the basic lines of authority, responsibility, and accountability for itself, its architect, and its construction manager and has drafted all of the relevant contracts it intends to utilize, it is critical that the owner make a final careful review of these contracts with its attorney not only to ensure internal consistency but to ensure that the overall allocation of duties and risks are consistent with the owner's desires and all parties' understandings. Finally, with respect to the duties and risks imposed upon various parties, lines of authority, responsibility, and accountability must be as clear as possible in order to avoid confusion during the decision, design, and particularly, the delivery of the project. Since during the course of delivery of a complex construction project literally thousands of questions arise, it is imperative that all parties understand from the inception which of them will have the authority, responsibility, and accountability for various kinds of issues. In order for an owner to ensure competent contract administration and effective, economical project delivery, parties must know to whom to turn with questions that must receive prompt, authoritative decisions.

Concurrent with deciding the basic contractual relationships to utilize in defining the duties and risks of the respective parties, with the assistance of its architect and construction manager, the owner must decide how to deliver the project. The owner must make four general decisions with respect to delivery: first, whether delivery will be in the traditional, sequential mode or in a phased construction mode; second, whether delivery will be accomplished by a general contractor or by multiple prime contractors; third, whether the basic construction contract will be a lump sum, unit price or a variation of a cost-plus contract, for example, a cost-plus contract with a guaranteed maximum price; and fourth, whether to accept bids for the contracts, to negotiate them, or to use a combination of methods.

Traditionally, owners utilized a sequential process of construction. As noted in Chapter 1, however, during the late 1960s and 1970s several disadvantages of the traditional, sequential approach to construction became apparent: a failure to incorporate at the decision and design phases expert construction cost analysis of materials, methods, and labor; increased time in that design does not start until decision is completed, and delivery does not start until design is completed; and problems that develop with utilization of a general contractor bidding a lump sum contract. Generally, the sequential approach takes more time, and in an inflationary economy more time means greater expense. Once an owner decides to utilize phased construction on a complex project, however, the owner virtually commits itself to utilizing some form of construction management. In phased construction some party must assist the owner's architect in developing bid packages and in scheduling and coordinating the work of multiple prime contractors as the project progresses. Although theoretically a traditional general contractor could be used with respect to phased construction, it is more common to utilize a construction manager and multiple prime contractors, one of which may be a general construction contractor. Perhaps the most significant ramification of using phased construction is that it is virtually impossible to utilize a lump sum contract, since the project is not fully designed once delivery starts. Thus, an owner's decision to utilize phased construction essentially commits it to use some form of cost-plus contract, often with a guaranteed maximum price. A decision to use phased construction does not mean that contracts cannot be bid. On the contrary, the use of phased construction with a construction manager is frequently very effective with respect to obtaining economical construction since bids may be taken from multiple prime contractors as discrete bid packages are developed. Although to some extent an oversimplification, probably the central tension inherent in an owner's decision to utilize either sequential or phased construction is that with sequential construction an owner can obtain a lump sum contract and be relatively secure in knowing the ultimate construction cost prior to delivery, whereas with phased construction an owner commences delivery without knowing the ultimate project cost, although it may know the guaranteed maximum price. Utilization of phased construction, however, frequently achieves the greatest project cost savings. On a complex project, once an owner decides to utilize phased construction, multiple prime contractors, and negotiated cost-plus contracts with guaranteed maximum prices, the owner probably can benefit from retaining a construction manager. Similarly, once an owner makes those decisions, the services an owner needs from a construction manager become more clear.

Thus, during the decision phase, aside from retaining an architect, the owner must make fundamental decisions with respect to whether to retain a construction manager and what delivery method to utilize. Once an owner makes these decisions, the owner may proceed to negotiate its contracts with its architect and its construction manager and to analyze the best method of project delivery. The remainder of this chapter will be devoted to an analysis of the kinds of services an owner may obtain from a construction manager, with a brief analysis of allocation of duties and risks relating to each such service. It is assumed that the owner has decided to use an architect for the basic design of the project, and a professional construction manager. It is also assumed that the owner will enter into separate cost-plus-guaranteed-maximum-price contracts with either a general contractor or multiple prime contractors for construction of the actual project. This analysis generally will refer to three standard construction management contracts developed during the last few years. They are the PBS/GSA's *Standard Construction Management Contract,* revised April 15, 1975, the American Institute of Architects, Document B801, *Standard Form of Agreement between Owner and Construction Manager,* revised June 1980 (reproduced in Chapter 4), and the Associated General Contractors' Document 8, *Standard Form of Agreement between Owner and Construction Manager* (Guaranteed Maximum Price Option), revised July 1980 (reproduced in Chapter 4).

CONTRACT PROVISIONS PERTAINING TO THE CONSTRUCTION MANAGER'S SERVICES

Design Services

Prior to the evolution of the construction management concept, architects were the principal parties participating with the owner in the development of the project concept and design. As noted in Chapter 1, owners and architects began experiencing the need for tested expertise with respect to the construction cost of materials, method, and labor during the decision and, in particular, the design phases. Thus, the owner–construction manager contract should contain specific provisions setting forth the construction manager's duties to the owner with respect to the review of the project concept and design. It is important to include services relating to the project concept as well as the project design.

The owner–construction manager contract should provide that during the decision and design phases the construction manager thoroughly familiarize itself with evolving architectural and structural drawings and specifications for the project. During the delivery phase this familiarization should extend to shop drawings. This familiarization should commence with the construction manager's thorough review and advice with respect to the project site. The first component of such advice should be an analysis of the construction labor market in the project locale, including an analysis of the availability and cost of labor and potential labor problems during the expected project delivery time. One factor for an owner to consider in retaining a construction manager is the construction manager's experience and the reliability of its advice as a result of having performed construction management services in the project locale.

The construction manager should also provide advice with respect to site use. The particular location of the site and the location of the project upon the site may have significant ramifications with respect to the delivery of labor and materials to the site. Suppose, for example, that an architect designed a project on a site where one side consisted of railroad tracks on high embankments and an adjacent side consisted of elevated approaches to a controlled-access highway. Suppose further that the normal, efficient flow of construction required the excavating and foundation contractors to commence by constructing retaining walls next to the railroad and highway approachs. Upon an early examination of the site the construction manager might well realize from its experience that significant problems are frequently encountered in constructing such retaining walls. Moreover, the construction manager might be able to predict, given the tight quarters with respect to two adjacent sides of the project, that the delivery of labor and materials would be significantly hampered during the course of the project because of various local government regulations affecting the routing of traffic, heavy equipment, and deliveries. The construction manager might be able to recommend a different location of the project on the site or a revised flow of construction to accommodate anticipated problems. At a minimum, the construction manager ought to be able to advise the owner and its architect about anticipated cost factors relating to the site selection and the project site placement.

Similarly, the construction manager ought to be able to provide advice with respect to the actual design of the building, in particular, the cost of alternative designs. Suppose the architect has designed a particular kind of foundation for the project on the hypothetical site referred to above. The construction manager might review that design and provide its advice with respect to its prior experience at constructing such foundations in such a location in certain types of soil. The construction manager might recommend to the architect possible alternatives and their respective costs.

Architects frequently will not be knowledgeable about the actual construction cost and performance of various major project subsystems and possible alternatives. Thus, current standard construction management contracts provide for the construction manager to make recommendations on the construction cost of alternative project subsystem components, in order that the owner and architect may consider various options. With respect to the basic project structural system an owner and architect may be uncertain whether to proceed with steel, poured-in-place concrete, or prestressed, precast concrete. The construction manager ought to be able to make recommendations to the owner and architect regarding the cost of these alternatives. In addition, the construction manager ought to be able to make recommendations with respect to sequencing of construction, such as whether the project exterior skin should be installed while the frame is being erected or afterward. The construction manager should review the expertise and reputation of construction firms likely to be bidding or to be considered with respect to each option. If cost and design factors are relatively equal, the construction manager's recommendations about the reputations of likely bidders and their ability to construct the project on time in a quality manner might make the difference between a successful project and a disaster. In addition, such recommendations are useful in determining which prime contractors should be bonded.

Along with project subsystems, the construction manager should recommend alternatives with respect to the availability, cost, and installation requirements of various critical materials. An architect might well specify the use of a particular kind of architectural spandrel glass. The construction manager's experience with that particular kind of glass might lead it to make alternative recommendations for a variety of reasons, for example, excessive cost, poor quality control, or difficulty of installation.

As the construction manager reviews evolving architectural and structural plans and specifications, it should provide recommendations relating to the development of construction sequencing and clear and discrete work scopes. Some designs may result in overlap of work scope with respect to various prime contractors. Some designs may result in a sequence of construction that requires too many trades in one area at one time. This is the peculiar kind of expertise that results largely from construction experience in which many architects are understandably lacking.

Finally, but perhaps most importantly, the construction manager ought to review the design for apparent defects. While the architect and its consulting engineers must bear the primary responsibility and liability for design defects, the owner ought to be able to rely on the construction manager for a general review of the design and identification of apparent defects. In addition, the construction manager ought to provide the owner and its architect with recommendations with respect to any ambiguities in the design documents which may lead to substantial change orders or disputes during project delivery. To the extent the construction manager is or was a general or prime contractor, it ought to be particularly adept at reviewing a set of architectural and structural plans and specifications and identifying potentially ambiguous areas which a contractor might attempt to utilize to gain additional compensation for its work. Early identification and resolution of these ambiguities provide significant cost savings to the owner and reduce the likelihood of expensive litigation and delays during the project delivery phase.

With respect to contractual allocation of risk, it is axiomatic that architects are liable in contract to those with whom they have contracted for breach of contract and possibly for negligent performance of the contract and are liable in tort to those with whom they have not contracted for negligence in providing defective plans and specifications.[1] Architects' traditional liability extends to those who incur personal injuries, for example, injured workers or ultimate consumers,[2] and to those who have sustained economic damages, for example, prime contractors and their sureties.[3] By analogy, to the extent a construction manager's advice results in design changes which prove defective, it may be liable in contract for breaching its contract with those with whom it has contracted and may be liable in tort to those same parties for the negligent performance of its contract. To those with whom it has not contracted, the construction manager will be liable in tort for negligent performance of its professional services to the extent it had a foreseeable duty to the plaintiff, and its negligence proximately caused the plaintiff's personal injuries or economic damages. Generally, the construction manager will be required to exercise the requisite skill and judgment of a similarly situated reasonable professional and will be held to a standard of ordinary care in that exercise. The construction manager's duties will probably be measured in large part by the specific terms of the owner–construction manager contract, various laws and regulations, and the construction manager's actual conduct on the project. In one recent decision, a court has suggested, at least by implication, that where the relevant contracts with the owner required the architect and construction manager jointly to participate in the design effort, they may be jointly and severally liable in contract to the owner if evidence is adduced at trial proving that the jointly developed plans and specifications were defective.[4] Thus, the owner, architect, and construction manager should be aware that although the architect will probably retain the primary duties and risk with respect to project design, the construction manager may be allocated some duty and consequent risk with respect to its design recommendations.

Long-Lead Procurement Items

The owner–construction management contract should require the construction manager to review the design for the purpose of identifying long-lead procurement items, for example, special subsystems, machinery, equipment, unusual materials or supplies, and even labor which must be procured in advance for efficient, timely project completion. Although somewhat analogous to the construction manager's design responsibilities, identification of and recommendations about long-lead procurement items can avoid numerous problems on construction projects. The identification of long-lead procurement items is a critical part of a successful project schedule, in particular, CPM schedules. The construction manager presumably has particular expertise at identifying such items and in facilitating their early purchase. This is especially important in phased construction inasmuch as drawings containing various long-lead procurement items evolve into the delivery stage, and sometimes are not complete prior to the award of prime contracts covering the scope of work in which such long-lead procurement items are included. Thus, it is critical for the construction manager to review evolving drawings, and in the event contracts are not yet awarded for long-lead procurement items which must be ordered, the construction manager must make recommendations with respect to the ordering and handling of such items. The construction manager's advice in this area directly relates to its services with respect to the identification of discrete bid packages, the award of contracts, contract administration, scheduling, coordination, and owner-purchased material.

The primary duty and risk for identification of long-lead procurement items is the construction manager's. To those with whom the construction manager has contracted, it may be liable for a breach of its contractual duties and for the negligent performance of them. To those with whom it has not contracted, it may be liable for negligent performance of its services as defined by its contract and its conduct.

Cost Estimates, Cost Accounting, and Preparation of Budgets

As noted in Chapter 1, the construction management concept evolved, in part, from owners' and architects' need for more accurate cost estimates for delivery of project design. The owner–construction manager contract should require the construction manager to evaluate the project concept and design during the decision and design phases with respect to the ultimate cost of delivery. During the delivery phase, the construction manager should evaluate the ongoing construction at periodical intervals to assess the extent to which the project is being constructed in accordance with budget. To the extent the project cost is exceeding budget, the construction manager should analyze why and should make recommendations with respect to achieving the budget.

Since cost estimates in evaluation of ongoing construction mean different things to different parties, it is prudent for the owner carefully to explore with its architect and construction manager the precise nature and extent of the cost estimate services the construction manager intends to perform in order to evaluate project cost.

The GSA construction management (CM) contract, for example, requires the construction manager to implement and utilize the PBS construction management control system (CMCS) during the course of the project. The GSA CM contract requires that the construction manager utilize the IBM 360 computer system, models 50 or higher, with an operating system to execute the CMCS computer programs and generate computer reports. The GSA CM contract does not permit any substitution with respect to the program evaluation or the computer system. Having generated its own computer system with programs for project cost evaluation, the GSA CM contract then requires that with respect to cost estimating the construction manager use those subprograms of the CMCS system. Because PBS/GSA generated its program, and because it requires the construction manager to make cost estimates and budget evaluations based on its subprograms, PBS/GSA is contractually able to bind a construction manager to a very specific set of requirements with regard to cost estimating and budget maintenance.

Simply providing in the owner–construction manager contract that the construction manager has to provide cost estimates, budget evaluations, and recommendations as to how to maintain the project budget is insufficient. At one extreme such cost estimate recommendations could consist simply of a review of the bids obtained for the work and a cursory analysis of whether payment requisitions, change orders, and extra work orders appear to be exceeding the guaranteed maximum price. At another extreme, the owner might be expecting cost estimate services in the nature of the CMCS program. Thus, it is critical for the owner and construction manager initially to understand and agree upon the nature and extent of the construction manager's cost estimate and budget evaluation services, including agreement upon a specific kind of computer program if one is to be utilized.

Once the precise nature and extent of the cost estimate services are agreed upon, the owner and construction manager should agree on the phases at which such information should be supplied. During the decision phase, the construction manager should prepare preliminary, rough project cost estimates based on general market experience for the area, volume, and basic type of project construction.

As the design phase evolves, the construction manager should prepare increasingly more detailed project cost estimates based on current labor and material costs as they apply to the specific equipment and quantities of material shown on the increasingly detailed drawings prepared by the architect. As discrete bid packages are developed and issued for bidding or negotiation, the construction manager should further refine its project cost estimates based on prices obtained from prime contractors which will be performing the work. Concurrent with reviewing the design and obtaining and reviewing bids for work, the construction manager should refine and make alternative recommendations as to different designs for different bid packages to provide maximum project cost savings to the owner.

During the delivery phase, the construction manager should be required to evaluate shop drawings for possible savings and should revise and refine the project cost estimate by, among other things, incorporating the actual project cost including approved change orders and extra work orders. Additionally, the construction manager should alert the owner to possible claims from prime contractors relating to unapproved change orders and extra work orders, acceleration and delay problems, and labor problems. Most importantly, the construction manager must provide the owner with an analysis of the relationship between prior project cost estimates and actual project costs. The construction manager should identify and delineate variances between estimated and actual project costs. The construction manager should be required to make an analysis of the reasons for such variances and make recommendations to the owner for achieving budget.

Depending upon the owner's accounting and supervisory capabilities, the owner may want to require the construction manager to maintain all project cost accounting records. In particular, the owner may want the construction manager to be responsible for reviewing all payment requisitions, including the cost of change orders and extra work orders. If so, the owner will want the construction manager to be responsible for obtaining and maintaining all supporting documentation from the prime contractors and for checking the accuracy of such documents.

With respect to allocation of duties and risks, traditionally only the architect incurred liability with respect to estimating construction cost. (Of course, general contractors always took the ultimate risk by submitting a bid for a fixed-price contract. Unlike the architect, however, the general contractor had the opportunity to make an entrepreneurial profit.) The AIA CM contract, Document B801, transfers the obligations for estimating project cost from the architect to the construction manager.[5] The critical issue in determining whether the cost estimator will be liable for its estimate is whether it guarantees or warrants the actual cost of construction. In the absence of such a contractual guaranty or warranty, and in the absence of contractual language making an accurate estimate a condition of payment, courts have usually found architects entitled to their fee regardless of whether bids are received at or below the construction estimate.[6] Where a fixed cost limit is agreed to or a warranty or guaranty is made and the actual project cost substantially or unreasonably exceeds the cost estimate, however, courts have held that architects have failed to comply with a condition precedent of the owner's duty to pay their fee.[7] Where a guaranteed or warranted cost estimate is exceeded as a result of the owner changing plans or specifications to meet changed project concepts, however, the architect is often able to recover its fee.[8] It should be stressed that even in the absence of a warranty or a guaranty of a cost estimate, an architect or construction manager may still be denied compensation for intentional misrepresentation or negligent preparation of cost estimates.[9]

The GSA and AIA construction management contracts provide only for the construction manager to estimate costs: these contracts do not provide for warranties, guaranties, or fixed-cost limits. It is important to recognize that both the GSA CM contract and the AIA CM contract assume the professional management concept. This highlights a critical point about contractual risk allocation with regard to project cost estimates. Traditionally, the general contractor took the ultimate project cost estimate risk since it submitted its bid in the form of a lump sum fixed-price contract. If its price was too low, its liability was a loss on the project. If it accurately estimated, it would make a reasonable entrepreneurial profit. Under the professional management concept, however, the construction manager offers its expertise as a professional service for a fixed fee as the owner's agent and does not guarantee the cost, timeliness, or quality of construction. Thus, in utilizing the professional management concept, the owner must realize that the construction manager is relinquishing its right to make an entrepreneurial profit on the project, and correspondingly for a fixed fee may not want to take the contractual risk of guaranteeing or warranting the project cost or the timeliness or quality of construction. If the owner prefers that the project cost be warranted or guaranteed by the construction manager, perhaps an adjustment to the fee can be negotiated or the owner should consider using contractor construction management.

Contract Document Assistance

The owner–construction manager contract should require the construction manager to review evolving architectural and structural drawings and specifications and to make recommendations to the owner and architect with respect to the development and delineation of discrete bid packages. The construction manager should be required to make recommendations to the owner and architect regarding work scope divisions with respect to the following factors: traditional work scope divisions within the industry, particularly as they affect potential labor jurisdictional disputes; project costs; construction sequences; potential work interfacing problems; potential trade interference problems; and any unusual labor or material requirements. The construction manager should be required to minimize work interfacing and trade interference problems by making recommendations with respect to different work divisions. The construction manager should make sure that each bid package identifies clearly and distinctly the work that is included and excluded for that particular contract. The construction manager should carefully assess all of the discrete bid packages to ensure that the entire project work is included. These recommendations are particularly important to avoid work scope disputes, interfacing and interference disputes, and jurisdictional labor disputes when delivery commences. Finally, the construction manager should make recommendations to the owner with regard to the development of such discrete bid packages to facilitate compliance with applicable government regulations, for example, the award of a requisite number of minority and Small Business Administration contracts.

Once the owner, architect, and construction manager agree on the precise scope of separate bid packages, the construction manager should be required to review the basic form of contract documents proposed by the owner and architect and make recommendations with respect to specific provisions in the general conditions. Regardless of whether the prime contracts are to be bid or to be negotiated, the construction manager should remain involved with each development in the process. On a complex construction project there will be numerous bulletins as the architectural and structural plans and specifications evolve, and during negotiations, work scope may be altered significantly as prices are received from different prime contractors with suggestions for project economies. As the bidding and negotiations proceed, it is imperative that the construction manager ensure that all project work is included in the prime contracts with particular attention to the factors outlined above.

With respect to allocation of duties and risks, the architect and construction manager probably have equal responsibility for the development and delineation of discrete bid packages. The construction manager, however, may have primary responsibility for a review of those bid packages with respect to the kinds of factors outlined above.

Bid, Negotiations, and Contract Award Assistance

The owner–construction manager contract should require the construction manager to provide the owner with substantial assistance with respect to the bidding, negotiating, and awarding of multiple prime contracts.

During the decision and design phases the construction manager should be required to analyze the construction market and the project locale with respect to potential prime contractors in critical construction areas, their competence, and their ability adequately to complete the owner's prospective project. Concurrent with such an analysis, the construction manager should develop prospective bidder interest in the project. The successful development of bidder interest in the project usually inures to the benefit of the owner. The greater number of prime contractors bidding with respect to discrete bid packages tends to result in more competitive prices and the generation of design alternatives which, upon review by the construction manager and architect, may save the owner substantial money. Concurrent with this task the construction manager should develop prequalification criteria for bidders. As discrete bid packages are prepared for ultimate bidding, the construction manager should be required to develop bid schedules and documents to assist the owner, architect, and construction manager in evaluating bids.

During the design and delivery phases, as discrete bid packages are finalized and issued to prospective prequalified bidders, the construction manager should be required to monitor all aspects of the issuance of bulletins, clarifications, and interpretations of the bid documents. Once bids are received, the construction manager should assist the owner and architect in analyzing the responsiveness of the bids, alternative designs, and overall costs. The construction manager should develop a negotiating strategy with respect to each bid package and bidder. Among the owner, architect, and construction manager, the construction manager will most likely have the greatest experience at analyzing a bid and developing the most effective negotiating strategy to lower the cost while maintaining quality. The construction manager should hold and organize bid conferences and pre-award conferences. During these conferences the construction manager should fully familiarize the bidders with critical aspects of the project, for example, scheduling mechanisms and coordination duties.

Once the owner, architect, and construction manager select specific prime contractors for discrete bid packages, the construction manager should monitor the final execution of contract documents and ensure that the work scopes included within the contracts are those to which the parties agreed. As multiple prime contracts are executed, the construction manager should be required to issue relevant notices to proceed. After multiple prime contracts are executed and notices to proceed are issued, the delivery phase commences.

With respect to the allocation of duties and risks for bidding, negotiation, and award assistance, the construction manager will have primary responsibility for the duties discussed above. In order to ascertain the precise extent of those duties, it is important for the owner and construction manager to delineate as clearly as possible in the owner–construction manager contract what is expected of the construction manager. With respect to development of bidder interest, for example, the owner–construction manager contract might require the monthly submittal of a report in a specified form by the construction manager scheduling those potential multiple prime contractors the construction manager contacted and those to which the construction manager sent its prequalification form, with an analysis of each one with respect to its qualifications and ability adequately to complete the project.

General Contract Administration Services

Once the multiple prime contracts are executed and notices to proceed are issued, the project enters the delivery phase. The owner–construction manager contract should require the construction manager to provide overall contract administration during the delivery phase. Generally, each multiple prime contract the owner executes will consist of two sections: first, a section with respect to each particular prime contractor, its work scope, and the architectural and structural drawings and specifications relevant thereto; and second, a section including the general conditions which will be similar for all multiple prime contracts on the project. In the general conditions to each multiple prime contract, the construction manager will have specific duties as agent of the owner. Thus, the owner–construction manager contract must require that the construction manager provide contract administration in accordance with its duties set forth in the general conditions to the multiple prime contracts. The AIA CM contract provides, for example:

> Unless otherwise provided in this Agreement and incorporated in the Contract Documents, the Construction Manager in cooperation with the Architect shall provide administration of the Contracts for Construction as set forth below and in the 1980 Edition of AIA Document A201/CM, General Conditions of the Contract for Construction, Construction Management Edition.[10]

It is critical to efficient administration of project delivery that the general conditions for each multiple prime contract be identical, that the construction manager's duties set forth in the general conditions be consistent with its duties set forth in the owner–construction manager contract, and that there be no conflict between the authority and duty of the architect and the construction manager. In the general conditions for separate multiple prime contracts, it is particularly important that the scope of the construction manager's authority as agent for the owner be clearly defined in contrast to that of the architect. On complex construction projects utilizing phased construction and multiple prime contracts, it is common to have literally hundreds of change orders and extra work orders. Depending upon the negotiating strengths of various prime contractors, the method for arriving at the price of change orders or extra work orders may be different. The general conditions must make clear to what extent the construction manager and architect, respectively, have authority to commit the owner to prices for various change orders and extra work orders.

In addition to performing overall contract administration in accordance with the general conditions for multiple prime contracts, the construction manager should provide other critical contract administration services. Perhaps the single most important negotiating point for the owner with respect to its construction manager is which of the construction manager's key personnel will be in charge of the owner's project. It is natural in marketing itself for the construction management firm to use its most effective, personable employees. From the owner's point of view, however, the critical question is whether these employees will be in charge of the owner's project; and what "in charge" means to the construction management firm with respect to time commitments and presence at the owner's project site. As with any firm of professionals, some of the construction manager's employees will be more effective, competent, and appealing to the owner than others. Similarly, regardless of the reputation of the construction management firm, if a relatively new employee with relatively little experience is placed in charge of a complex project, the owner is not going to be served as well as if a more senior, experienced professional were in charge of its project. While neither the owner nor the construction management firm can guarantee the presence of a particular person on the project (the person may die or decide to leave the employment of the construction management firm), the owner should bargain for the right to have one of three or four key personnel in charge of its project. Prior to executing the owner–construction management contract, the owner should insist on meeting and interviewing several prospective members of the construction management firm who in its opinion are capable of being in charge of its project. The owner should establish a priority list with the construction manager, and the construction management contract should provide that in the event the owner's first selection leaves the employment of the construction management firm (or other specified conditions), the owner will be able to obtain its second choice for the duration of the project. Similarly, if the owner is unlucky enough to have a second choice leave during the course of the project, the owner will be entitled to its third choice.

A second important consideration for the owner is carefully to define in the owner–construction manager contract precisely what "in charge" of the project means for the construction manager's project administrator. Without such a careful definition, the construction management firm may assume that its person in charge of the project will maintain an office at the construction management firm and only visit the project site on a weekly basis. On the other hand, the owner may well expect and want the person in charge of its project to have an office at the project site and to spend a 40-hour workweek or more at the site. During their contract negotiations the owner and construction manager should carefully consider different options within this range and the cost of each option. Both parties must be aware of the fact that on a complex, phased, multiple prime contraction project, overall contract administration and, in particular, scheduling and coordination by the construction manager are critical to the success of the project. Consequently, the success of the project may depend substantially upon the competence of the particular person from the construction management firm in charge on a daily basis at the project site.

With respect to general contract administration the construction manager should also be required to establish a format for and run weekly job progress meetings. The construction manager should be required to develop with the owner and its architect a general agenda for each such meeting which should include, for example, a review of the minutes of the prior week's meeting; a review with each active prime contractor of its general work progress and any work problems; a careful review of the scheduling status of the project to determine whether any of the prime contractors are behind schedule or interfering with other prime contractors' work and whether any of the prime contractors are asserting acceleration or delay claims as as result of the conduct of any other prime contractor, the owner, the construction manager, or the architect; a comprehensive listing for each prime contractor of any and all time extension requests, change orders, and extra work orders; and a careful description by each prime contractor of its intended work schedule, work areas, and personnel requirements for the upcoming week. The construction manager should also type and distribute minutes of weekly job progress meetings.

The construction manager should be required to have at its project trailer complete final copies of all multiple prime contract construction documents, including the latest version of approved architectural, structural, and shop drawings and specifications. In addition to possessing an entire set of drawings and specifications for the project for each prime contractor, the construction manager should keep binders for each prime contractor containing all relevant change order and extra work order requests, time extension requests, claims, and supporting documents. The construction manager should maintain detailed current logs with respect to the submittal of all shop drawings, catalog cuts, and samples for each multiple prime contractor. Finally, and probably most importantly, the construction manager must maintain at its project trailer an entire copy of the current project CPM schedule, including all computer printouts and network logic diagrams. These must be readily accessible by all multiple prime contractors, and the construction manager should make it a point at each weekly job progress meeting to invite a review of the current revised CPM schedule by each prime contractor. This function is critical in keeping the project on schedule and in providing the owner and its architect with relevant recommendations in the event a particular multiple prime contractor begins to slip behind schedule. Moreover, in the event litigation develops involving acceleration and/or delay claims, it will be critical from the owner's point of view to have had such information readily available to multiple prime contractors.

With respect to allocation of duties and risks, the construction manager may primarily be liable for the performance of its contractual duties set forth in the general conditions to the multiple prime contracts and set forth in the owner–construction management contract relating to overall contract administration. As has been assumed in this chapter, if the owner selects professional construction management, the architect is primarily responsible for the design of the project. Once the delivery phase commences with the execution of multiple prime contracts and the issuance of notices to proceed, the construction manager assumes primary liability for overall contract administration. The construction manager will be liable to the owner for a breach of its contractual terms as set forth in the owner–construction management contract and in the general conditions for the multiple prime contracts. To those with whom the construction manager has not contracted, it may well be liable in tort for the negligent performance of its duties, which will be established by the terms of the two aforesaid contracts and its conduct on the project site.

Scheduling and Coordination Services

If, as has been assumed in this chapter, an owner decides to utilize phased construction and multiple prime contracts, the owner has a critical need for the construction manager to provide overall scheduling and coordination services. As noted in Chapter 1, the traditional general contractor in sequential construction theoretically earned its markup on each of its subcontractors' work by providing overall contract scheduling and coordination services to the owner. The traditional general contractor guaranteed the cost, timeliness, and quality of construction. In phased construction utilizing multiple prime contracts, the owner theoretically saves the general contractor's markup on the work of each of the trade contractors. Each prime contractor guarantees only the cost, timeliness, and quality of its own construction. Thus, the owner and construction manager must decide which party will assume overall project scheduling and coordination responsibility. The ability effectively to schedule and to coordinate project delivery will be critical to the success of the project and to project litigation which arises during and after delivery.

When a project is substantially delayed by actions of either the owner or one or more of the multiple prime contractors, the strategy of the delayed prime contractors often is to sue the owner asserting that the owner had the duty to schedule and to coordinate all of the multiple prime contractors. In the event one of the multiple prime contractors delays the project, the other prime contractors will allege that the owner had the duty either to force the offending prime contractor to maintain the project CPM schedule or to terminate it and hire another prime contractor in order to maintain the project schedule. Thus, in negotiating their contract, the owner and construction manager must decide which party will take primary responsibility and the consequent risk for project scheduling and coordination. Whichever party assumes that responsibility should also possess the requisite power to enforce it.

The owner should consult its architect and construction manager with regard to what kind of schedule will be most effective for the owner during project delivery and, in the event of delays, litigation. On complex construction projects the owner, architect, and construction manager may decide to hire a scheduling expert to provide and revise a sophisticated CPM or PERT schedule. In the event the owner makes such a decision, it will be important to decide whether the consulting CPM expert contracts with the owner, the architect, or the construction manager. Regardless of whether the construction manager or an independent consulting CPM expert provides the scheduling service, it is imperative that the owner fully understand the nature and extent of such a CPM before making its selection. The owner may well want to consider alternative computer programs and then contractually bind the construction manager or the CPM expert to the use of a specific program.

Along with the use of a specific program, the owner should negotiate for a periodic review and revising of the program. The owner should require that the party responsible for CPM scheduling meet frequently with the prime contractors to obtain relevant data to ensure an accurate CPM schedule. It is imperative that the owner and construction manager make sure the multiple prime contracts contain provisions requiring the multiple prime contractors to provide all relevant information to the CPM scheduler.

Once a particular type of CPM schedule is agreed upon, it should be specifically referenced in the owner–construction manager contract, and the construction manager should be required to make periodic reports to the owner with respect to variances between the original CPM and revisions reflecting the current project status, particularly to the extent it is different from that originally anticipated. Such reports should contain an analysis of the reasons for any such variations. If the owner has retained the ultimate responsibility for coordinating the project, the construction manager's report should also consist of recommendations to the owner as to what steps should be taken with respect to offending prime contractors. If the construction manager has retained the ultimate responsibility for coordination of the project, the report should summarize the actions the construction manager has taken on the owner's behalf to require multiple prime contractors to maintain schedule. In either event, the report should contain an analysis of prospective acceleration or delay claims that might be asserted by prime contractors.

The construction manager should be required to provide general project coordination. In addition to scheduling, such coordination services would include a review of all interfacing, interference, and potential labor disputes among multiple prime contractors. In addition, the construction manager, as noted above, should be required to develop an agenda for and to run weekly project meetings and distribute the minutes.

With respect to contractual allocation of duties and risks, the critical question is which party will have the ultimate responsibility for scheduling and coordination and the power to enforce it. In the traditional construction approach, the general contractor has the duty to construct the project in accordance with the time requirements of its contract with the owner and, therefore, to schedule and coordinate the work of its subcontractors. The general contractor is liable to the owner if it fails to construct the project in a timely manner by failing to schedule and to coordinate properly the work of its subcontractors. Similarly, the general contractor is liable to its subcontractors for causing them delay by failing to schedule and to coordinate adequately the project.[11]

With the advent of phased and fast track construction, multiple prime contracts became a necessity. When an owner utilizes multiple prime contractors, the critical question is which party will be responsible for scheduling and coordinating their work. Since the owner saves the general contractor's markup for, among other things, management services, the owner normally has the duty.[12]

Since a construction manager offers an owner expertise in the area of scheduling and coordination, the construction manager's liability is most likely to be a function of the extent to which an owner delegates that duty and provides the power to enforce it.[13] Under the GSA CM contract and the AIA CM contract where the construction manager acts in a professional capacity, it assumes the duty of advising the owner on scheduling and coordination, leaving the ultimate responsibility for scheduling and coordination with the owner, which retains the requisite power to enforce compliance. When the construction manager guarantees a maximum price, however, the construction manager frequently retains the requisite authority to enforce its scheduling and coordinating duties and, presumably, incurs liability for breaches of those duties.

Under the GSA CM contract, the construction manager ensures that the construction will be completed as early as possible, but not later than the scheduled completion date.[14] The construction manager is responsible for the development of a comprehensive schedule for the project.[15] The ultimate power to enforce the construction manager's schedule and coordination directives, however, is retained by GSA's contracting officer. The relevant provision is as follows:

> The Construction Manager shall coordinate and provide general direction of the work of the separate construction contractors; he shall inspect the work performed by the separate contractors to ensure conformity with requirements of their respective contract. In the event any differences arise between the Construction Manager and any separate construction contractor, the Construction Manager shall inform the Contracting Officer promptly in writing, giving both the details of pertinent facts and applicable contract provisions and his recommendation as to action to be taken by the Contracting Officer. Promptly after receipt of the Contracting Officer's written interpretation, the Construction Manager shall transmit it to the separate construction contractor.[16]

Thus, the GSA retains the ultimate power to terminate and default a contractor for failure to comply with a construction manager's schedule and advice with respect to coordination. Consequently, as agent for a disclosed principal which has ultimate responsibility for scheduling and coordination, pursuant to a GSA CM contract, the construction manager probably will not incur any liability for scheduling and coordination to multiple prime contractors.

On GSA and government contracts, the government has been held responsible for scheduling and coordination of multiple prime contractors.[17] To date, only two decisions have included a discussion of the construction manager's role. In each decision the Board of Contract Appeals reasoned that providing the construction manager adequately advised the government regarding the project status, it would not incur liability.[18]

In summary, under the GSA CM contract a construction manager probably will not be liable to either prime contractors or GSA for a failure to schedule and to coordinate if it diligently performs its duty to establish a schedule and to keep GSA's contracting officer advised as to multiple prime contractor's compliance with that schedule. Indeed, this standard is articulated in the GSA CM contract:

> A Construction Manager will not be deemed to have failed to meet his contractual undertakings and will not be held responsible for. . .
> (2) Time overrun, provided the Construction Manager timely developed and maintained a master schedule reflecting all activities having significant impact on scheduling, prepared accurate, timely up-dates or revisions to the schedule as required by the contract, has taken all reasonable measures to phase the construction work, to eliminate access and availability constraints, to anticipate problems and to eliminate or minimize their adverse impact on the completion of the construction within the time specified.[19]

There appear to be no decisions on GSA actions against its construction manager for a breach of the GSA CM contract. Two actions were recalled by PBS/GSA's former assistant commissioner for construction management; however, both were settled.[20] Presumably, a construction manager would be liable to GSA in the event it failed to establish a schedule and to advise GSA's contracting officer of prime contractor's failures to comply with the schedule or its coordination directives or if it failed to transmit orders of the contracting officer in a timely manner.

In 1980, the GSA issued a directive that it would no longer use construction management except in rare instances.[21] Although PBS/GSA had not, as of this writing, issued any public statement with respect to its reasons for reducing its use of the construction management system, PBS/GSA's former assistant commissioner for construction management indicated in an interview that PBS/GSA's problem with the construction management system set forth in the GSA CM contract was precisely that the construction manager had no direct authority to compel compliance with its schedule for and coordination directives to the prime contractors, and GSA's contracting officer frequently was too removed from the project to effectively enforce his powers.[22] Consequently, GSA's directive no longer to use construction management except in rare instances is attributable to the failure of PBS/GSA's own contracting officer and project manager system. Ironically, in 1970 the PBS study severely criticized that very problem.[23]

AIA Document B801, *Standard Form of Agreement between Owner and Construction Manager,* requires the construction manager to establish a schedule for the performance of the work and to coordinate the work of the prime contractors.[24] In AIA Document A201/CM, *General Conditions of the Contract for Construction* (reproduced in Chapter 4), however, the owner is given the right to use prime contractors, the duty to coordinate the work of prime contractors,[25] and the requisite power to stop and carry out the work in the event a prime contractor is not complying with the construction schedule or the coordination directives of the owner's agent, the construction manager.[26] Since these contract provisions are similar to those of the GSA CM contract, apparently the owner will be liable for scheduling and coordination of multiple prime contractors. Additionally, the construction manager as agent for a disclosed principal with limited authority may not be liable to prime contractors if diligent in performing its duties.

To date, only three courts arguably have ruled on the construction manager's liability for scheduling and coordination and consequent delay to a prime contractor. In *John E. Green Plumbing & Heating Co. v. Turner Construction Co.,*[27] the plumbing and mechanical prime contractor entered into a contract with the owner to construct the mechanical portion of a hospital in Detroit. The owner contracted with Turner to provide construction management for the project. The mechanical prime had no contract with the construction manager.

The mechanical prime instituted an action against the construction manager alleging substantial delay damages as a result of the construction manager's conduct: first, the construction manager tortiously interfered with its contract with the owner; and second, the construction manager negligently performed its scheduling and coordination duties.

The construction manager argued that the no damage for delay clause in the contract between the owner and the mechanical prime precluded the action. In addition, the construction manager defended on the ground that there was no privity of contract between the construction manager and the mechanical prime.

The court denied the construction manager's summary judgment motion on the tortious interference count, but granted the motion on negligence.[28] The court reasoned that Michigan law permitted a party to pierce a no damage for delay clause when the party attempting to enforce it had intentionally interfered with the contractor's duties. The court reasoned that by definition tortious interference implies that the defendant intentionally interfered with the contractor's performance, and therefore, such allegations were sufficient to defeat the construction manager's summary judgment motion.[29] On the negligence count, however, the court reasoned that since negligence does not include intentional conduct, the construction manager was entitled to summary judgment, obtaining the benefit of the owner's no damage for delay clause.[30]

Unfortunately, this case is not enlightening due to the lack of legal analysis of the principles involved. The court apparently implicitly ruled that the construction manager's defense of lack of privity was not sufficient to defeat either tort count. Additionally, although the court acknowledged that the mechanical prime's suit was in tort, it nevertheless assumed without discussion that the construction manager would be able to defend by availing itself of the no damage for delay clause contained in the owner's contract with the mechanical prime. Finally, the decision does not set forth the relevant contract provisions with respect to whether the owner or construction manager had the duty and power to enforce coordination.

In the consolidated cases known as *Edwin J. Dobson, Jr., Inc. v. Rutgers State University*,[31] the construction manager–general prime, plumbing prime, and electrical prime all sued the owner of a construction project for, *inter alia,* failing to coordinate the work of the primes, alleging active interference on the part of the owner in order to pierce a no damage for delay clause.

Essentially, the owner protected itself with a comprehensive no damage for delay clause. The owner retained the right to withhold retainage from each prime, including withholding retainage to cover claims by separate primes against the owner or other primes. The owner retained the right to stop the work and default a prime, but gave the duty of scheduling and coordinating all the primes to the construction manager–general prime, which had duties primarily of a construction manager, but performed some of the general trade work itself. One clause required the construction manager "to enforce" the schedule for completion of the work; however, the contract did not grant the construction manager any enforcement mechanisms.

The owner retained as an independent consultant a CPM expert to aid the construction manager and other primes in establishing an effective CPM chart to assure timely completion of the work. The owner retained an architect to interpret plans and specifications and advise it on certain questions including, but not limited to, various primes' claims against other primes in order to ascertain the amount of retainage that should be withheld from one prime to protect the owner and other primes.

Finally, the contract contained numerous mutual responsibility and coordination clauses in which each prime agreed to coordinate its work with others, to abide by the construction schedule of the construction manager, to pay delay damages as a result of its delay, and to indemnify the owner in the event other primes sued the owner as a result of delay.

The court reasoned that since a substantial portion of the damages sought by the primes was delay damages as a result of delayed completion of the project, the critical question was which party had the duty to coordinate the work of the primes to ensure timely completion. The court held that although the owner did retain the right to stop the work, to default a prime, and to withhold retainage from a prime contractor as a result of its delay of the project and for damages caused by it to other primes, the owner had, in fact, delegated the duty to coordinate the scheduling and progress of the work to the construction manager.[32]

The court relied heavily on the clauses in the contract with the construction manager requiring it to coordinate the work and a clause in the specifications relating to scheduling and coordination which provided the construction manager with the duty "to enforce the combined schedule" for completing the project upon which all the primes agreed with the advice of the CPM expert.[33]

In responding to the other prime's argument that the construction manager had not been given *any power to enforce the schedule* since the owner retained the right to withhold funds and to terminate a prime and, thus, that the owner retained the duty to coordinate,[34] the court reasoned that "the effective means of enforcing the contract is not the ability to withhold funds or to terminate, but to sue and to enforce according to remedies provided by law.[35]

Thus, the court recognized the importance of the primes' argument that the party charged with coordination must have appropriate power to enforce that coordination.[36] Since the court ruled that although the owner possessed the power to terminate and withhold retainage, the construction manager had the duty to coordinate, the court required a legal theory that would provide the construction manager with a right to enforce its duty to coordinate. The court reasoned that the existence of a cause of action against the other primes would be a sufficient enforcement mechanism.[37]

The court proceeded to articulate essentially two theories upon which a construction manager–general prime could assert a cause of action against another prime: first, an apparently direct contractual cause of action based upon reciprocal coordination agreements; and second, a third-party beneficiary cause of action. The court arrived at the first conclusion by reasoning in the following manner. First, the court found that eventually all the primes attempted to follow a particular CPM schedule with relatively complete data which had been developed from an arrow network diagram prepared by and with the CPM expert. Most significantly, however, the court found that the CPM schedule was developed by the construction manager and the CPM expert with each prime's extensive input and analysis. The court maintained that at a given point in time all of the primes essentially agreed to follow this particular CPM schedule, whether entirely fair or not.

According to the court, that agreement combined with the mutual responsibility clauses pursuant to which each prime agreed to pay other primes' delay damages and the owner's right to withhold retainage for the payment of those damages gave "rise to a contract right among themselves, for they are the only ones involved in the process [of developing the agreed-upon and enforceable schedule]."[38] Although the court did not specifically explain whether this was a direct contract action based upon an oral agreement to abide by a jointly developed schedule or a direct contract action based on reciprocal mutual responsibility clauses and agreements to abide by a schedule in separate prime contracts, the court did explain:

> . . . that where contractors among themselves agree that segments of the work to complete a project shall be done by certain dates in order to complete a project for an owner by a stated date or within a stated time, then the contrac-

tor which fails to perform on time is liable in damages for breach of contract, whether the contractor have [sic] overall supervision [citations omitted] or the contractor is responsible only for a portion of the work [citations omitted].[39]

The court then held that the construction manager–general prime also had a third-party beneficiary cause of action against the other primes. In so holding, the court principally relied upon the following clause:

> *In case the Contractor . . . shall unnecessarily delay . . . the work of the Owner or other Contractors, by not properly cooperating with them or by not affording them sufficient opportunity or facility to perform work as may be specified, the Contractor shall, in that case, pay all costs and expenses incurred by such parties due to any such delays* and he hereby authorizes the Owner to deduct the amount of such costs and expense from any monies due or to become due the Contractor under this Contract. . . . [based upon Architect's recommendations].[40]

The court noted that since the owner was protected by a no damage for delay clause, the primes obviously intended to relieve the owner from liability for delay damage by having each contractor who caused delay pay damages to the injured primes, with the owner possessing a mechanism, the withholding of retainage, to assure compliance with the promise. After reciting but without analyzing general axioms regarding donee and creditor beneficiaries, the court reasoned that since each prime agreed to pay the costs of its delay to another prime, there existed a promise the performance of which was to be rendered directly to the third party who was intended by the promisee or owner to be benefited thereby. Thus, the court apparently considered a prime under this clause to be a donee beneficiary of other primes' contracts with the owner.[41]

The court concluded that the construction manager did possess a cause of action against other primes for their delay either on a direct contract action as a result of reciprocal agreements to comply with an agreed-upon schedule of work for timely completion or as a third-party beneficiary of the separate prime contracts.[42] Thus, the court supported its holding that the construction manager retained the duty to coordinate and not the owner. The court concluded that the construction manager was responsible to other primes for their delay damages.[43]

Thus, at least one court has held that an owner which attempts to impose an ultimate duty to schedule and to coordinate upon a construction manager–general prime may be successful. In such a case the construction manager will be liable to other prime contractors for its failure to enforce its scheduling and coordination duties, and will, at least in New Jersey, possess a cause of action against the delaying primes.

Supervision and Inspection Services

On any construction project unless the owner has competent personnel to conduct supervision and inspection responsibilities, the owner needs either an architect or a construction manager to supervise and inspect the multiple prime contractors' work. The traditional elements of architectural practice are design of plans and specifications and superintendence of the construction project.[44] In recent years the proliferation of litigation against architects has resulted largely from architects' supervisory duties.[45] Under prior versions of the AIA owner-architect and owner–general contractor agreements where the architect had the duty to supervise construction, and had the power to stop project work, architects frequently were found liable to, among others, general contractors for failure to perform their supervision responsibilities.[46] In reaching such decisions, courts relied upon the architects' duties set forth in their agreement with owners and owners' agreements with general contractors.[47] In response to their expanding liability, architects through the AIA have altered standard AIA contract documents to mitigate the effect of such decisions.[48] The most salient alterations have been the change of the architects' duty to supervise to a duty of periodic observations, a deletion of the architects' independent right to stop the project work, and the reinforcement of an express disclaimer of the architects' responsibility for the construction means and methods employed by the general contractor.[49] It is now relatively infrequent to find an architect or professional construction manager who is willing to undertake such responsibilities, as they carry a liability for an end result over which they may not have sufficient control. A leading commentary has noted, however, that courts have not been particularly sympathetic to these changes when important areas of responsibility are disclaimed by all parties.[50]

As noted in Chapter 1, as architects increasingly have been unwilling to assume supervisory and inspection duties for owners, a primary area of expertise which a construction manager offers an owner is contract supervision and inspection services. Either an architect or a general contractor operating as a construction manager has the expertise to perform such functions. Thus, the owner–construction manager contract should require the construction manager to supervise and inspect the multiple prime contractors' work.

The owner and construction manager should negotiate specifically the extent of the construction manager's supervisory and inspection duties with particular attention to the meaning of terms such as "supervision," "observation," and "inspection." The GSA CM contract provides that the construction manager is responsible for supervision and inspection, but its responsibility only extends to advising the GSA contracting officer in writing of the details and pertinent facts relating to its supervision and inspection of prime contractors' work. The ultimate responsibility for supervision and inspection along with the right or duty to stop project work remains with GSA.[51]

Similarly, the AIA CM contract merely requires general supervision and inspection responsibilities from the construction manager.[52] In other words, the construction manager's duties consist only of determining in general that each multiple prime contractor's work is being performed in accordance with its contract documents. The construction manager is only required to "endeavor to guard the owner against defects and deficiencies in the work."[53] In addition, the AIA CM contract contains the express disclaimer of supervision and inspection responsibilities that the AIA owner-architect contract now contains—that the construction manager shall not be responsible for the construction means, methods, techniques, sequences, and procedures employed by multiple prime contractors in the performance of their contracts and that the construction manager shall not be responsible for the failure of any multiple prime contractor to carry out work in accordance with its contract.[54] Moreover, in the AIA construction management edition of *General Conditions of the Contract for Construction,* the owner, as opposed to the construction manager, retains the right to stop the project work.[55] In other words, the AIA document attempts to distinguish between a supervisory function which involves observation of the work and one which could involve a duty to superintend the work.

The owner, however, may well desire that its construction manager assume more responsibility than simple on-site observation for its supervision and inspection services. If so, the owner should probably give the construction manager greater powers, for example, the right to stop the project work and the right to reject work performed by multiple prime contractors. The owner should be prepared, however, to confront the construction manager's perspective. The construction manager will assert that when it agreed to act as a professional construction manager for a fee, it relinquished the opportunity to make an entrepreneurial profit on the project. As a result, the construction manager will assert that only multiple prime contractors should retain ultimate liability for the actual means, methods, techniques, and construction of project work.

One of the few court decisions to date specifically ruling on a construction manager's liability for supervision is enlightening. In *First National Bank of Akron v. William F. Cann, et al.,*[56] the owner (the bank) entered into a contract with Cann, a licensed architect, and into a construction management contract with BEC to expand and renovate an existing bank. The construction manager, BEC, was a corporation engaged in the business of assisting lending institutions in making improvements to their physical facilities, providing assistance in the selection of an architect, and providing construction management.[57] Unknown to the owner, the architect, Cann, was president of the construction management firm, and all architectural fees paid to him were ultimately turned over to the construction management company. Cann was paid a salary in his capacity as president of the

construction manager but received no separate compensation as an architect. Thus, as the court concluded, the construction manager was both a
preliminary consultant and the construction manager, and its president,
Cann, was the architect.[58]

Sometime after construction was completed, the granite panels on the
face of the bank failed. After receiving an unsatisfactory response from
both the architect and the construction manager, the bank engaged an
engineering firm to assess the problem. Ultimately, it was determined that
the granite panels had not been installed in accordance with the plans and
specifications for their installation; rather, substantial amounts of cutting
of structural steel and welding had occurred with respect to the parts which
had been designed to hold the granite to the retaining wall of the bank
prior to the installation of the granite.

The bank sued the architect and the construction manager on the
grounds that they were jointly and severally liable to the bank in both
contract and tort for, *inter alia,* inadequate supervision and inspection.

After analyzing the architect's and the construction manager's contracts
with the bank and emphasizing the fact that although the bank was not
aware of it, the architect was an employee of the construction manager, the
court concluded that the architect and construction manager jointly bound
themselves under contract to the owner for the construction of the project.[59] Although the architect's contract contained various disclaimers and
exculpatory clauses attempting to limit the architect's liability for supervision, the court held:

> Admittedly, the contract documents are explicit on the provision that exhaus-
> tive, continuous on-site inspections are not required. That exhaustive, continu-
> ous on-site inspections were not required, however, does not allow the architect
> to close his eyes on the construction site, refrain from engaging in any inspec-
> tion procedure whatsoever, and then disclaim liability for construction defects
> that even the most perfunctory monitoring would have prevented. Even the
> most general supervision would have placed Cann [architect] on notice as to
> the activities of laborers engaging in unauthorized, extensive welding and
> cutting of structural steel.
>
>   Turning to principles of contract interpretation, the Court construes Cann's
> [architect's] duty under article 3.4.3 to include inspections and monitoring of a
> nature that would have uncovered the vast majority of defective conditions on
> the south wall [the failing granite exterior wall].[60]

An additional factor in the court's holding was the architect's employment
relationship with the construction manager and the potential conflict. As a
result of that relationship the architect had a higher duty than usual to the
bank to ensure that the work was proceeding as planned.[61]

The court analyzed the terms of the construction management agreement which required the construction manager to act "promptly to establish a work program under which the project will proceed with all possible
speed consistent with reasonable cause, good workmenship and safety."[62]

The construction management contract required the construction manager to secure approval from the owner for any changes in the work as originally designed. Finally, the construction manager had certain duties with respect to supervision.

The court concluded that the construction manager materially breached its contract by permitting substantial cutting and welding to occur on the structural support system for the granite without obtaining any authorization from the owner and without adequately supervising the changes.[63] Thus, the court found the architect and the construction manager jointly and severally liable in contract for breach of their respective supervision responsibilities in their contracts.

With respect to the owner's tort claims against the architect and construction manager, the court concluded, in essence, that since there were contracts among the relevant parties and since pursuant to those contracts the parties had attempted to allocate their risks and the damages that would flow from the breach thereof, no tort cause of action existed.[64]

Thus, in negotiating their contracts, the owner, its architect and its construction manager should carefully define and allocate duties and risks with respect to supervision and inspection. Ultimate liability for supervision and inspection will be determined based upon duties set forth in the respective contracts and actual conduct at the project site.

Review and Certification of Progress Payments

Construction industry analysts know that prime contractors frequently engage in "front-end loading" and occasionally divert funds received as progress payments from owners to uses other than the payment of subcontractors and suppliers which have supplied labor and material to the project.[65] Unless an owner has competent personnel to review payment requisitions, inspect multiple prime contractors' work, and certify that payment may safely be made in accordance with the requisition, the owner needs a party to be responsible to it to make such a review and certification. Much as with supervision and inspection duties, prior versions of AIA contracts allocated the risk for review and certification of payment requisitions upon the architect.[66] When the architect breached that duty or negligently performed it, the owner, general contractor, its surety, and subcontractors often sustained economic damages, and courts began finding architects liable to parties other than the owner for breach of such duties.[67] Acting much as they did upon being found liable for supervision and inspection duties, architects responded through the AIA by modifying their duties set forth in the AIA *General Conditions of the Contract for Construction*. An express disclaimer was added which provides that by issuing a certificate for payment the architect does not represent that it has made an exhaustive or continuous on-site inspection to check the quality or quantity of project work by multiple prime contractors.[68] An additional disclaimer provides that the architect's certificate does not constitute a representation by the architect that it has made any analysis of the multiple prime contractor's use of the payments.[69]

The AIA CM contract provides that the construction manager is required to develop and implement procedures for reviewing and processing payment requisitions by multiple prime contractors. The construction manager is required to make recommendations with respect to payment and certification to the owner and its architect. The construction manager is provided, however, with the same disclaimers added by the AIA to the most recent edition of *General Conditions of the Contract for Construction.*

Thus, the owner and the construction manager must carefully negotiate this issue and the precise language of the provisions requiring the construction manager to review and certify payment requisitions. From the owner's point of view, of all parties in the construction process the construction manager probably has the most expertise and is probably in the best position to make such a certification. The owner may well desire the construction manager to assume greater responsibility for its review and certification of payment requisitions than it would expect of the architect in a traditional construction project.

Change Order Assistance

On any complex construction project where phased construction and multiple prime contracts are utilized, hundreds of changes occur. For each change that occurs there are four significant ramifications: first, a time factor; second, a cost factor; third, an interfacement factor; and fourth, an interference factor. Thus, unless the owner has competent personnel with the time to review and monitor change orders, the owner needs a party to perform this service for it. Both the GSA CM contract and the AIA CM contract require the construction manager to identify and recommend to the owner and its architect desirable changes to the multiple prime contracts, and to review and evaluate requests for change orders from multiple prime contractors. In addition, these contracts require the construction manager to assist the owner in the negotiation of the time and cost factors of such change orders. The GSA CM contract also provides that the construction manager must monitor the work associated with such change orders to assess the actual time and cost factors.

In negotiating provisions of the owner–construction manager contract relating to change order assistance, it is critical that the owner realize in advance that identification and recommendation of a particular change to a multiple prime contract is only one small portion of the kind of assistance the owner needs. Once the necessity for a change is identified as a result of changing architectural plans and specifications as phased construction evolves, the construction manager's identification of a change to a multiple prime contract, or a change request from a multiple prime contractor, the owner needs prompt and competent advice on the impact of each such change on project time, cost, interfacement, and interference factors.

One of the most difficult areas of contract administration on phased, multiple prime construction projects is the negotiation of the cost of change order work. It is imperative that the owner, its architect, and its construction manager carefully review the provisions in the general conditions for each multiple prime contract relating to the payment for change order work. Frequently, such clauses contain alternatives. One alternative is that the owner, through its construction manager, and the prime contractor agree on a lump sum price. A second alternative is that the owner, through its construction manager, and the multiple prime contractor agree on a cost plus a guaranteed maximum price for the change order work. A third alternative is simply that the multiple prime contractor completes the work on a cost-plus-a-fixed-fee basis. Regardless of which option is utilized, it is critical for the owner whether the provision requires the multiple prime contractor to proceed with the change order work in advance of an agreement on the price or even in the event of a dispute with respect to the price. The owner, its architect, and its construction manager should carefully consider these provisions and agree in advance on the one best suited for delivery of the owner's contract. Once the payment method is selected and inserted in the general conditions for the multiple prime contracts, the owner–construction manager contract should require that the construction manager be the owner's authorized agent to negotiate and monitor such change orders. In particular, in the event multiple prime contractors are likely to be performing change order work on a cost-plus basis, the construction manager should be required carefully to monitor the time and materials utilized by the multiple prime contractors. The construction manager should be required to obtain and evaluate all supporting documentation.

In addition, the owner–construction manager contract should require the construction manager to evaluate the impact of each such change order on the work of other multiple prime contractors, particularly from an interfacement and interference viewpoint. All too often, the owner considers the cost of a particular change to be simply the cost of the work of the prime contractor directly involved. Often, however, the change requires other multiple prime contractors to remove some of their work or to alter their work sequence in such a way as to create additional cost for them. Indeed, as a result of such interfacement and interference issues, other prime contractors may assert substantial acceleration or delay claims against the offending prime contractor, the construction manager, and the owner. Thus, the construction manager should be required to advise all possibly affected prime contractors of each impending change and to seek from them an advance evaluation of the change impact on their work from time, cost, interfacement, and interference viewpoints.

On a complex, phased, multiple prime construction project the construction manager's services with respect to the identification, evaluation, and negotiation of change orders can be critical to project success. It is imperative that the owner fully discuss with its architect and construction manager the precise scope of each of their duties with respect to change orders, and delineate those duties as specifically as possible in the relevant contracts.

Substantial and Final Completion Services

The owner–construction manager contract should require the construction manager to assist the owner and architect in determining when the project or a designated portion of it is substantially complete. The construction manager should be required to summarize each multiple prime contractor's work and to prepare a punch list indicating incomplete or unsatisfactory items. After the preparation of the initial punch list, the construction manager should monitor each multiple prime contractor's completion of the punch list. The construction manager should obtain from the multiple prime contractors all the requisite affidavits, mechanic's lien releases, general releases, operating manuals, guarantees, required extra or spare parts, and, perhaps most importantly, as-built drawings in the possession of multiple prime contractors.

As multiple prime contractors request final inspection of their work, the construction manager should establish inspection dates and conduct the inspections along with the owner and its architect. In the event multiple prime contracts require the multiple prime contractors to train the owner's personnel in the operation and maintenance of particular equipment, the construction manager should ensure that the owner's personnel have received competent training and are prepared to continue the operation and maintenance of the equipment.

Finally, the construction manager should be required to evaluate each multiple prime contractor's final payment requisition, negotiate all remaining change orders and extra work orders, and arrange for the owner to make final payment.

Coordination of Owner-Purchased Materials, Systems, and Equipment

Frequently, the owner will be purchasing specialized materials, systems, and equipment for installation in the project. Unless the owner intends to administer the procurement, delivery, storage, and installation coordination of such items, the owner should require the construction manager to assume those duties. Needless to say, it can be extremely costly, aside from being embarrassing, for the owner to delay the project by failing to arrange for the timely procurement and delivery of its special materials, systems, and equipment. Probably the most important component of this service is the identification and purchasing of long-lead procurement items. Additionally, the construction manager should be required to monitor the fabrication of such specialty, owner-purchased items, to ensure compliance with the project CPM schedule.

Government Liaison

During the course of project delivery on a phased, multiple prime construction project, by necessity the owner must have numerous contacts with federal, state, and local government agencies. On a federal level, the owner may monitor its prime contractors' compliance with various federal requirements, for example, regulations relating to equal employment opportunity, minority business enterprises, Small Business Administration contracts, and the Occupational Safety and Health Act (OSHA). Depending upon the nature of the project, there may well be various state requirements. During the past several years, for example, several Altanic City casinos have been constructed on a phased, mutiple prime contract basis. The New Jersey Casino Control Act contains a myriad of complex provisions with which owners have had to comply. Similarly, local government agencies have regulations affecting traffic and pedestrian control, site delivery of materials, building permits, special certificates relating to elevators and electricity, and temporary and final certificates of occupancy. The owner may want to require the construction manager to act as its government liaison. If so, the owner–construction manager contract specifically should address the extent to which the construction manager is required to identify applicable regulations and to monitor compliance therewith.

Labor Assistance

The owner–construction manager contract should require the construction manager to analyze the project locale with respect to the availability of labor during the expected project delivery time and the labor cost. In addition, the construction manager should carefully evaluate each multiple prime contractor's work scope to minimize labor jurisdictional disputes and union–open shop issues. The construction manager should advise the owner with respect to expected major unions' rate increases during anticipated project delivery time. Finally, the construction manager should be prepared to identify and monitor multiple prime contractors' compliance with government regulations affecting labor, for example, equal employment opportunity regulations, minority business enterprise regulations, Small Business Administration regulations, and Occupational Safety and Health Act regulations.

Safety Services

Owing to the well-known physical hazards of working in the construction industry, safety has become of paramount importance to owners, architects, construction managers, and contractors working on a project. Federal, state, and local governments all have various regulations affecting safety at a construction site. Therefore, the owner must decide which party is going to be responsible for the identification, evaluation, review, and monitoring of multiple prime contractors' safety programs on the project. Prior to the emergence of the construction manager within the construction industry, architects were generally responsible for safety. Architects were traditionally subject to suit only by those with whom they had contracted. Perhaps the single, most significant legal development with respect to architects' liability on construction projects was the abolition of the privity defense to suits by injured third parties.[70] The architects' privity defense initially was eroded in decisions resulting from cases brought by injured third-party plaintiffs alleging architects' failures to maintain safe construction sites.[71] Federal regulations, principally the Occupational Safety and Health Act and the Consumer Product Safety Act, have imposed complicated requirements affecting the construction industry.[72] As with duties relating to supervision and inspection and certification of payment requisitions, once architects found themselves increasingly liable for safety-related duties, architects through the AIA have attempted to withdraw from those duties and consequent liability.

Both the GSA CM contract and AIA CM contract place the entire responsibility for safety on the construction manager.[73] Therefore, unless the owner intends to assume safety duties for itself, in the owner–construction manager contract the owner should require the construction manager to handle all safety-related duties with respect to the project. The owner and construction manager should recognize, however, that such allocation of duties and risks places liability squarely on the construction manager. The majority of reported decisions construing a construction manager's liability in any respect in the construction context involve the construction manager's liability for injuries to persons at the project site. From these decisions it is now clear that a construction manager is subject to OSHA as administrator and coordinator of all phases of construction including the safety program even though none of its employees is engaged in actual, physical construction work.[74] The construction manager may also be liable for personal injuries sustained at the project site, depending upon its duty to the injured person, negligent performance of that duty, and the proximate cause of the injuries.[75] The construction manager's liability will be determined by the terms of its contract and its conduct at the project site.[76]

Project Security Services

Upon the commencement of project delivery, the owner must make a decision on how to provide for project security. Generally, the owner will contract with a security agency to guard the project during the course of construction. Nevertheless, the owner may consider providing in the owner–construction manager contract for the construction manager to develop a security program with the security agency and to supervise and monitor the enforcement of it. In particular, if a project encounters serious disputes between the owner and some of its prime contractors, and prime contractors threaten to abandon the job or the owner threatens to terminate the prime contractors, careful plans must be developed and precautions taken with respect to guarding the project and maintaining all items on the project site and within the possession, custody, and control of the owner to which the owner has title. In fact, it is prudent for the owner and its attorney to set forth provisions in the general conditions for each multiple prime contract with respect to title to any items to which title may be disputed in the event the threat of abandonment or termination arises during project delivery. Although it is difficult for owners, architects, construction managers, and multiple prime contractors to imagine at the decision, design, and initial delivery phases of a project that goodwill could deteriorate to the point where such issues might develop, those with experience in litigation in the construction industry all too often have encountered cases in which parties took direct, physical action to enforce their asserted rights supported by guards or local police. While no contract provisions can preclude a party from taking such action, well-drafted provisions delineating parties' rights under such circumstances can provide substantial assistance to owners and their counsel in obtaining prompt and comprehensive relief in the event such actions occur.

General Condition Work

The owner–construction manager contract should provide for some party to complete general condition work on the project, for example, general project cleanup. Frequently, the construction manager will provide such general condition work with its own forces. The owner–construction manager contract should contain provisions specifically delineating the work scope and should provide for the incorporation by reference of the general conditions. The owner should recognize that to a minimal extent this places the construction manager in a conflict position with the owner. Indeed, under a technically pure form of professional construction management, a construction manager would not even perform general condition work.

Additional Miscellaneous Services

The owner and construction manager may wish to negotiate an addendum to their contract setting forth additional miscellaneous services the owner may require of the construction manager and an appropriate adjustment of the construction manager's fee. Such additional services may include the following: services related to investigations, appraisals, or evaluations of existing conditions, facilities, or equipment; verification of the accuracy of existing drawings or other information furnished to multiple prime contractors; services relating to the negotiation of construction contracts modifying specific tenant or rental areas of the project; evaluation and assistance in negotiating insurance claims as a result of losses with respect to the project; recruiting and training of owner's maintenance personnel; expert witness services in the event of project litigation; and additional services as a result of the owner's termination of the project or default of a prime contractor. Such an addendum may include a general provision with respect to other services the owner may desire which are not specifically included in the contract.

OTHER KEY PROVISIONS OF THE OWNER–CONSTRUCTION MANAGER CONTRACT

The owner–construction manager contract should address other key areas. First, it should address the owner's responsibilities to the construction manager. Second, it should address the construction manager's fee and provide a definition of construction costs, including direct personnel expense and reimbursable costs, to the extent the construction manager's fee is related to the construction cost. Third, it should contain provisions relating to the payment of the construction manager's fee. Fourth, it should contain provisions relating to the construction manager's insurance and bonding. Fifth, it should contain provisions relating to the handling of disputes among the owner, architect, construction manager, and multiple prime contractors. Sixth, it should contain provisions relating to termination of the owner–construction manager contract. Finally, it is recommended that the owner and construction manager negotiate specific provisions relating to the handling of any conflicts which may develop between the owner and construction manager during the course of the project.

FOOTNOTES

[1] 5 Am. Jur. 2d *Architects*, sec. 8, p. 670 (1962); 19 Am. Jur. Trials *Architectural Malpractice*, sec. 17, p. 267 (1972).

[2] See, e.g., *Hanna v. Fletcher,* 231 F. 2d 469 (D.C. Cir.), *cert. denied sub nom. Gichner Ironworks v. Hanna,* 351 U.S. 989 (1956); *Paxton v. Alameda County,* 119 Cal. App.2d 393, 259 P.2d 934 (1953); *Laukkanen v. Jewel Tea Co.,* 78 Ill. App.2d 153, 222 N.E.2d 584 (1966); *Miller v. DeWitt,* 37 Ill.2d 273, 226 N.E.2d 630 (1967); *Inman v. Binghamton Hous. Auth.,* 3 N.Y.2d 137, 164 N.Y.S.2d 699, 143 N.E.2d 895 (1957).

[3] See generally Annot., *Tort Liability of Project Architect for Economic Damages Suffered by Contractor,* 65 A.L.R. 3d 249, (1975). See e.g., *Owen v. Dodd,* 431 F. Supp. 1239 (N.D. Miss. 1977); *A. R. Moyer, Inc. v. Graham,* 285 So.2d 397 (Fla. 1973); *C. H. Leavell & Co. v. Glantz Contracting Corp.,* 322 F. Supp. 779 (D.C. La. 1971).

[4] *First National Bank of Akron v. William F. Cann, et al.,* 503 F. Supp. 419 (N.D. Ohio 1980) *aff'd.,* No. 80-3484 (6th Cir. Jan. 28, 1982).

[5] Compare American Institute of Architects, Document B141, *Standard Form of Agreement between Owner and Architect,* July 1977, with American Institute of Architects, Document B141/CM, *Standard Form of Agreement between Owner and Architect,* construction management ed., June 1980.

[6] See, e.g., *Kurz v. Quincy Post No. 37, American Legion,* 5 Ill. App.3d 412, 283 N.E.2d 8 (1972); *Cobb v. Thomas,* 565 S.W.2d 281, Tex. Civ. App.-Tyler 1978, *writ ref'd n.r.e.* (Tex. Civ. App. 1978); Annot., *Building Contract-Cost Limitation,* 20 A.L.R.3d 778 (1968).

[7] See, e.g., *Stanley Consultants, Inc. v. H. Kalicak Constr. Co.,* 383 F. Supp. 315 (E.D. Mo. 1974); *Torres v. Jarmon,* 501 S.W.2d 369 (Tex. Civ. App. 1973); 20 A.L.R. 3d, sec. 4 (1968); 5 Am. Jur. 2d *Architects,* sec. 17 (1962).

[8] See, e.g., *Jay Dee Shoes, Inc. v. Ostroff,* 191 Md. 87, 59 A.2d 738 (1948).

[9] See, e.g., *Stanley Consultants, Inc. v. H. Kalicak Constr. Co.,* 383 F. Supp. 315 (E.D. Mo. 1974) *(dictum).*

[10] American Institute of Architects, Document B801, *Standard Form of Agreement between Owner and Construction Manager,* rev. ed., Apr. 15, 1975, art. 1.2.1.

[11] See, e.g., *Guerini Stone Co. v. P. J. Carlin Constr. Co.,* 248 U.S. 334 (1919); *Johnson v. Fenestra,* 305 F.2d 179 (3d Cir. 1962).

[12] See, e.g., *Hoel-Steffen Constr. Co. v. United States,* 197 Ct.Cl. 561, 456 F.2d 760 (1972); *Baldwin-Lima-Hamilton Corp. v. United States,* 434 F.2d 1371 (Ct. Cl. 1970); *Natkin & Co. v. George A. Fuller Co.,* 347 F. Supp. 17, 35–36 (W.D. Mo. 1972); *Paccon, Inc. v. United States,* 185 Ct.Cl. 24, 399 F.2d 162 (1968); *L. L. Hall Constr. Co. v. United States,* 117 Ct.Cl. 870, 878, 379, F.2d 559, 564 (1966); *Hoffman v. United States,* 166 Ct.Cl. 39, 340 F.2d 645 (1964); *J. A. Jones Constr. Co. v. Dover,* 372 A.2d 540 (Del. Super.), *appeal dismissed,* 377 A.2d 1 (Del. 1977); *Forest Electric Corp. v. State of New York,* 30 A.D.2d 905, 292 N.Y.S.2d 589 (1968); *Pennsylvania v. S. J. Groves & Sons Co.,* 20 Pa. Commw. Cts. 526, 343, A.2d 72 (1975); *Gasparini Escav. Co. v. Pennsylvania Turnpike Comm'n,* 409 Pa. 465, 187 A.2d 157 (1963); *Henry Shenk Co. v. Erie County,* 319 Pa. 100, 178 A. 662 (1935). Contra, *Edwin J. Dobson, Jr., Inc. v. Rutgers State Univer.,* 157 N.J. Super. 357, 384 A.2d 1121 (1978) *aff'd,* 180 N.J. Super. 350, 434 A.2d 1125 (1981) (hereinafter cited as *Dobson*). See also an interesting line of cases construing New York's law requiring multiple prime contracts on public projects and precluding the owner from delegating the duty to coordinate to any other party. *General Building Con. v. Syracuse,* 40 A.D.2d 584, 334 N.Y.S.2d 730 (1972), *aff'd,* 32 N.Y.2d 780, 344 N.Y.S.2d 961, 298 N.E.2d 122 (1973); *Forest Elect. Corp. v. State of New York,* 30 A.D.2d 905, 292 N.Y.S.2d 589 (1968).

[13] See, e.g., *Dobson,* note 12.

[14] General Services Administration, Public Building Service, *Standard Construction Management Contract,* rev. ed., Apr. 15, 1975, art. 3(a)(2), p.6.

[15] Ibid., art. 4, pp. 7–12.

[16] Ibid., art. 14, p. 19.

[17] See note 12.

[18] *Casson Constr. Co. v. United States,* (1978) Contract Appeals Decisions (CCH), par. 13,032, p. 63,559 (GSBCA No. 4884, Feb. 10, 1978); *Pierce Assoc., Inc. v. United States* (1977) Contract Appeals Decisions (CCH), par. 12,746 p. 61,937 (GSBCA No. 4163, Aug. 18, 1977).

[19] General Services Administration, Public Building Service, *Standard Construction Management Contract,* art. 3(c)(2), p. 6.

²⁰ Interview with David R. Dibner, Assistant Commissioner for Design and Construction, General Services Administration, Washington, D.C., by telephone, Dec. 22, 1980.

²¹ General Services Administration, Public Building Service, Order No. ADM/P/5450.39B, Oct. 20, 1980.

²² Dibner interview.

²³ General Services Administration, Public Building Service, *Construction Contracting Systems, A Report on the Systems used by PBS and Other Organizations,* Mar. 17, 1970, pp. 2-33–2-37 and 4-30–4-31.

²⁴ American Institute of Architects, Document B801, *Standard Form of Agreement between Owner and Construction Manager,* arts. 1.2.2 and 1.2.2.2.

²⁵ American Institute of Architects, Document A201/CM, *General Conditions of the Contract for Construction,* construction management ed., June 1980, art. 6.1.3.

²⁶ Ibid., arts. 3.3.1 and 3.4.3.

²⁷ *John E. Green Plumbing & Heating Co. v. Turner Constr. Co.,* 500 F. Supp. 910 (E.D. Mich. 1980). See also *R. S. Noonan, Inc. v. Morrison-Knudsen Co.,* 522 E. Supp. 1186 (E.D. La. 1981).

²⁸ *Green v. Turner,* p. 913.

²⁹ Ibid., p. 911.

³⁰ Ibid., p. 913.

³¹ *Dobson,* note 12.

³² Ibid., p. 397, 384 A.2d, p. 1142.

³³ Ibid., p. 393, 384 A.2d, p. 1142 (emphasis omitted).

³⁴ Ibid., p. 407, 384 A.2d, p. 1146.

³⁵ The court also ruled that neither the CPM expert nor the architect was an agent of the owner with respect to coordination. The court reasoned that the CPM expert was an independent contractor similar to an independent testing laboratory and that it was merely to furnish advice and its expertise as a tool to the primes in developing an effective schedule pursuant to which the contract completion date could be met. Similarly, the court ruled that the architect was not an agent of the owner and had no duty to coordinate. The court relied heavily on the architect's "neutral role" in advising the owner on what amounts, if any, to withhold from primes' retainages as a result of the owner's claims and other primes' claims. Moreover, the court found that the architect was under a duty to advise the owner and the other parties as to the progress and status of the work, but that he had no power to enforce compliance with the schedule for completion agreed upon by the primes. Ibid., pp. 390, 406–407, 384 A.2d 1137–1138, 1146.

³⁶ Apparently part of the construction manager/general prime's strategy in not asserting a third-party beneficiary cause of action against the other primes was to support its position that it had no power to enforce its alleged duty to coordinate the project and, consequently, that the owner, in fact, retained the effective obligation to coordinate and the potential liability for failure so to do.

³⁷ Ibid., p. 407, 384 A.2d, p. 1146.

³⁸ Ibid., p. 403, 384 A.2d, p. 1144.

³⁹ Ibid., p. 406–407, 384 A.2d, p. 1146.

⁴⁰ Ibid., p. 408, 384 A.2d, p. 1147 (emphasis in original).

⁴¹ *Restatement of Contracts,* sec. 133, pp. 151–152 (1932).

⁴² *Dobson,* note 12, p. 412, 384 A.2d, p. 1149.

⁴³ *Dobson* was affirmed in 1981 by the New Jersey Superior Court, Appellate Division. In an additional decision, *R. S. Noonan, Inc. v. Morrison-Knudsen Co.,* 522 F. Supp. 1186 (E.D. La. 1981), the court found an owner and construction manager liable to a prime contractor for failing to coordinate the use of the project site to afford the prime contractor adequate access.

⁴⁴ 5 Am. Jur. 2d *Architects,* sec. 1, p. 662 (1962).

[45] G. Wayne Murphy, "Liability of the Architect to the Construction Surety before and after the 1976 Edition of AIA Document A201," speech delivered to the American Bar Association Section of Insurance, Negligence and Compensation Law, Fidelity & Surety Law Committee, Mid-Winter Meeting, January 1978, New York, N.Y.; Note, "Architectural Malpractice: A Contract-Based Approach."

[46] Harvard note. *Harvard Law Review*, vol. 92, p. 1086 (hereinafter cited as Harvard note).

[47] Ibid.

[48] Murphy, op. cit., p. 13; Harvard note, p. 1086.

[49] Harvard note, p. 1086.

[50] Ibid.

[51] General Services Administration, Public Building Services, *Standard Construction Management Contract*, art. 14, p. 19.

[52] American Institute of Architects, Document B801, *Standard Form of Agreement between Owner and Construction Manager*, arts. 1.2.7 and 1.2.7.1.

[53] Ibid.

[54] Ibid., art. 1.2.7.1.

[55] American Institute of Architects, Document 201/CM, *General Conditions of the Contract for Construction*, art. 3.3.1.

[56] *First National Bank of Akron v. William F. Cann, et al.*, 503 F. Supp. 419 (N.D. Ohio 1980) *aff'd.*, No. 80-3484 (6th Cir. Jan. 28, 1982).

[57] Ibid., p. 423.

[58] Ibid.

[59] Ibid., p. 435.

[60] Ibid., p. 436.

[61] Ibid., p. 437.

[62] Ibid., p. 438.

[63] Ibid., p. 437.

[64] Ibid., p. 439.

[65] *Murphy*, op. cit., p. 16.

[66] Ibid., p. 17.

[67] See, e.g., *Aetna Insurance Co. v. Hellmuth, Obata & Kassabaum, Inc.*, 392 F.2d 472 (8th Cir. 1968); *Hall v. Union Indemnity Co.*, 61 F.2d 85 (8th Cir.), *cert. denied*, 287 U.S. 663 (1932); *Peerless Ins. Co. v. Cerny & Assocs.*, 199 F. Supp. 951 (D. Minn. 1961); *Calandro Dev., Inc. v. R. N. Butler Contractors, Inc.*, 249 So.2d 254 (La. 1971); *State for the Use of National Surety Corp. v. Mulvaney*, 221 Miss. 190, 72 So.2d 424 (1954); *Westerhold v. Carroll*, 419 S.W.2d 73 (Mo. 1967). See also Annot., *Liability of Architect or Engineer for Improper Issuance of Certificate*, 43 A.L.R. 2d 1227. But compare *Blecick v. School District*, 2 Ariz. App. 115, 406 P.2d 750 (1965).

[68] American Institute of Architects, Document A201, *General Conditions of the Contract for Construction*, August 1976, art. 9.4.2.

[69] Ibid.

[70] Harvard note. p. 1081.

[71] See, e.g., *Miller v. DeWitt*, 37 Ill. 2d 273, 226 N.E.2d 630 (1967); *Inman v. Binghamton Hous. Auth.*, 3 N.Y.2d 137, 143 N.E.2d 895, 164 N.Y.S.2d 699 (1957).

[72] Occupational Safety and Health Act, 29 U.S.C. 651, *et seq.*; Consumer Product Safety Act, 15 U.S.C., sec. 2051, *et seq.*

[73] General Services Administration, Public Building Service, *Standard Construction Management Contract*, art. 17; American Institute of Architects, Document B801, *Standard Form of Agreement between Owner and Construction Manager*, art. 1.2.4.

[74] *Bechtel Power Corp. v. Secretary of Labor*, 548 F.2d 248 (8th Cir. 1977).

⁷⁵ See, e.g., *DiSalvatore v. United States,* 456 F. Supp. 1079 (E.D. Pa. 1978); *Everette v. Alyeska Pipeline Serv. Co.,* 614 P.2d 1341 (Alaska 1980); *Hammond v. Bechtel, Inc.,* 606 P.2d 1269 (Alaska 1980); *Mackey v. Campbell Constr. Co.,* 101 Cal. App.3d 774, 162 Cal. Rptr. 64 (1980); *Cumberbatch v. Bd. of Trustees,* 382 A.2d 1383 (Del. Super. 1978); *Lemmer v. IDS Properties, Inc.,* No. 50327 and 50463 (Minn., Oct. 24, 1980).

⁷⁶ Ibid.

CHAPTER THREE

CONFLICTS IN
THE CONSTRUCTION
MANAGEMENT PROCESS

The previous chapter has suggested many services and responsibilities that a construction manager may undertake for the owner. At the same time, there are pitfalls in the process which must be guarded against. Construction management will have the same types of conflict among owner, architect, contractor, and subcontractors as in traditional construction—disputes over delays, additional costs, unforeseen conditions, bid errors, poor workmanship, etc. However, construction management also creates the opportunity for some substantial conflicts of interest to arise. What these are and how they can be dealt with are discussed below.

The relationship between an architect or a professional construction manager and the client is, in essence, a fiduciary relationship existing between an agent and a disclosed principal. As defined in *Black's Law Dictionary*, a fiduciary is "a person having a duty, created by his undertaking, to act primarily for another's benefit in matters connected with such undertaking." A fiduciary relationship is one "founded on trust or confidence reposed by one person in the integrity and fidelity of another. . . . It exists where there is special confidence reposed in one who in equity and good conscience is bound to act in good faith and with due regard to interests of one reposing the confidence."

A fiduciary relationship may exist as a matter of common law, such as between an attorney and client, or as a matter of statutory or administrative law, such as where a specific charge to serve the client's interests is found in architectural and engineering licensing laws or regulations. This special relationship may also arise as a matter of contract: for instance, both Associated General Contractors construction management agreement forms (reproduced in Chapter 4) provide in the first paragraph that "the construction manager accepts the relationship of trust and confidence established between him and the owner under this Agreement." AIA Document B801, *Standard Form of Agreement between Owner and Construction Manager* (reproduced in Chapter 4), and the AGC forms continue with language to the effect that the construction manager covenants with the owner to furnish his best skill and judgment in furthering the interests of the owner. Such language is similar to that which has historically been contained in AIA's cost-plus construction contract (AIA Document A111, *Standard Form of Agreement between Owner and Contractor,* reproduced in Chapter 4), whereby the contractor, by virtue of having a cost-plus contract, is charged with using resources in the best overall interests of the owner. Finally, a fiduciary relationship may arise by implication out of dealings between the parties which establish an agency relationship.

The agent acting for a disclosed principal will have the power to bind the principal to commitments made by the agent, to the extent of the agent's apparent authority in the eyes of third parties. The agent who discharges his or her duties faithfully and in a nonnegligent manner will not be liable to the client or to third parties for actions taken by the agent on behalf of the principal. Because the fiduciary relationship is one in which the principal has placed a degree of trust and reliance in the agent, it demands, in return, the utmost good faith, honesty, and loyalty on the part of the agent. The degree of loyalty involved in different kinds of relationships will vary with the level of trust and confidence placed by the principal in, and the extent of authority of, the agent or quasi-agent. In some cases, the agent is bound to act in the manner that best serves the principal and must avoid taking any financial advantage; in others, the agent is permitted to act in any number of ways to serve his or her own ends, so long as the interests of the principal are not prejudiced. The former situation is that the trustee, while a contractor working under a cost-plus arrangement falls into the latter. Regardless of the degree of confidence and loyalty required, conflicts of interest can arise.

Generally, conflicts of interest fall into two categories: those that can be cured by disclosure and those that are incurable (and therefore, intolerable). Incurable conflicts are condemned because the existence of a conflicting loyalty, no matter how slight, is inconsistent with the degree of trust placed in the agent by the principal. For example, it is entirely wrong for a judge who is a stockholder of a company to sit on a case involving that company. A person charged with a public trust should not be in a position to benefit from his or her decisions, lest those decisions become suspect. A curable conflict, so far as the judge is concerned, could be one where the judge previously owned stock in the company, or had the company as a client while practicing law, or had otherwise been employed by the company. Here, disclosure to the litigants, in and of itself, would probably be sufficient to dispel any suggestion of a conflict—particularly if there is no objection from the litigants. The judge may have a bias, having had a previous relationship with the company, and that bias may be so strong as to be objectionable; but this conflict does not rise to the level of the one previously cited because the judge will have nothing tangible to gain or lose by his or her decisions in the case.

Conflicts of interest will arise in the construction management process where one party has been entrusted with the responsibility for making judgments—including judgments on behalf of another and judgments of an impartial nature—and where the party stands to receive a tangible benefit, directly or indirectly, as a result of such decisions.

PERCENTAGE FEES AND COST-SAVINGS INCENTIVES

One of the most longstanding potential conflicts of interest for the architect, or, more recently, for the contractor (as part of a cost-plus construction contract) or for the construction manager is found in percentage fees. Although this fee arrangement satisfies the test of a conflict of interest, the percentage-of-cost basis for paying a fee has seldom been strongly criticized as a conflict of interest, having wide acceptance in other professions and pursuits such as real estate brokerage and interior design. A percentage fee has only an indirect relationship to the amount of work required; it assumes that a more expensive project requires more services than a less expensive one. The percentage fee provides a windfall, not related to the amount of work done, if the cost of the project escalates. Less likely, but also occurring on occasion, is a fee penalty where the project cost comes in below expectations. The architect designing a building on a percentage-fee basis will be in a position to secure extra compensation on account of its design decisions: the choice of more expensive equipment and materials will be rewarded, and any conscious effort to minimize costs will be penalized. A construction manager working on a percentage-fee basis is in a similar position: efforts to economize are rewarded by a reduced fee, and there will be a "bonus" if construction costs are high. Is this the type of a conflict that requires disclosure? Generally not. Because the percentage-fee arrangement is a historical one, its benefits and pitfalls are well known and easily perceived. Moreover, it is frequently the client who suggests or chooses to use a percentage fee. There is generally no "secret" which is withheld by not disclosing the conflicts inherent in such a fee arrangement.

How can the conflicts inherent in the use of a percentage fee be avoided? The client can take some solace in that there are probably few design professionals who take advantage of a percentage fee by running up construction costs in order to increase their own fees. Design professionals do not make their fortunes on a single project, and generally, they look to their clients for future business. Overly costly projects do not win such repeat work and often result in premature termination of the designer's services. Ultimately, running up costs is self-defeating for the design professional.

Construction managers and contractors who are not familiar with percentage-fee work, however, may find themselves tempted by the opportunity to neglect economies to their own benefit. And whereas the architect on a 6 percent fee stands to gain $6000 for every $100,000 overrun, the cost-plus contractor with a typical 10 percent fee not only gains $10,000 in additional compensation, it also does an extra $100,000 of business. The client who has any concern over "gliding the lily" by architect, construction manager, or contractor should consider a fixed fee, or a maximum dollar limit on a percentage fee (based on an estimated maximum cost of construction) to ensure that abuses are not rewarded by higher fees. Such a fee limitation will not, in itself, encourage economy and efficiency—it will only limit ultimate costs. Shared-savings provisions, which provide a positive incentive, are frequently found in construction contracts, but they are extremely rare in professional services agreements. Design professionals are expected by their clients to use their best efforts on behalf of their clients to contain costs—and most expect it of themselves without the need for additional encouragement—while contractors, in general, are not expected to take any measures to benefit the project owner unless some extra consideration is provided in return. As long as the contractor is within any guaranteed maximum price which it has given, its obligation under the contract to keep costs at a minimum will be difficult to enforce. This illustrates the difference between a fiduciary relationship and one which is, at best, a quasi-fiduciary one: an extra incentive for saving costs is inconsistent with the basic assumptions of a fiduciary relationship.

It is likely that a construction manager who has a background as a contractor will expect to receive a bonus from any cost savings achieved in the project. This can create conflicts in reverse and can become the basis for additional disputes with the project designers: such a construction manager may have an interest in reducing the quality of the project in an attempt to reduce the initial construction costs (and to share in cost savings). The architect or engineer may consider some changes or substitutions as being uneconomical over the life of the building, when operating, maintenance, and replacement costs are taken into account. In such cases, the project owner will have to become a referee and determine whether its interests are best served by lower initial costs or lower long-term costs. If the construction manager has participated in the early design of the project (before establishing a base cost for the purpose of a shared-savings provision), the client might well question whether the construction manager's suggestions for cost savings could have been raised earlier. The client can avoid doubts as to the motives of the construction manager by establishing a true fiduciary relationship at the outset and maintaining that relationship throughout construction untainted by the prospect of personal gain or loss as a result of decisions or recommendations.

COST-REIMBURSABLE CONTRACTS

The second area of conflict can be found in the cost portion of any cost-plus contract. For professionals, construction managers and contractors alike, there may be some benefit to having a larger volume of work, even if the fee portion of the contract is fixed. The consultant or contractor who has been provided by the client with a cost-plus contract also has something of a fiduciary obligation to the client and should not needlessly or deliberately run up costs—particularly, it should not take advantage of the contract to keep otherwise idle forces busy.

Most cost-plus contracts are written so that they will be subject to an upper limit of cost, sometimes referred to in construction contracts as a GMP or guaranteed maximum price. Although the GMP may set an upper limit on the price to be paid by the client, this price will likely be higher than what might be obtained by competitive bids. What incentive, if any, is there for the contractor or consultant to keep costs significantly under the maximum limit provided in the contract?

Some contracts attempt to control costs within the upper limit by establishing strict definitions of what are to be reimbursed as cost items. Any costs (normally overhead items) incurred by the contractor or consultant outside of the allowable costs must be absorbed in the fee portion of the compensation. Such descriptions of allowable costs can be found in the forms of agreement published by AIA and AGC. Allowable costs should specifically exclude costs arising out of the negligence of the contractor or consultant, in order to ensure some fiscal responsibility. As discussed previously, incentive clauses allowing the contractor or consultant to share in any savings if costs are kept below a certain dollar level can create conflicts, and their potential benefit must be viewed in light of those conflicts. Historically, professional consultants have not been subject to strict definitions of cost, nor are they offered incentive provisions, as their fees are small compared with the costs of construction.

One example of a conflict of interest arising out of a cost-plus contract would be where the contract provides for reimbursement of rental costs of equipment leased for use on the project, but requires that the capital or ownership costs of the contractor's own equipment be absorbed in the fee as a part of the contractor's overhead—a typical arrangement. Many contractors, for business and tax reasons, purchase equipment through one corporate identity and lease it to themselves in another. Under a cost-plus contract, this leasing arrangement allows the contractor to recover a cost (as rent) that otherwise (as capital) would have had to be absorbed as an overhead item. The contractor has three choices—lease from an unrelated outside party, lease from itself, or purchase. Would there be a conflict of interest if the rental charges between the contractor and its leasing arm were comparable to rates in the marketplace? After all, the client would not be paying any more than if the equipment were leased from a third party.

The test of a conflict of interest in a fiduciary relationship is not whether the client is being placed at a disadvantage. The test is whether the contractor has a financial interest in the outcome of a decision which he is expected to make on behalf of the client or some third party. Even if the contractor rents the scaffolding, formwork, or truck to himself at no more than the market rate, he still has a conflict of interest. The conflict exists because his obligation is to serve the client faithfully as an agent and to advance the client's interests where possible—not simply to avoid cheating the client. At the very least, disclosure by the contractor that he intends to lease equipment from his own subsidiary is required. And then, it might be appropriate to reduce the allowable rental charges to an amount that represents the actual costs of ownership (taking into account financing costs and depreciation), so that the contractor is not, in essence, making a profit as a result of this decision to lease equipment to himself.

A cost-plus contract can provide other opportunities for conflicts of interest. Although the contractor will have a duty to use the most efficient means and to protect the interests of the client as if they were its own, it may be faced with conflicting allegiances. For example, is a cost-plus contractor properly serving its client's interests by subcontracting work out on a cost-plus basis when it could obtain fixed prices? A contractor who has a relationship with a particular subcontractor or supplier on more than one project is in the position to reward that subcontractor or supplier on the cost-plus project, at the client's expense, in exchange for favorable prices on another project where the contractor is providing a fixed price.

This is the type of conflict that can be protected against, but perhaps not cured, by a requirement that the contractor disclose whether any subcontractors or suppliers on the project are also working for it on other projects at the same time, or whether the contractor has any unsettled claims from or indebtedness to any subcontractor or supplier to be used on the project. Also, a requirement that the contractor obtain competitive subbids for portions of the work, with final award of subcontracts subject to the client's approval, can go a long way toward avoiding such conflicts.

Although the owner is inviting conflicts of interest if it engages a contractor or a construction manager to perform construction work on a cost-plus basis, the architect/engineer (A/E) and the professional construction manager acting strictly as agent for the owner can be faced with similar (but less severe) conflicts when working on a cost-plus basis. They generally are trusted to take no more time than is necessary on the project, consistent with their normal professional approach to their assignments, while at the same time, they are expected to spend all the time required to do a satisfactory job. The A/E or the professional construction manager may have idle employees that could be "loaded on" to the project, they may be leasing space from a controlled or related entity, and they may well have the same consultants on several projects. The same type of conflicts may arise; similar remedies may be utilized to cure or avoid such conflicts.

REVIEW OF DESIGN DOCUMENTS

Some commentators have suggested that a third conflict of interest in construction management arises when the review of the architect or engineer's drawings for completeness, economy, and constructability is done by a construction manager who is the same as or is owned or controlled by the architect or engineer. The suggestion is made that, when acting as the construction manager, the A/E will not be likely to identify errors, much less require changes in the design if doing so would cost any time or money or would expose any negligence on the part of the A/E. Therefore, this conflict prevents the architect–construction manager from serving the best interests of the client.

This argument, unfortunately, is based on a faulty premise—one which can and should be avoided in developing contracts. This faulty premise is that the construction manager is undertaking to review the design drawings as if it were an independent, third-party construction manager. Obviously, an "independent" review is not a discrete service that a construction manager who is also the architect or engineer will be providing or should be expected to provide. The client has selected an A/E who is also providing construction management services; the A/E has an original obligation to design the project in a way which will be efficient and economical to construct, and presumably has the ability to do so. If the client desires to have an independent check of the design documents, that service must be obtained from an independent party.

Construction management involves any number of different skills, capabilities, and services—and whether a conflict of interest arises for an architect or engineer who also provides construction management services to the client depends on whether the particular services expected of the construction manager require an independent third party. The suggestion that an A/E has a conflict of interest when acting as a construction manager assumes that the construction manager *necessarily* will be undertaking an independent review of the A/E's design documents. Such a "conflict of interest" is similar to the "conflict" that arises when a set of construction contract documents requires the contractor, on substantial completion of the work, to develop a punch list of items to be completed or corrected. Clearly, the client is not expecting the contractor to provide the same independent advice as is provided by the architect or engineer who independently develops a punch list. One cannot be expected to give totally independent evaluation of one's own work in the first place; thus, there is no conflict unless the client places the consultant or contractor in that position.

It goes without saying, however, that the client should be fully aware of any financial relationships between the A/E and the construction manager.

DECISIONS ON DISPUTES

A more persuasive argument can be made as to the existence of conflicts of interest in a situation common to both traditional construction and the construction management process; this involves the circumstance where the designer (in a traditional capacity or where acting as construction manager) is called upon to decide a claim based on an alleged error, omission, or inconsistency in its own drawings and specifications. The conflict is clear: if an impartial, objective judgment were to sustain the contractor's claim, it would also tend to set the blame on the designer, who might be liable to the client for any extra costs awarded to the contractor. Various commentators have long maintained that an A/E cannot be impartial when judging claims which call its own work into question.

This apparent conflict could reach abusive levels if the A/E were to exercise the power to order minor changes in the work, or to approve substitutions in order to reduce contract requirements for the contractor, in exchange for abandoning a claim which would clearly merit a cost increase. "Horse trading" on contract requirements will occur to a certain extent on any project, and can simply represent a fine tuning of the design during the construction process where contract requirements in some areas, which turn out to be needless or noncritical, are used to balance additional requirements elsewhere. If done in a manner which is properly authorized, documented, and aboveboard, this can serve the client's best interests. In a properly administered contract, the client will receive notice of all such changes.

However, where the changes are being authorized informally between the architect and contractor, the opportunity arises for the architect to overlook the contractor's poor workmanship in exchange for the contractor's not pursuing a claim for an extra because the architect made an error or omission. This type of abuse can be encouraged by the project owner who insists that there be no change orders and who makes it clear that the A/E will be expected to pay for any extra costs resulting from an error or omission. When such a client refuses to perceive that designing a building is not an exact science and holds its consultants to a standard requiring a perfect result, it is more likely that imperfect judgments will never be brought to its attention.

Even where the A/E is simply called upon to decide impartially a dispute between owner and contractor, there can be an apparent conflict of interest: that is, the A/E is being called upon to decide impartially a dispute between its client and the contractor while being compensated by one of the parties. Although this has been a traditional function of the project designer, contractors have long argued that the benefit of any doubt will always go to the client and that the contractor is at a disadvantage. Although the architect's decisions are subject to arbitration or may become the subject of litigation if the contractor pursues the question, an arbitrator, like a court reviewing the arbitrator's decision, is entitled to accept the judgment made by the architect unless it is based on a mistake or fraud. Sometimes, decisions by an architect which are clearly erroneous are upheld because of the narrow grounds afforded for overturning a decision.

Architects and engineers will rarely, if ever, acknowledge or agree that deciding disputes between contractors and their clients involves a conflict of interest. In practice, they are accustomed to calling the shots as they see them, with a primary interest in achieving what is best for the project itself rather than what may be best for client or contractor. The general acceptance by owners and contractors of this role for architects and engineers, combined with the quasi-judicial immunity afforded to A/E's by the courts, and the availability of arbitration and litigation as appeal mechanisms, all combine to make this role an impartial and effective one in practice. Perhaps because the project owner is ultimately paying all costs of the project in any event, it makes no difference that the owner and contractor do not equally contribute to the A/E's fee for dispute-settling services.

The practical absence of a conflict in this situation is not hard to explain. The interest that the architect or engineer has in keeping on good terms with the client (which could be in conflict with an obligation to decide a dispute impartially) is likely to be counterbalanced by an interest in maintaining the respect and cooperation of the contractor through the duration of the project, and likely on other projects in the future. The architect may have some bias, of course—for instance, the architect may be especially beholden to a long-standing client or may have a close relationship with the builder. The wise client will take pains to determine what allegiances the project referee may have.

Where the architect is called upon to make a decision involving the adequacy of its own documents or services, the test of a conflict of interest is met, although indirectly so. For instance, assume that the contractor makes a claim for additional costs, based on an allegation that the architect has failed to process and approve shop drawings within a reasonable time. If this allegation is true, the architect (regardless of whether it has used due diligence in processing those shop drawings) should grant the contractor the extra costs. But by so ruling, the architect is exposed to the owner's claim against the architect for the amount of those extras. If the architect has been negligent, it may well be responsible for the extra costs incurred. However, if the architect has used due diligence, it should not have any liability to the owner (the client) notwithstanding the delay.

A second example might involve an error or omission in the construction documents. Assume that the architect has omitted any requirement for reinforcing steel in a concrete slab. During construction, the architect discovers the omission and issues a correction to the drawings. The contractor demands a change order reflecting the additional cost for the reinforcing and costs for delay. The architect, reasoning that the owner would have had to pay extra for the reinforcing in any event, approves the extra cost for the reinforcing but denies any delay costs.

In both situations described, the architect is faced with a potential conflict between an obligation to render an impartial decision and its own pecuniary interests. If the architect decides against the contractor, the decision will likely be appealed to arbitration, resulting in even greater costs to the owner. A decision made in bad faith could eliminate any quasi-judicial immunity and render the architect liable to both the owner for the costs of defending that decision and to the contractor for the costs of challenging it. On the other hand, if the architect decides against the owner and grants the contractor an extra, it risks a great embarrassment at the least, and possibly the confidence of the client and a claim for the deficiency as well. If the architect's conduct was not negligent but nevertheless caused extra costs, liability for those costs may be avoided, but the architect will certainly not have strengthened its relationship with the client. And if the architect's conduct was negligent, it will be difficult for the architect to admit this to the client without jeopardizing the architect's liability insurance coverage. Thus, even disclosure of the conflict involves a conflict.

The architect faced with judging the adequacy of its own product or services should not be rendered impotent thereby. The client should be advised by the architect that a decision against the contractor might be challenged on the grounds that it was made in good faith. After advising the client, the architect should then proceed to render, insofar as possible, a decision without partiality to either the client or the contractor. If the architect was in fact negligent, this fact will likely come out sooner or later, and there will be nothing to gain and even more to lose by allowing potential exposure to govern the architect's decision. In the final analysis, then, there are counterbalancing factors which minimize this apparent conflict of interest. In fact, the courts have failed to establish any precedent that an architect is incapable of serving as arbiter of disputes between the owner and contractor simply because the architect's own performance has been questioned.

If an architect has taken the way out of trading some of the contractor's deficiencies off against its own, the architect may have attempted to rationalize this approach. The architect may honestly believe that the project owner is not being placed under any disadvantage; in fact, the owner may be getting more, for less money, than if a higher initial contract price was paid for a project which included the omitted items. What the architect here overlooks (as did the contractor who, in the earlier example, leased equipment to itself) is that the fiduciary duty to the client goes beyond keeping the client from being disadvantaged. The architect owes the client the opportunity to make the decisions on what should be left in or out of the project and what deficiencies will be excused or overlooked.

The construction manager, depending on the role that is provided for it, will be faced with similar conflicts. Normally, the professional construction manager will have the primary responsibility for estimating, scheduling, coordinating, and inspecting construction. It will also have responsibility for processing payment applications from contractors and either approving them for payment or recommending approval for payment to the architect. In addition, the construction manager will normally have some responsibility for reviewing claims for additional cost, requests for change orders, and the like. The professional construction manager can find itself in exactly the same position as the architect in the previous examples. Assume, for instance, that a contractor on the project makes a claim for delay damages that is based on an alleged failure by the construction manager to schedule or coordinate the work properly. The professional construction manager must then investigate, evaluate, and make a recommendation (if not a decision) on the claim. It has the same choices as did the architect: approve the extra and face a claim from the project owner; deny a meritorious claim and risk being overturned in a subsequent arbitration, after an even greater expense to the owner; or attempt to resolve the claim (while avoiding any liability) by accepting lower-quality work or otherwise waiving contract requirements. The best course of action for the construction manager is also the same—to admit one's fallibility, give the contractor what is properly due, and either hope for the project owner's understanding or be prepared to face the consequences.

The important point for the project owner to understand is that the professional construction manager, like the architect, is providing services on behalf of the client which involve a high degree of informed judgment. Their decisions will not always turn out to be correct, and some may cost the client additional money. If the professional A/E and construction manager were expected to bear the financial risk of every act, judgment, decision, error, or omission they made, they would not be agents of the owner working for a fee; they would be in an entrepreneurial position which would require much higher rewards for success.

CONSTRUCTION MANAGER ACTING AS CONTRACTOR

The most significant conflict of interest in construction management is found where the construction manager undertakes, in providing services on the project, to protect the project owner's interests while at the same time providing construction labor and materials for the project on which the construction manager is deriving a profit. Although there appears to be a growing trend toward the use of professional construction management services, contractor construction management is currently the most prevalent approach. The potential conflicts of interest involved are clear and classic.

Consider a contractor–construction manager whose normal mode of operation is to "broker" all the construction work on the project. The construction manager negotiates prices for all the subcontracted work and gives the project owner a guaranteed maximum price for the project based on these negotiated subcontract prices. The guaranteed maximum price is subject to any number of exceptions—essentially, whenever one of the subcontractors is awarded an extra, the GMP is increased by a like amount. Some of the subcontract items are handled under separate, direct contracts between the owner and subcontractors (acting as prime contractors) whose work is being coordinated by the construction manager. These subcontract items may even be assigned back to the construction manager by the project owner. As to other subcontract items, the construction manager will act as the general contractor.

Should the construction manager have the discretion to determine what work is handled under one of the multiple prime contracts with the owner, and what construction work, if any, will be performed by the construction manager or under contract to the construction manager? Should the construction manager be entitled to perform any portions of the work it desires, so long as the cost does not exceed the price quoted by the lowest bona fide bidder? If a subcontractor defaults, should the construction manager be required to complete the work within the original GMP? What should the role of the construction manager be in reviewing and approving payments to subcontractors, and who, if anyone, will review and approve payments to the construction manager for the construction work it performs and for work performed by the construction manager's subcontractors?

Consider the first questions. Is there an opportunity for a conflict of interest when the construction manager is given the option to assign work to the project owner or to undertake it itself for the same price? Undoubtedly, the existence of that option removes any assurances that the bids provided by others were bona fide. Presumably, the construction manager will only be undertaking work that it knows can be performed for less than the price quoted by the outside contractor; how can the owner be assured that the construction manager has gotten the best price for the owner from the outside contractor? In addition to having a vastly better knowledge of the project, the construction manager also has the ability to manipulate project costs to ensure that the portions it undertakes will be profitable to it. The construction manager's management decisions on behalf of the owner will directly affect its profit or loss; the construction manager has the opportunity to steer profitable construction work to itself, should it choose.

If there are portions of the work on which a bid cannot be obtained, the construction manager normally will perform those portions on a cost-plus basis. When the construction manager itself has had the responsibility for preparing bid packages and soliciting bids, however, can the owner even be sure whether a desire on the construction manager's part to perform part of the work with its own forces (which provides it not only a certain volume of work but also a predictable profit) had anything to do with the apparent inability to secure an outside contractor?

Then there arise questions of inspection of the work. Is the owner confident that the contractor–construction manager will perform its own work (or the work that it has subcontracted out on its own account) satisfactorily? Can the contractor–construction manager be expected to apply the same standards in inspecting its own work as it does in inspecting the work of others?

When it comes to reviewing payment requests, the construction manager will collect payment applications from the separate prime contractors and from its own subcontractors, as well as prepare its own payment request. Can the contractor–construction manager truly be expected to approve, on behalf of the owner, payments to itself or its own subcontractors?

Can the contractor–construction manager be expected to fairly decide disputes between its own subcontractors and the separate prime contractors working directly for the owner? If the construction manager has given a guaranteed maximum price but has the opportunity to control quality of the work, will the construction manager allow quality to suffer in order to keep from exceeding its GMP?

These conflicts are not a type that can be cured by disclosure. They can be controlled, to a certain extent, by contract language clearly defining allowable costs in cost-plus work performed by the construction manager acting as a contractor, by requiring proper documentation of all costs, and by strictly defining the conditions under which the construction manager may perform work itself.

It may be to the owner's advantage to limit the extent of work which the construction manager may do itself. If the construction manager also does business as a general contractor, the owner might prohibit the construction manager from taking on any work except "general conditions" items normally performed by a general contractor's own forces.

The important point in terms of conflict of interest is the same as was made earlier with respect to an A/E providing an "independent" construction document review while acting as a construction manager. Construction management services are by no means standardized; they involve the specific functions that the client contracts for. Where a conflict of interest exists, the client should not expect that such service can be provided strictly on the client's behalf or with the best interests of the client in mind. A construction manager who is reviewing and approving the work of others can be expected to do so conscientiously. When that same construction manager is "reviewing" and "approving" its own work, however, it is in a position no different from that of the contractor who submits a punch list to the client. The client should be under no delusion that the construction manager's approval of its own work is anything other than a representation that the work required has been performed; it certainly is no independent verification.

The construction manager who also performs work on the site is essentially no different from a general contractor, working on a cost-plus basis, who is also consulting during the design phases of the project. Such a construction manager is being given something of a carte blanche by the project owner, and the owner must therefore have a great amount of trust and confidence in the construction manager. In fact, the construction manager will usually have enough power over the project that no amount of contractual limitations or protections for the client will be an adequate substitute for the integrity of a construction manager who will fairly and honestly pursue the client's interests along with its own.

But there are some basic decisions for the project owner to make regarding the construction manager–cost-plus contractor. The client should decide whether it is willing to accept the construction manager's applications for payment on its own work at face value, or, if not, whether it wants to follow a traditional procedure of having the architect or engineer inspect the work done by the contractor–construction manager and review and approve its payment requests. Where the construction manager is expected to perform construction work, the project owner will be well advised to consider use of a traditional cost-plus contract to protect its interests in other respects as well. Although it may appear to be self-defeating or a duplication of effort to have a construction manager overseeing the work of a number of subcontractors and separate primes, and then to have another entity providing the same oversight with respect to the work performed by the construction manager, the owner may consider it desirable to have this control over the potential conflicts of interest peculiar to the contractor–construction manager.

In sum, a conflict of interest does not necessarily arise when a construction manager is performing part of the work. The conflict arises out of particular responsibilities undertaken by the construction manager on the owner's behalf which can be inconsistent with its undertaking to perform part of the work—such as the approval of its own work as conforming to contract requirements and the approval of payments to itself.

The owner may have the opinion that if the construction manager can be trusted to manage the project, the construction manager will also be conscientious enough to perform part of the work properly. If that degree of confidence in the construction manager exists, the project owner will feel no need to protect its own interests vis-à-vis the construction manager. If the owner would not be willing to have a cost-plus contractor perform the project without any independent oversight, however, it would be well advised to condition the contractor–construction manager's contract by the types of checks and balances found in a traditional cost-plus contract.

CHAPTER FOUR

STANDARD INDUSTRY CONSTRUCTION MANAGEMENT CONTRACT FORMS

The construction industry, not unlike other industries, has developed standard forms of construction contracts which have eliminated much of the need for custom-tailored agreements for most projects. The standard construction contract documents published by The American Institute of Architects have been well known in the industry for many years and form the basis for probably the majority of building construction contracts entered into in the United States. The National Society of Professional Engineers began publishing form contracts, similar to those of AIA, some twenty years ago, and those documents are now issued under the aegis of several engineering societies. Finally, the Associated General Contractors of America, in addition to endorsing various of the construction contract forms published by AIA, has developed contract forms to meet situations not addressed by the AIA forms.

THE STANDARD INDUSTRY FORMS

In the mid-seventies, the AIA, NSPE, and AGC were all confronted with the phenomenon of construction management, and responded in various ways to the needs of their members for information and guidance. All three of the organizations developed construction management contract forms reflecting the particular perspective of that organization, if not a recommended approach to construction management. The AIA and AGC originally attempted to develop a single owner–construction manager agreement form that could be endorsed by both organizations; however, different policy approaches eventually led to the creation of separate agreement forms, both of which are now in their second editions. The NSPE form was developed in conjunction with the American Consulting Engineers Council, independent of the AIA and AGC forms.

The following forms will be referred to as the AIA, AGC, and NSPE/ACEC forms, respectively, and will be analyzed in depth:

AIA Document B801, *Standard Form of Agreement between Owner and Construction Manager,* 1980

AGC Document 8, *Standard Form of Agreement between Owner and Construction Manager (Guaranteed Maximum Price Option),* 1980

NSPE/ACEC Publication 1910-15, *Standard Form of Agreement between Owner and Project Manager for Professional Services,* 1977

In addition to the above forms, the AIA and AGC have issued related contract documents to cover the relationships among other parties in the construction process. These documents are complementary to the standard forms mentioned above and are designed to be used in conjunction with them. Thus, the AIA and AGC have each published a family of internally consistent construction management documents. These families, reproduced within, are as follows:

AIA family of construction management documents

Document A101/CM	*Standard Form of Agreement between Owner and Contractor*
Document A201/CM	*General Conditions of the Contract for Construction*
Document A311/CM	*Performance Bond and Labor and Material Payment Bond*
Document B141/CM	*Standard Form of Agreement between Owner and Architect*
Document B801	*Standard Form of Agreement between Owner and Construction Manager*

Document G701/CM	*Change Order*
Document G722	*Projected Application and Project Certificate for Payment*
Document G723	*Project Application Summary*

AGC family of construction management documents

Document 8	*Standard Form of Agreement between Owner and Construction Manager (Guaranteed Maximum Price Option)*
Document 8a	*Amendment to Owner–Construction Manager Contract (Guaranteed Maximum Price)*
Document 8b	*General Conditions for Trade Contractors under Construction Management Agreements*
Document 8d	*Standard Form of Agreement between Owner and Construction Manager (Owner Awards All Trade Contracts)*
Document 8f	*Change Order*

For comparative purposes, AIA Document A111, *Standard Form of Agreement between Owner and Contractor,* is also reproduced within.

ADVANTAGES OF USING AN INDUSTRY FORM DOCUMENT

The AIA, AGC, NSPE, and ACEC are organizations which represent a wealth of experience, both in construction and in the drafting of construction contract forms. Although the various forms, to a greater or lesser extent, will reflect and be responsive to the needs of architects, contractors, or engineers, respectively, they do not ignore the needs of the owner. Nor can the forms be cited as examples of overreaching by a party with superior economic or bargaining power; after all, the objective of the organizations is to promote widespread use of their documents, which will depend on general acceptance by construction owners.

The industry forms, at the least, provide a reasonable format for an agreement and cover the most important subject areas within their texts. In addition, in those subject areas the industry form provisions are in specific language which has usually been carefully considered and is able to be understood by the construction manager. The language used in the industry forms is usually derived from or is similar to the language found in the traditional construction contract forms and has a long history of usage in the industry and, in many cases, of judicial interpretation. The question for the user of the industry forms is not usually the adequacy of the language, but whether the provisions reflect the responsibilities which the parties are willing to undertake.

The most valuable aspect of the industry forms is that they provide a baseline of what is considered to be a reasonable balance of responsibilities by the authoring organization, which will normally be acceptable to both parties. The forms represent a standard of practice in the industry with which specific variations can be compared during contract negotiations. Because architects, engineers, and contractors can be expected to be familiar with the forms, the owner can reasonably expect that any construction manager whom it engages will be willing to provide services in accordance with at least one of the industry forms. The existence of the forms thereby provides a level of certainty to the owner who is wondering what the construction manager should be doing for it; the forms also provide a basis to compare specific proposals from prospective construction managers.

Finally, the industry forms have been developed to interface with the related agreements between owner and architect and between owner and contractor and with the general conditions which will apply to the performance of the construction itself. Because all these relationships are interrelated, the use of industry documents in which all these relationships have previously been thought through can save the owner much time and expense in drafting specialized agreements covering all the parties involved in the construction project. Few owners have, or can afford to buy, the level of experience or consideration which compares to that invested in these forms. The owner can avoid the uncertainty of having substantial resistance by one party to entering into a series of relationships in which that party's position has not been adequately protected, or the need to go back and amend agreements previously entered into in order to satisfy the last party to enter the relationship. The simple fact is that few architects, engineers, or contractors can argue with the owner's choice to use one of the standard industry forms.

The industry forms presented herein are most often used without extensive modification. They are used at the suggestion—sometimes the insistence—of owners, as well as of engineers, architects, and contractors. The parties can easily do far worse for themselves by not using a standard document.

DRAWBACKS TO USING AN INDUSTRY FORM DOCUMENT

There are situations where none of the standard forms of agreement comes anywhere close to the type of agreement which the owner desires to have with the construction manager. Sometimes, this will be a matter of the level of detail embodied in the agreement, rather than of any desire to deviate from the basic relationships described in one or another of the standard industry documents. In other cases, despite general agreement with the relationships envisaged by the industry forms, the terms of the agreement will not be seen by either the owner or the construction manager as providing sufficient protections to its interests. Finally, there may be instances where the owner does not wish to enter into a relationship anything like those outlined in the standard forms, such as a turnkey operation. In such a case, the owner will usually have already selected a construction manager, and the drafting of an agreement must be preceded by an exhaustive discussion of the duties and responsibilities not only of owner and construction manager, but also of the others who will or have already become involved in the project.

A decision not to use the industry forms should be based on a knowledge of what protection those forms provide for the owner and what relationships they describe and enough knowledge of the construction management market to know whether alternative approaches are realistic and available. An owner should be satisfied that its project is sufficiently unique that the need for a departure from the standard forms justifies the additional uncertainties which will emerge, and protect itself against those uncertainties accordingly.

INTRODUCTION TO THE STANDARD FORMS: CONCEPTUAL DIFFERENCES

There are three well-defined conceptual approaches to construction management, each of which is represented by one of the standard industry documents. In one approach, represented by the AIA documents, the construction manager is an *independent professional* providing management services to the owner and has no contractual or organizational relationship to either the architect or engineer responsible for design of the project, nor any relationship to the multiple, separate prime contractors used to construct the project. This approach is referred to in this book as *professional construction management*. The entire AIA family of construction management documents is based on this approach, where the construction manager is an independent, separate entity responsible directly to the owner. Because the construction manager coordinates rather than controls the performance of the prime contractors in the various trades, it does not undertake to guarantee their work, nor to guarantee the cost of the project to the owner.

Another conceptual approach is reflected in the AGC documents. The construction manager will perform similar services, but will also be given a greater amount of control over the project—in some cases, performing portions of the construction work with its own forces, or letting out work to subcontractors as a general contractor. Under this approach, the construction manager may be in a position to offer a guaranteed maximum price for the project and to provide some guarantee of performance. This approach to construction management currently appears to be the most popular, because of the appeal to the owner of single-point responsibility for construction. Because of the similarity of this approach to the use of a traditional general contractor, it will be referred to as *contractor construction management*.

The third approach is the converse of the second, and is illustrated by the NSPE/ACEC project management agreement. Essentially, the design professional (architect or engineer) will perform construction management services in addition to traditional design and construction administration. This approach is probably the oldest approach to construction management, having been utilized historically whenever the owner elected to use multiple, separate prime contractors instead of a general contractor with multiple subcontractors. Under this approach, which may be called *architectural construction management,* the owner is again spared the necessity of having to deal with yet another party. In contrast to the contractor construction management approach, the construction phase responsibilities undertaken by the architect or engineer acting as construction manager are limited and are similar to those of the independent professional construction manager.

In would not be unfair to say that the AGC forms promote an approach to construction management designed to preserve a position for the general contractor and that the NSPE/ACEC project management agreement attempts to preserve or promote a strong role for the design professional during construction. The AIA form does not side with either the design professional or the contractor (a similar approach, in fact, is offered in AGC Document 8d), but this neutral approach is at the cost of adding another party to the construction team. While the AGC forms minimize the role of the design professional during construction, both the AIA and the NSPE/ACEC forms eliminate the need for the general contractor entirely. It is no surpise, then, that the approaches to construction management and the forms themselves are the subject of heated discussion and debate in the industry; potentially severe economic consequences to one or another segment of the industry could result from widespread use of one form or another.

CHOOSING A FORM OF AGREEMENT

After deciding to use the services of a construction manager, the first choice the owner must make—regardless of whether any standard forms are to be used—will be which approach to use. Once that decision is made, the selection of forms becomes relatively easy.

Architectural Construction
Management Approach

The NSPE/ACEC project manager agreement form, Publication 1910-15, is most suitable for use where the architect or engineer not only designs the project, but also manages it from a point in time before design through construction. (The NSPE/ACEC form may also be modified to cover the circumstance of the *professional construction manager,* by deleting services and responsibilities for which a design professional other than the construction manager has been or is to be placed under contract.)

As an alternative to using the NSPE/ACEC form, the AIA construction management form, Document B801 (discussed in the next section of this chapter), may be used where the architect or engineer also performs construction management services, as a supplement to the architect's or engineer's basic contract for services (such as AIA Document B141, owner-architect agreement, or NSPE Publication 1910-1, owner-engineer agreement). If the B801 document is used to supplement another agreement, care should be taken to eliminate, in both documents, any duplications or discrepancies in services and conditions which are based on an assumption that the construction manager will be an independent third-party professional.

To modify the AIA construction management agreement in article 1 and generally, phrases such as "with the Architect's assistance" and "in cooperation with the Architect" should be eliminated, as well as the term "Architect" in phrases such as "Advise the Owner and the Architect." Other modifications to article 1 should not be necessary; where an overlap with the owner-architect agreement occurs, these provisions most often simply expand on the requirements of the owner-architect agreement and do not create any real inconsistencies. The additional services provisions, paragraph 1.3, may be deleted as duplicative of language in the owner-architect agreement, as may the following provisions: article 2, "The Owner's Responsibilities" (with the exception of paragraph 2.10); article 3, "Construction Cost"; article 5, "Direct Personnel Expense"; article 7, "Payments to the Construction Manager" (with the exception of paragraph 7.4.2); article 8, "Construction Manager's Accounting Records"; article 9, "Arbitration"; article 10, "Termination of Agreement"; article 11, "Miscellaneous Provisions"; article 12, "Successors and Assigns"; paragraph 13.1; and article 15, "Basis of Compensation" (unless the compensation for construction management services is to be identified separately).

In addition to the modifications to AIA Document B801 outlined above, modifications to the owner-contractor agreement (AIA Document A101/CM) and to the general conditions (AIA Document A201/CM) will be necessary to eliminate the identification of the construction manager as a separate entity. This will normally be accomplished through supplementary conditions; a simple statement that the architect and construction manager are one and the same and that the general conditions should be interpreted accordingly may be sufficient.

The choice between the NSPE/ACEC form and a modified AIA form supplementing a standard owner-architect or owner-engineer agreement will often depend on the type of project (engineering or architectural), the identity of the construction manager (most architects will prefer to work under a modified B801 form, while engineers may prefer the NSPE/ACEC form), the desire for specificity in the agreement (the AIA form is considerably more detailed in describing the construction manager's responsibilities), and the form's flexibility and application to the overall project relationship (for instance, the simpler NSPE/ACEC form can be matched more easily with nonstandard forms of general conditions, whereas the AIA form shares much language with the AIA general conditions and will require extensive amendments if a nonstandard form of general conditions is used on the project).

STANDARD
FORM OF AGREEMENT
BETWEEN OWNER AND PROJECT MANAGER
FOR
PROFESSIONAL SERVICES

Jointly Issued by

PROFESSIONAL ENGINEERS IN PRIVATE PRACTICE

A practice division of the

NATIONAL SOCIETY OF PROFESSIONAL ENGINEERS

AMERICAN CONSULTING ENGINEERS COUNCIL

© 1977 National Society of Professional Engineers
2029 K Street, N.W., Washington, D.C. 20006

American Consulting Engineers Council
1155 15th Street, N.W., Washington, D.C. 20005

NSPE/PEPP-ACEC (1977 Edition)
Pub. #1910-15

COMMENTARY FOR PROJECT MANAGER AGREEMENT

This Project Manager Agreement was prepared by the Joint Contract Documents Committee of the two professional societies in response to numerous requests for a carefully thought-out document that would combine most of the customary functions of a professional engineer with those of a construction manager. The professional engineering services described in the attached Agreement are, for the most part, quite similar to those in the Standard Form of Agreement between Owner and Engineer for Professional Services (Doc. 1910-1, 1974 Ed.) issued by NSPE/PEPP which is currently in the process of its periodic review by the Joint Contract Documents Committee of NSPE/PEPP and ACEC. Construction management services may be variously described and include a broad spectrum of functions, most of which appear in many construction management contracts, but many are peculiar to the job or firm involved. The attached Agreement contemplates the combination of these design and management functions from the start of the relationship with the Owner and the continuation thereof until completion of construction. While carefully coordinating all aspects of his services with the Owner, the Project Manager will take over for the Owner many administrative and coordinating functions. He will, in effect, take charge of the Project from beginning to end in order to provide special expertise and relieve the Owner of duties and responsibilities which the Owner is neither qualified to undertake nor for which he will have a continuing need. However, under the attached Agreement, the Project Manager will not perform any of the functions of the contractors. He may assist in purchasing on the Owner's behalf, coordinate the work of separate contractors and assist in expediting various aspects of the work; but neither in his capacity as a professional engineer nor as a construction manager will he be involved in or assume responsibility for the means, methods, techniques, sequences, or procedures of construction or the safety precautions or programs incident thereto.

The matter of insurability of the Project Manager's services has been considered and discussed with certain carriers providing liability coverage against negligent errors and omissions in furnishing professional engineering services. Certain policies provide coverage for design professionals who offer services as project managers. The coverage of other policies extends only to the performance by a Project Manager of those services which are considered "customary" for engineers (whatever that may mean), with no coverage being provided for those project management services that are not considered customary. In either case, all coverage is subject to exclusions and specific policy language. It is important, therefore, before entering into any agreement to furnish project management services, to discuss the matter of coverage with competent insurance counsel. In some cases a solution may be found through a special endorsement of the professional liability policy, plus an endorsement of the Project Manager's comprehensive general liability policy. As a minimum, coordination of both coverages is desirable. Note should also be made that most professional liability policies exclude coverage in respect of estimating, scheduling and time of performance which are functions usually undertaken by a Project Manager. Because of the Project Manager's deep involvement in a Project, there is the temptation to provide assurance to the Owner as to the Project's performance. One should bear in mind, however, that professional liability insurance does not extend to express warranties or guarantees of performance, and accordingly, contractual provisions to that effect should be avoided in all cases.

Recent court decisions indicate the need for caution in respect of the possible treatment of the Project Manager as an employer under OSHA on the theory that the cumulative effect of all construction management functions at the job site can become so extensive that the Project Manager is, in reality, an integral part of the total construction effort and thus engaged in construction work at the site. Suffice to say that the greater the degree of control, direction or supervision exercised by the Project Manager over the construction process, the greater will be his chance of being considered part of the construction team with the consequent exposure to liability and regulation as such. Accordingly, it is important that the Project Manager adhere carefully in practice to the statement contained in paragraph 1.6.3 that he will not be responsible for the means, methods, techniques, sequences or procedures of construction selected by contractors or for safety precautions and programs incident to the work of contractors.

The attached Agreement contains several blank spaces to be filled in. In paragraph 1.1.1 the blank will be filled in after consultation with the Owner as to the extent of the services to be rendered. Depending upon how this blank is filled in, the description of basic services may have to be modified. This conclusion will also determine how the percentages of payment are to be completed in paragraph 5.2.2 when compensation is on the basis of a lump-sum or percentage of construction costs.

It should be noted that in all phases of basic services, the Project Manager assumes responsibility for estimating (see particularly paragraphs 1.3.3 and 1.4.5). This contemplates a greater degree of detail and accuracy than when furnishing "statements of probable construction costs" which is the term used in the Standard Form of Agreement between Owner and Engineer for Professional Services.

No reference is made in the description of the construction phase services to the form of general conditions which are to be attached to the construction contracts and which will describe services to be furnished by the Project Manager as a representative of the Owner during construction. The Joint Contract Documents Committee intends that the Standard General Conditions of the Construction Contract (NSPE Doc. 1910-8, 1974 Ed. currently under revision) will be used with modification by Supplementary Conditions to cover the additional responsibilities of the Project Manager. Suggestions for the language of such modifications are under review by the Joint Contract Documents Committee.

Because the timeliness of performance is so important in relationships with one's client, in the determination of fees and in the estimates of construction costs, blank spaces have been left in paragraphs 4.2, 4.3, 4.4 and 4.10 to be filled in after consideration of the Project's complexity. In addition, one should examine the other provisions of Section 4 as to timing for relevancy to the particular Project.

Section 5 contemplates the selection of a method of payment, filling in the pertinent blanks and crossing out the inapplicable paragraphs. Paragraph 5.1.1.7 recognizes that none of the suggested methods may be wholly applicable, in which case another method or combination of methods is to be set out in an attachment to the Agreement. Care should be taken in preparing any such attachment that the terms common to both the Agreement and the attachment are used with precisely the same meaning.

Because there is no uniform understanding about what persons are intended by the term "principals" of the Engineer, paragraphs 5.3.2 and 5.3.3 require their identification and a statement of hourly rates. Paragraph 5.3.3 also provides for a possible agreement on the use of a percentage figure as to the amount of the customary and statutory benefits to be included in payroll costs; many firms have found this to be a convenient way of handling the amount rather than going through a detailed computation in support of each monthly statement. Where the services will be furnished over a period of many months, inclusion of a provision for adjusting the percentage figure is suggested.

Arbitration provisions are included in paragraph 6.4, and are recommended. However, arbitration provisions are not enforceable in all jurisdictions. Some firms and their insurance carriers have strong objections to arbitration as a method of solving disputes. Consultation with legal counsel is suggested where appropriate. Paragraph 6.4 may be crossed out in its entirety.

This Agreement represents the first attempt by the Joint Contract Documents Committee to develop a document of this sort. It is hoped that as the document is used, comments and suggestions will be forwarded to the Committee for its refinement and improvement prior to publication of the next edition.

TABLE OF CONTENTS

Page No.

STANDARD FORM OF AGREEMENT
BETWEEN
OWNER AND PROJECT MANAGER
FOR
PROFESSIONAL ENGINEERING PROJECT MANAGEMENT SERVICES

THIS IS AN AGREEMENT made as of _____ day of _____

IN THE YEAR Nineteen Hundred and _____ by and

between _____

_____ (hereinafter

called OWNER) and _____

_____ (hereinafter called

PROJECT MANAGER). OWNER intends to engage PROJECT MANAGER to perform professional project management services in connection with a project known as _____

_____

which is referred to herein as the Project and is described in more detail as follows:

OWNER and PROJECT MANAGER in consideration of their mutual covenants herein agree in respect of the performance of professional project management services by PROJECT MANAGER and the payment for those services by OWNER, as set forth below.

SECTION 1—BASIC SERVICES OF PROJECT MANAGER

1.1. General.

1.1.1. PROJECT MANAGER's basic professional services shall be rendered in .... phases as described below during which he shall provide design services including normal civil, structural, mechanical and electrical professional engineering services and normal architectural services incidental thereto and construction management services and other related services, all as are set forth below.

1.2. Basic Project Development Phase.

After written authorization to proceed, PROJECT MANAGER shall:

1.2.1. Consult with OWNER to determine his requirements for the Project and review available data.

1.2.2. Advise OWNER as to the necessity of his providing or obtaining data or services from others, such as core borings, probings and subsurface explorations, hydrographic surveys, laboratory tests and inspection of samples, materials and equipment, with appropriate professional interpretations thereof; environmental assessments and impact statements, and property, boundary, easement, right-of-way topographic and utility surveys; and when authorized by OWNER act as OWNER's representative in connection with obtaining such data and services from others.

1.2.3. Provide analyses of OWNER's needs and perform planning surveys, site evaluations and comparative studies of prospective sites and solutions, assess local availability of materials, research and review state and local building codes and ordinances and requirements of proposed funding agencies.

1.2.4. Provide general economic analyses of OWNER's requirements applicable to various alternatives.

1.2.5. Prepare a Report with appropriate exhibits indicating the considerations involved and the alternative solutions available to OWNER, and setting forth PROJECT MANAGER's findings and recommendations with respect to the Project scope, the initial Project budget and time for completion and containing PROJECT MANAGER's plan for coordination and supervision of design and for coordination of construction.

1.2.6. Provide assistance to OWNER in determining the requirements of regulatory authorities, including those having jurisdiction over the anticipated environmental impact of the Project, and assistance in obtaining tentative approvals where appropriate.

1.2.7. Furnish five copies of the Report and present and review it in person with OWNER.

1.3. Preliminary Design Phase.

After written authorization to proceed with the Preliminary Design Phase, PROJECT MANAGER shall:

1.3.1. Provide preliminary design documents consisting of preliminary drawings, design criteria, flow sheets, material and equipment lists, outline specifications and preliminary construction schedules.

1.3.2. Make recommendations concerning construction contract arrangements and scheduling that will be advantageous to OWNER in terms of cost and timing, including the advantages of letting the work on the basis of one or more prime contracts, negotiated versus bid prime contracts, cost incentive or cost plus contracts, guaranteed maximum price and liquidated damages provisions, and approval of subcontractors.

1.3.3. Based on the information contained in the preliminary design documents, submit an estimate of cost for the Project which shall include Construction Cost, contingencies, compensation for all professionals and consultants, costs of land, rights-of-way and compensation for damages to properties and shall also include interest and financing charges as estimated by OWNER's financial consultant. All of the foregoing are collectively referred to herein as "Project Costs." Develop with OWNER a revised Project budget.

1.3.4. Furnish five copies of the above preliminary design documents and present and review them in person with OWNER.

1.4. Final Design Phase.

After written authorization to proceed with the Final Design Phase, PROJECT MANAGER shall:

1.4.1. On the basis of the accepted preliminary design

documents prepare for incorporation in the Contract Documents, final drawings to show and describe the character and scope of the work to be performed by contractors on the Project (hereinafter called "Drawings"), and Specifications.

1.4.2. Analyze conditions anticipated during Construction Period, prepare final design and construction schedules, that will be advantageous to OWNER in terms of cost and timing.

1.4.3. Advise OWNER concerning preparation of purchase orders and making purchases in OWNER's name, prepare purchase specifications, obtain and analyze vendors' proposals for items of material and equipment to be prepurchased, assist in the expediting of delivery of vendors' drawings and review such drawings for compliance with the information given in the purchase order and for coordination with the rest of the Project.

1.4.4. Furnish to OWNER such documents and design data as may be required for, and assist in the preparation of, the required documents so that OWNER may apply for approval of such regulatory authorities as have jurisdiction over design criteria applicable to the Project and its environmental impact, and assist in obtaining such approvals by participating in submittals to and negotiations with appropriate authorities.

1.4.5. Monitor any adjustments to his latest estimate of Project Cost caused by changes in scope, design requirements or Construction Costs and periodically advise OWNER of his conclusions in respect thereof; and furnish finalized Drawings and Specifications with a revised estimate of Project Cost.

1.4.6. Prepare for review and approval by OWNER, his legal counsel and other advisors contract agreement forms, general conditions and supplementary conditions, and (where appropriate) bid forms, invitations to bid and instructions to bidders, and assist in the preparation of other related documents.

1.4.7. Furnish five copies of these documents and of the Drawings and Specifications, and present and review them in person with OWNER.

1.5. Bidding or Negotiating Phase.

After written authorization to proceed with the Bidding or Negotiating Phase, PROJECT MANAGER shall:

1.5.1. As representative of OWNER, obtain bids or proposals for each separate prime contract for construction and furnishing materials and equipment and make recommendations as to the award of the same.

1.5.2. After consultation with OWNER, determine the acceptability of subcontractors and other persons and organizations proposed by the prime contractor(s) for those portions of the work as to which such acceptability is required by the Contract Documents.

1.5.3. Determine the acceptability of substitute materials and equipment prepared by the apparent low bidder when substitution before award of the Contract is allowed by the bidding documents.

1.6. Construction Phase.

During the Construction Phase, the PROJECT MANAGER shall:

1.6.1. Consult with and advise OWNER and act as his representative in the administration of the construction contracts. All of OWNER's instructions to contractors will be issued through PROJECT MANAGER. The prime contractors with whom OWNER enters into contracts for construction of, and furnishing of materials and equipment for, the Project are hereinafter referred to as "Contractor(s)."

1.6.2. Report to OWNER at appropriate intervals on progress, costs and cost control, and advise Contractor(s) of OWNER's decisions on such matters.

1.6.3. Provide full time resident project representation by resident project staff being present at the site at all times when the construction work is in progress in order to provide through experienced project management observation of the progress and quality of the construction work, to determine in general if it is proceeding in accordance with the Contract Documents, and to guard OWNER against defects and deficiencies in the work of Contractor(s). The resident project staff will direct its efforts toward providing greater protection for OWNER that the completed Project will conform to the Contract Documents, but neither PROJECT MANAGER nor his staff shall be responsible for the means, methods, techniques, or procedures of construction selected by Contractor(s) or for safety precautions and programs incident to the work of Contractor(s) or for any failure of Contractor(s) to comply with any laws, ordinances, rules or regulations applicable to the construction work or for any failure of Contractor(s) to perform the construction work in accordance with the Contract Documents.

1.6.4. Provide major material and equipment purchasing services for OWNER acting solely as his agent; all purchase orders for such materials and equipment shall be on OWNER's purchase order forms; and PROJECT MANAGER shall not be deemed to have, nor shall he assume, title to any materials or equipment purchased on behalf of OWNER.

1.6.5. Expedite and coordinate the receipt of major material and equipment deliveries, inspect materials and equipment as received for damage and compliance with the Contract Documents, report when delivery is later than scheduled, schedule the arrival of vendor's erection or assembly supervisors, inspect installed material and equipment for mechanical, electrical, piping and instrument connections, for correct rotation and lubrication and readiness for delivery to OWNER's operating personnel.

1.6.6. Coordinate the sequence of operations and other relationships among separate Contractor(s) and maintain liaison between them, the engineering design staff and OWNER.

1.6.7. Review and approve Shop Drawings (as that term is defined in the aforesaid Standard General Conditions) and samples, the results of tests and inspections and other data which each Contractor is required to submit, but only for conformance with the design concept of the Project and compliance with the information given in the Contract Documents; determine the acceptability of substitute materials and equipment proposed by Contractor(s); and receive and review (for general content as required by the Specifications) maintenance and operating instructions, schedules, guarantees, bonds and certificates of inspection which are to be assembled by Contractor(s) in accordance with the Contract Documents.

1.6.8. Issue all instructions of OWNER to Contractor(s); prepare change orders as required; he may, as OWNER's representative, require special inspection or testing; he shall act as interpreter of the requirements of the Contract Documents and judge of the performance thereunder by the parties thereto and shall make decisions on all claims of OWNER and Contractor(s) relating to the execution and progress of the work and all other matters and questions related thereto; but PROJECT MANAGER shall not be liable for the results of any such interpretations or decisions rendered by him in good faith.

1.6.9. Based on his on-site observations as a professional experienced and qualified in project management and on his review of Contractor(s) applications for payment and the accompanying data and schedules, determine the amounts owing to Contractor(s) and recommend in writing payments to Contractor(s) in such amounts; such recommendations of payment will constitute a representation to OWNER, based on such observations and review, that the work has progressed to the point indicated and that, to the best of his knowledge, information and belief, the quality of the work is in accordance with the Contract Documents (subject to an evaluation of the work as a functioning Project upon Substantial Completion, to the results of any subsequent tests called for in the Contract Documents, and to any qualifications stated in his recommendation), but by recommending any payment PROJECT MANAGER will not be deemed to have represented that he has made any examination to determine how or for what purposes any Contractor has used the moneys paid on account of the Contract Price, or that title to any of Contractor(s) work, materials or equipment has passed to OWNER free and clear of any lien, claims, security interests or encumbrances.

1.6.10. Notify OWNER of all permanent work which does not conform to the Contract Documents, prepare a written report describing any apparent non-conforming permanent work and make recommendations to OWNER for its correction, and, when authorized by OWNER, instruct Contractor to carry out the acceptable corrective measures.

1.6.11. Conduct an inspection to determine if the Project is substantially complete and a final inspection to determine if the Project has been completed in accordance with the Contract Documents and if Contractor(s) have fulfilled all of their obligations thereunder so that PROJECT MANAGER may recommend, in writing, final payment to Contractor(s).

1.6.12. Prepare a set of reproducible record prints of Drawings showing those changes made during the construction process, based on marked-up prints, drawings and other data furnished to PROJECT MANAGER by Contractor(s) and which PROJECT MANAGER considers significant.

1.6.13. PROJECT MANAGER shall not be responsible for the acts or omissions of any Contractor(s), any subcontractor(s) or any of Contractor(s)' or subcontractor(s)' agents or employees or any other persons (except his own employees and agents) at the Project site or otherwise performing any of the work of the Project.

SECTION 2—ADDITIONAL SERVICES OF PROJECT MANAGER

If authorized in writing by OWNER, PROJECT MANAGER shall furnish or obtain from others Additional Services of the following types which are not included as Basic Services in Section 1; these will be paid for by OWNER as indicated in Section 5.

2.1. Preparation of applications and supporting documents for governmental grants, loans or advances in connection with the Project; preparation of environmental assessments and impact statements; and review and evaluation of the effect on the design requirements of the Project of any such statements and documents prepared by others.

2.2. Providing special feasibility studies, cash flow and economic evaluations, assisting in obtaining financing for the Project, evaluating processes available for licensing and assisting OWNER in obtaining process licensing.

2.3. Providing Value Engineering and documentation thereof during the course of engineering design.

2.4. Services to make measured drawings of or to investigate existing conditions or facilities, or to verify the accuracy of drawings or other information furnished by OWNER.

2.5. Services resulting from significant changes in general scope of the Project or its design including, but not limited to, changes in size, complexity, OWNER's schedule, character of construction or a method of financing; and revising previously accepted studies, reports, design documents or Contract Documents when such revisions are due to causes beyond PROJECT MANAGER's control.

2.6. Providing renderings or models for OWNER's use.

2.7. Preparing documents for alternate bids requested by OWNER for work which is not executed or documents for out-of-sequence work.

2.8. Investigations involving detailed consideration of operations, maintenance and overhead expenses; and the preparation of rate schedules, earnings and expense statements, feasibility studies, appraisals and valuations; detailed quantity surveys of materials, equipment and labor; and audits or inventories required in connection with construction performed by OWNER.

2.9. Furnishing the services of special consultants for other than the normal civil, structural, mechanical and electrical engineering and normal architectural design incidental thereto, such as consultants for interior design, selection of furniture and furnishings, communications, acoustics, kitchens and landscaping; and providing data or services of the types described in paragraph 1.2.2 when OWNER authorizes PROJECT MANAGER to provide such data and services in lieu of furnishing the same in accordance with paragraph 3.3.

Page 4 of _ _ _ _ pages

2.10. Providing any type of field surveys for design purposes, and engineering surveys and staking to enable Contractor(s) to proceed with their work; and providing other special field surveys.

2.11. Services resulting from the involvement of more separate prime contracts for construction or for equipment than are contemplated by paragraphs 5.1.1.2 or 5.1.1.4.

2.12. Services in connection with change orders to reflect changes requested by OWNER if the resulting change in compensation for Basic Services is not commensurate with the additional services rendered, and services resulting from significant delays, changes or price increases occurring as a direct or indirect result of material, equipment or energy shortages.

2.13. Services during out-of-town travel required of PROJECT MANAGER other than visits to the Project site as required by Section 1.

2.14. Additional or extended services during construction made necessary by (1) work damaged by fire or other cause during construction, (2) a significant amount of defective or neglected work of any Contractor, (3) prolonging of the contract time of any prime contract by more than sixty days, (4) acceleration of the work schedule involving services beyond normal working hours, (5) default by any Contractor, and (6) investigation, evaluation and negotiation of Contractor(s) claims for additional time or money.

2.15. Preparation of operating and maintenance manuals; protracted or extensive assistance in the utilization of any equipment or system (such as initial start-up, testing, adjusting and balancing); and training personnel for operation and maintenance.

2.16. Services after completion of the Construction Phase, such as inspections during any guarantee period and reporting observed discrepancies under guarantees called for in any contract for the Project.

2.17. Preparing to serve or serving as a consultant or witness for OWNER in any litigation, public hearing or other legal or administrative proceeding involving the Project.

2.18. Additional services in connection with the Project, including services normally furnished by OWNER and services not otherwise provided for in this Agreement.

SECTION 3—OWNER'S RESPONSIBILITIES

OWNER shall:

3.1. Provide all criteria and full information as to his requirements for the Project, including design objectives and constraints, space, capacity, and performance requirements, flexibility and expandability, and any budgetary limitations; and furnish copies of all design and construction standards which OWNER will require to be included in the Drawings and Specifications.

3.2. Assist PROJECT MANAGER by placing at his disposal all available information pertinent to the Project including previous reports and any other data relative to design and construction of the Project.

3.3. Furnish to PROJECT MANAGER as required by him for performance of his Basic Services data prepared by, or services of others, including without limitation data or services of the types described in paragraph 2.2.2; and furnish property descriptions and zoning, deed and other land use restrictions; all of which PROJECT MANAGER may rely upon in performing his services.

3.4. Provide engineering surveys and staking to enable Contractor(s) to proceed with their work.

3.5. Guarantee access to and make all provisions for PROJECT MANAGER to enter upon public and private property as required for PROJECT MANAGER to perform his services.

3.6. Examine all studies, reports, sketches, Drawings, Specifications, proposals and other documents presented by PROJECT MANAGER, obtain advice of an attorney, insurance counselor and other consultants as he deems appropriate for such examination and render in writing decisions pertaining thereto within a reasonable time so as not to delay the services of PROJECT MANAGER.

3.7. Cooperate in obtaining bids or proposals and related negotiations.

3.8. Provide such legal, accounting, independent cost estimating and insurance counseling services as may be required for the Project, and such auditing service as OWNER may require to ascertain how or for what purpose any contractor has used the moneys paid to him under the construction contract.

3.9. Designate in writing a person to act as OWNER's representative with respect to the work to be performed under this Agreement. Such person shall have complete authority to transmit instructions, receive information, interpret and define OWNER's policies and decisions.

3.10.   Give prompt written notice to PROJECT MANAGER whenever OWNER observes or otherwise becomes aware of any defect in the Project.

3.11.   Furnish approvals and permits from all governmental authorities having jurisdiction over the Project and such approvals and consents from others as may be necessary for completion of the Project.

3.12.   Furnish, or direct PROJECT MANAGER to provide, necessary Additional Services as stipulated in Section 2 of this Agreement or other services as required.

3.13.   Be responsible for the protection of survey stakes and other markers until Contractor(s) commence their work at site, and replace all stakes and markers damaged or moved prior to commencement of such work.

3.14.   Bear all costs incident to compliance with the requirements of this Section 3.

SECTION 4—PERIOD OF SERVICE

4.1.   The provisions of 4.2 through 4.8, inclusive, and the various rates of compensation for PROJECT MANAGER's services provided for elsewhere in this Agreement have been agreed to in anticipation of the orderly and continuous progress of the Project through completion of the Construction Phase. PROJECT MANAGER's obligation to render services hereunder will extend for a period which may reasonably be required for the design, award of contracts and construction of the Project including extra work and required extensions thereto.

4.2.   The services called for in the Basic Project Development Phase will be completed and the Report submitted within . . . . calendar days following the authorization to proceed with that phase of services.

4.3.   After acceptance by OWNER of the Report, indicating any specific modifications or changes in scope desired by OWNER, and upon written authorization from OWNER, PROJECT MANAGER shall proceed with the performance of the services called for in the Preliminary Design Phase, and shall submit preliminary design documents and a revised estimate of Project Cost within . . . . calendar days following the authorization to proceed with that phase of services.

4.4.   After acceptance by OWNER of the preliminary design documents and revised estimate of Project Cost, indicating any specific modifications or changes in scope desired by OWNER, and upon written authorization from OWNER, PROJECT MANAGER shall proceed with the performance of the services called for in the Final Design Phase, so as to deliver Contract Documents and a revised estimate of Project Cost for all authorized work on the Project within . . . . calendar days after the authorization to proceed with that phase of services.

4.5.   PROJECT MANAGER's services under the Basic Project Development Phase, Preliminary Design Phase and Final Design Phase shall each be considered complete at the earlier of (1) the date when the submissions for that phase have been accepted by OWNER or (2) thirty days after the date when such submissions are delivered to OWNER for final acceptance, plus such additional time as may be considered reasonable for obtaining approval of governmental authorities having jurisdiction over design criteria applicable to the Project.

4.6.   After acceptance by OWNER of the PROJECT MANAGER's Drawings, Specifications and other Final Design Phase documentation including his most recent estimate of Project Cost and upon written authorization to proceed, PROJECT MANAGER shall proceed with performance of the services called for in the Bidding or Negotiating Phase. This Phase shall terminate and the services to be rendered thereunder shall be considered complete upon commencement of the Construction Phase, or upon cessation of negotiations with prospective Contractor(s) (except as may otherwise be required to complete the services called for in paragraph 6.3.2.5).

4.7.   The Construction Phase will commence with the execution of the first prime contract to be executed for the work of the Project or any part thereof, and will terminate upon written approval by PROJECT MANAGER of final payment on the last prime contract to be completed. Construction Phase services may be rendered at different times in respect of separate prime contracts if the Project involves more than one prime contract.

4.8.   In the event that the work of the Project is to be performed under more than one prime contract, OWNER and PROJECT MANAGER shall, prior to commencement of the Final Design Phase, develop a schedule for performance of PROJECT MANAGER's services during the Final Design, Bidding or Negotiating and Construction Phases in order to sequence and coordinate properly the services applicable to the work under such separate contracts. This schedule is to be prepared whether or not the work under such contracts is to proceed concurrently and is to be attached as an exhibit to and made a part of this Agreement, and the provisions of paragraphs 4.4 through 4.6 inclusive, will be modified accordingly.

4.9.   If OWNER has requested significant modifications or changes in the scope of the Project, the time of performance of PROJECT MANAGER's services and his various rates of compensation provided for elsewhere in this Agreement shall be adjusted appropriately.

4.10.   If OWNER fails to give prompt written authorization to proceed with any phase of services after completion of the immediately preceding phase, or if the Construction Phase has not commenced within . . . . calendar days after completion of the Final Design Phase, PROJECT MANAGER may, after giving seven days' written notice to OWNER, suspend services under this Agreement.

4.11.   If PROJECT MANAGER's services for design or during construction of the Project are delayed or suspended in whole or in part by OWNER for more than three months for reasons beyond PROJECT MANAGER's control, PROJECT MANAGER shall on written demand to OWNER (but without termination of this Agreement) be paid as provided in paragraph 5.3.6 for services rendered. If such delay or suspension extends for more than one year for reasons beyond PROJECT MANAGER's control, or if PROJECT MANAGER for any reason is required to render services more than one year after Substantial Completion, the various rates of compensation provided for elsewhere in this Agreement shall be subject to renegotiation.

SECTION 5—PAYMENTS TO PROJECT MANAGER

5.1.   Methods of Payment for Services and Expenses of PROJECT MANAGER.

5.1.1.   *Basic Services.* OWNER shall pay PROJECT MANAGER for Basic Services rendered under Section 1 on one of the following bases (except as otherwise provided in paragraph 5.1.7):

5.1.1.1.   *Lump Sum.* If the work of the entire Project is awarded on the basis of one prime contract, a lump sum fee of $ . . . .; but, if the prime contract contains cost-plus or incentive savings provisions for the Contractor's basic compensation, a

lump sum fee of $ . . . .
or

5.1.1.2.   *Lump Sum.* If the work of the Project is awarded on the basis of not more than a total of . . . . separate prime contracts for construction and for equipment, a lump sum fee of $ . . . .; but, if the prime contracts contain cost-plus or incentive savings provisions for the Contractor's basic compensation, a lump sum fee of $ . . . .
or

5.1.1.3.   *Percentage.* If the work of the entire Project is awarded on the basis of one prime contract, ... % of the Construction Cost; but, if the prime contract contains cost-plus or incentive savings provisions for the Contractor's basic compensation, ... % of the Construction Cost.
or

5.1.1.4.   *Percentage.* If the work of the Project is awarded on the basis of not more than a total of . . . . separate prime contracts for construction and for equipment, ... % of the Construction Cost; but, if the prime contracts contain cost-plus or incentive savings provisions for Contractor's basic compensation, ... % of the Construction Cost.
or

5.1.1.5.   *Direct Labor Costs Times a Factor.* An amount based on Direct Labor Costs times a factor of . . . . for services rendered by principals and employees assigned to the Project.
or

5.1.1.6.   *Payroll Cost Times a Factor.* An amount based on Payroll Costs times a factor of . . . . for services rendered by principals and employees assigned to the Project.
or

5.1.1.7.   *Other Method.* (To be used in case none of the above methods of compensation is applicable.) (Refer to and attach schedule when applicable)

5.1.2.   *Additional Services.* OWNER shall pay PROJECT MANAGER for Additional Services rendered under Section 2 as follows:

5.1.2.1.   *General.* For Additional Services rendered under paragraphs 2.1 through 2.18, inclusive (except services covered by paragraph 2.9 and services as a consultant or witness under paragraph 2.17), on the basis of (Direct Labor Costs) (Payroll Costs) times a factor of . . . . for services rendered by principals and employees assigned to the Project.

5.1.2.2.   *Special Consultants.* For services and reimbursable expenses of special consultants employed by PROJECT MANAGER pursuant to paragraphs 2.9 or 2.18, the amount billed to PROJECT MANAGER therefor times a factor of . . . .

[Delete inapplicable paragraphs and initial]

Page 7 of _ _ _ _ pages

5.1.2.3. *Serving as a Witness.* For the services of the principals and employees as consultants or witnesses in any litigation, hearing or proceeding in accordance with paragraph 2.17 at rates to be agreed upon for each day or any portion thereof (but compensation for time spent in preparing to appear in any such litigation, hearing or proceeding will be on the basis provided in paragraph 5.1.2.1).

5.1.3. *Reimbursable Expenses.* In addition to payments provided for in paragraphs 5.1.1 and 5.1.2, OWNER shall pay PROJECT MANAGER the actual costs of all Reimbursable Expenses incurred in connection with all Basic and Additional Services.

5.1.4. As used in this paragraph 5.1, the terms "Construction Cost," "Direct Labor Costs," "Payroll Costs" and "Reimbursable Expenses" will have the meanings assigned to them in paragraphs 5.3.1 through 5.3.4, inclusive.

5.2. Times of Payment.

5.2.1. PROJECT MANAGER shall submit monthly statements for Basic and Additional Services rendered and for Reimbursable Expenses incurred. When compensation is on the basis of a lump sum or percentage of Construction Cost the statements will be based upon PROJECT MANAGER's estimate of the proportion of the total services actually completed at the time of billing. Otherwise, these monthly statements will be based upon PROJECT MANAGER's Direct Payroll Cost times a factor, as applicable. OWNER shall make prompt monthly payments in response to PROJECT MANAGER's monthly statements.

5.2.2. Where compensation for Basic Services is on the basis of a lump sum or percentage of Construction Cost, OWNER shall, upon conclusion of each phase of Basic Services, pay such additional amount, if any, as may be necessary to bring total compensation paid on account of such phase to the following percentages of total compensation for all phases of Basic Services:

Phase	*Insert Actual Percentage and Initial in Margin*
Basic Project Development	. . . .%
Preliminary Design	. . . .%
Final Design	. . . .%
Bidding or Negotiating	. . . .%
Construction	. . . .%
TOTAL	100%

5.2.3. Payments for Basic Services in accordance with paragraph 5.1.1.7 shall be made at the following times:
(Refer to and attach schedule when applicable)

5.3. General.

5.3.1. The construction cost of the entire Project (herein referred to as "Construction Cost") means the total cost of the entire Project to OWNER, but it will not include PROJECT MANAGER's compensation and expenses, the cost of land, rights-of-way, or compensation for or damages to properties, unless this Agreement so specifies; nor will it include OWNER's legal, accounting, insurance counseling or auditing services, or interest and financing charges incurred in connection with the Project. When Construction Cost is used as a basis for payment it will be based on one of the following sources with precedence in the order listed:

5.3.1.1. For work constructed, the total Construction Cost of all work performed as designed or specified by PROJECT MANAGER.

5.3.1.2. For work designed but not constructed, the lowest bona fide bid received from a qualified bidder for such work; or if the work is not bid, the lowest bona fide negotiated proposal for such work.

5.3.1.3. For work designed but for which no such bid or proposal is received, PROJECT MANAGER's most recent estimate of Construction Cost.

Labor furnished by OWNER for the Project will be included in the Construction Cost at current market rates including a reasonable allowance for overhead and profit. Materials and equipment furnished by OWNER will be included at current market prices except used materials and equipment will be included as if purchased new for the Project. No deduction is to be made from PROJECT MANAGER's compensation on account of any penalty, liquidated damages, or other amounts withheld from payments to Contractor(s).

5.3.2. Direct Labor Costs used as a basis for payment mean salaries and wages paid to all personnel engaged directly on the Project, including, but not limited to, engineers, architects, surveymen, designers, draftsmen, specification writers, estimators, other technical personnel, stenographers, typists and clerks; but does not include indirect payroll related costs or fringe benefits. For the purposes of this Agreement the principals of PROJECT MANAGER and their hourly direct labor costs are:

_____
_____
_____
_____

5.3.3. Payroll Costs used as a basis for payment mean

the salaries and wages paid to all personnel engaged directly on the Project, including, but not limited to, engineers, architects, surveymen, designers, draftsmen, specification writers, estimators, other technical personnel, stenographers, typists and clerks; plus the cost of customary and statutory benefits including, but not limited to, social security contributions, unemployment, excise and payroll taxes, workmen's compensation, health and retirement benefits, sick leave, vacation and holiday pay applicable thereto. For the purposes of this Agreement, the principals of **PROJECT MANAGER** and their hourly payroll costs are:

_____
_____
_____
_____
_____

The amount of customary and statutory benefits of all other personnel will be considered equal to ....
% of salaries and wages.

5.3.4. Reimbursable Expenses mean the actual expenses incurred directly or indirectly in connection with the Project for: transportation and subsistence incidental thereto; obtaining bids or proposals from Contractor(s); furnishing and maintaining field office facilities; subsistence and transportation of all resident project personnel; toll telephone calls and telegrams; reproduction of reports, Drawings and Specifications and similar Project-related items in addition to those required under Section 1; computer time including an appropriate charge for previously established programs; and if authorized in advance by **OWNER**, overtime work requiring higher than regular rates.

5.3.5. If **OWNER** fails to make any payment due **PROJECT MANAGER** for services and expenses within sixty days after receipt of **PROJECT MANAGER**'s bill therefor, the amounts due **PROJECT MANAGER** shall include a charge at the rate of 1% per month from said sixtieth day, and in addition **PROJECT MANAGER** may, after giving seven days' written notice to **OWNER**, suspend services under this Agreement until he has been paid in full all amounts due him for services and expenses.

5.3.6. If this Agreement is terminated by **OWNER** upon the completion of any phase of the Basic Services, progress payments due **PROJECT MANAGER** for services rendered through such phase shall constitute total payment for such services. If this Agreement is terminated by **OWNER** during any phase of the Basic Services, **PROJECT MANAGER** will be paid for services rendered during that phase on the basis of payroll costs times a factor of . . . . for services rendered during

that phase to date of termination by principals and employees assigned to the Project. In the event of any termination, **PROJECT MANAGER** will be paid for all unpaid Reimbursable Expenses, plus all termination expenses. Termination expenses mean Reimbursable Expenses directly attributable to termination, which shall include an amount computed as a percentage of total compensation for Basic Services earned by **PROJECT MANAGER** to the date of termination, as follows:

20% if termination occurs after commencement of the Preliminary Design Phase but prior to commencement of the Final Design Phase; or

10% if termination occurs after commencement of the Final Design Phase.

SECTION 6—GENERAL CONSIDERATIONS

6.1. Termination.

This Agreement may be terminated by either party upon seven days' written notice in the event of substantial failure by the other party to perform in accordance with the terms hereof through no fault of the terminating party.

6.2. Reuse of Documents.

All documents including Drawings and Specifications furnished by **PROJECT MANAGER** pursuant to this Agreement are instruments of his services in respect of the Project. They are not intended or represented to be suitable for reuse by **OWNER** or others on extensions of the Project or on any other project. Any reuse without specific written verification or adaptation by **PROJECT MANAGER** will be at **OWNER**'s expense and without liability or legal exposure to **PROJECT MANAGER**, and **OWNER** shall indemnify and hold harmless **PROJECT MANAGER** from all claims, damages, losses and expenses including attorneys' fees arising out of or resulting therefrom. Any such verification or adaptation will entitle **PROJECT MANAGER** to further compensation at rates to be agreed upon by **OWNER** and **PROJECT MANAGER**.

6.3. Estimates of Cost.

6.3.1. Since **PROJECT MANAGER** has no control over the cost of labor, materials or equipment or over Contractor(s)' methods of determining prices, or over competitive bidding or market conditions, his estimates of Project Cost and Construction Cost provided for herein are to be made on the basis of his experience and qualifications and represent his best judgment as a professional experienced and qualified in project management and familiar with the construction industry, but PROJECT MANAGER cannot and does not guarantee that proposals, bids or the actual Project and Construction Costs will not vary from estimates of cost prepared by him. If prior to the Bidding or Negotiating Phase OWNER wishes greater assurance as to the Construction Cost he shall employ an independent cost estimator as provided in paragraph 3.8.

6.3.2. If a Construction Cost limit is established as a condition to this Agreement, the following will apply:

6.3.2.1. The acceptance by OWNER at any time during the Basic Services of a revised estimate of Project Cost in excess of the then established cost limit will constitute an increase in the Construction Cost limit to the extent indicated in such revised estimate.

6.3.2.2. Any Construction Cost limit established by this Agreement will include a bidding contingency of ten percent unless another amount is agreed upon in writing.

6.3.2.3. PROJECT MANAGER will be permitted to determine what materials, equipment, component systems and types of construction are to be included in the Drawings and Specifications and to make reasonable adjustments in the scope of the Project to bring it within the cost limit.

6.3.2.4. If the Bidding or Negotiating Phase has not commenced within six months of the completion of the Final Design Phase, the established Construction Cost limit will not be effective or binding on PROJECT MANAGER, and OWNER shall consent to an adjustment in such cost limit commensurate with any applicable change in the general level of prices in the construction industry between the date of completion of the Final Design Phase and the date on which proposals or bids are sought.

6.3.2.5. If the lowest bona fide proposal or bid exceeds the established Construction Cost limit, OWNER shall (1) give written approval to increase such cost limit, (2) authorize negotiating or rebidding the Project within a reasonable time, or (3) cooperate in revising the Project scope or quality. In the case of (3), PROJECT MANAGER shall, without additional charge, modify the Contract Documents as necessary to bring the Construction Cost within the cost limit. The providing of such service will be the limit of PROJECT MANAGER's responsibility in this regard and, having done so, PROJECT MANAGER shall be entitled to payment for his services in accordance with this Agreement.

6.4. Arbitration.

6.4.1. All claims, counter-claims, disputes and other matters in question between the parties hereto arising out of or relating to this Agreement or the breach thereof will be decided by arbitration in accordance with the Construction Industry Arbitration Rules of the American Arbitration Association then obtaining, subject to the limitations stated in paragraphs 6.4.3 and 6.4.4 below. This Agreement so to arbitrate and any other agreement or consent to arbitrate entered into in accordance therewith as provided below, will be specifically enforceable under the prevailing law of any court having jurisdiction.

6.4.2. Notice of demand for arbitration must be filed in writing with the other parties to this Agreement and with the American Arbitration Association. The demand must be made within a reasonable time after the claim, dispute or other matter in question has arisen. In no event may the demand for arbitration be made after the time when institution of legal or equitable proceedings based on such claim, dispute or other matter in question would be barred by the applicable statute of limitations.

6.4.3. All demands for arbitration and all answering statements thereto which include any monetary claim must contain a statement that the total sum or value in controversy as alleged by the party making such demand or answering statement is not more than $200,000 (exclusive of interest and costs). The arbitrators will not have jurisdiction, power or authority to consider, or make findings (except in denial of their own jurisdiction) concerning any claim, counter-claim, dispute or other matter in question where the amount in controversy thereof is more than $200,000 (exclusive of interest and costs) or to render a monetary award in response thereto against any party which totals more than $200,000 (exclusive of interest and costs).

6.4.4. No arbitration arising out of, or relating to, this Agreement may include, by consolidation, joinder or in any other manner, any additional party not a party to this Agreement.

6.4.5. By written consent signed by all the parties to this Agreement and containing a specific reference hereto, the limitations and restrictions contained in paragraphs 6.4.3 and 6.4.4 may be waived in whole or in part as to any claim, counter-claim, dispute or other matter specifically described in such consent. No consent to arbitration in respect of a specifically described claim, counter-claim, dispute or other matter in question will constitute consent to arbitration any other claim, counter-claim, dispute or other matter in question which is not specifically described in such consent or in which the sum or value in controversy exceeds $200,000 (exclusive of interest and costs) or which is with any party not specifically described therein.

6.4.6. The award rendered by the arbitrators will be final, not subject to appeal and judgment may be entered upon it in any court having jurisdiction thereof.

6.5. Confidentiality.

PROJECT MANAGER shall not disclose or permit the disclosure of any confidential information except to his agents, employees and other consultants who need such confidential information in order to properly perform their duties relative to this Agreement.

6.6. Publicity.

No information relative to Project shall be released by PROJECT MANAGER for publication, advertising or for any other purpose without prior approval of OWNER.

6.7. Accounting Records.

Records of PROJECT MANAGER's Direct Labor Costs, Payroll Costs, contracted professional services from others, Reimbursable Expenses pertaining to the Project, and records of payments and claims between OWNER and Contractor(s), shall be kept on a generally recognized accounting basis and shall be available to OWNER or his authorized representative, at mutually convenient times and shall be retained for three years after final payment or abandonment of the Project

unless OWNER otherwise instructs PROJECT MANAGER in writing.

6.8. Extent of Agreement.

This document together with the exhibits and schedules, which are identified herein and which are hereby made a part hereof and incorporated herein by reference, constitutes the entire agreement between OWNER and PROJECT MANAGER and supersedes all prior negotiations, representations or agreements, either written or oral. There are no conditions, agreements or representations between the parties except those expressed herein. This Agreement may be altered, amended or repealed only by a duly executed written instrument signed by both OWNER and PROJECT MANAGER.

6.9. Applicable Law.

This Agreement shall be governed by the law of the principal place of business of PROJECT MANAGER.

6.10. Successors and Assigns.

OWNER and PROJECT MANAGER each binds himself and his partners, successors, executors, administrators and assigns to the other party to this Agreement and to the partners, successors, executors, administrators and assigns of such other party, in respect to all covenants of this Agreement. Except as above, neither OWNER nor PROJECT MANAGER shall assign, sublet or transfer his interest in this Agreement without the written consent of the other; however, PROJECT MANAGER may employ others to assist him in carrying out his duties under this Agreement. Nothing herein shall be construed as giving any rights or benefits hereunder to anyone other than OWNER and PROJECT MANAGER.

SECTION 7—SPECIAL PROVISIONS

This Agreement is subject to the following special provisions in addition to those set forth above:

7.1.

IN WITNESS WHEREOF the parties hereto have signed this Agreement as of the day and year first above written.

OWNER: PROJECT MANAGER:

. . . .

. . . .

. . . .

Page 12 of _ _ _ _ pages

Professional Construction
Management Approach

Comparison of AIA and AGC forms.   Both the AIA owner–construction manager agreement, Document B801, and a derivative of the AGC owner–construction manager agreement, AGC Document 8d, are designed for use when the construction manager is an independent third party and the owner is awarding and entering into contracts for construction directly. Under neither the AIA nor the AGC document will the construction manager provide a guaranteed maximum price. (Although article 8 of AGC Document 8d does contain a reference to a "Guaranteed Maximum Price, if established," the AGC agreement itself makes no further mention of the concept.)

Because the AIA and AGC documents were originally developed as one draft document, there are few differences in the services offered under the two agreement forms. In AIA Document B801, the construction manager provides a "preliminary evaluation of the program and budget requirements" (par. 1.1.1) not present in the AGC document, while AGC Document 8d calls for the construction manager to "advise the Owner on site selection and participate in development of project feasibility studies." Under the AIA document the construction manager is required to develop a progress schedule for each of the contractors on the project (par. 1.1.5.3) and to make recommendations to the owner regarding allocations of responsibility for safety, temporary facilities, and "general conditions" support items—equipment, materials, and services for common use of the contractors (par. 1.1.5.1). Under the AIA document, any construction support activities must be governed by a separate agreement form. In the AGC document, the construction manager actually performs these services (par. 2.2.2), which are described in detail (par. 1.3.3).

The AIA document contains a few more activities for the construction manager during the design phase than does the AGC document, including an analysis of types and quantities of labor required for the project (par. 1.1.6), advising on methods of selecting contractors and awarding contracts (par. 1.1.5.2) (the AGC document assumes that this will be done through competitive bidding and award of fixed-price contracts), the development of prequalification criteria and the conduct of prebid conferences (par. 1.1.7), and the conduct of preaward conferences and recommendations on the suitability of proposed subcontractors and material suppliers (par. 1.1.8).

During the construction phase, the major differences in services between the AIA and AGC documents are that (1) the AGC's construction manager performs the "General Conditions" construction support items, as defined in paragraph 1.3.3, while the AIA's construction manager does so, if at all, only under a separate agreement; (2) the AIA's construction manager has the authority to reject work not in conformance with the contract documents and to require special inspection or testing of the work (par. 1.2.7), while the AGC's construction manager does not (par. 2.2.8.1); (3) the AIA's construction manager arranges for the delivery and storage of equipment (par. 1.2.12), a provision not found in the AGC document; and (4) the AGC's construction manager determines the dates of substantial completion (par. 2.2.12) and final completion (par. 2.2.14) of the work, while under the AIA agreement, these determinations are left to the architect (par. 1.2.14).

In terms of owner rights and responsibilities, the AIA document makes it clear that the owner retains the right to perform work with its own forces and to award separate contracts which will be outside the construction manager's management authority (par. 2.10). The AGC document requires the owner to demonstrate satisfactory financial strength to the construction manager, failing which the construction manager may stop the project (par. 3.11).

The AIA agreement may be made subject to a fixed limit of construction cost if the owner desires some assurances on cost (par. 3.4). Because the construction manager is not in a position to guarantee costs, however, the fixed limit is subject to many conditions, as in AIA's professional services agreements.

Compensation of the construction manager in the AIA and AGC documents differs in form, if not substance; the AIA document closely follows AIA's other professional services agreements, while the AGC document provides for reimbursement of the construction manager's costs plus payment of a fee as in the AGC cost-plus construction contract forms. Under the AGC agreement, the construction manager is given further economic protection than in the AIA document, interest is payable on amounts owing to the construction manager at the rate of the owner's construction loan (par. 9.4), and the construction manager can stop the project on seven days' notice if it has not been paid within seven days of the due date (par. 9.3). AIA Document B801, on the other hand, does not give the construction manager even the right to suspend performance if it has not been paid; its only remedy is to terminate the agreement if nonpayment has been a substantial breach.

Insurance and termination provisions of the AIA and AGC documents also reflect the institutional bias of the authors. Finally, there is a significant difference in the arbitration provisions of the two documents, in that consolidation of actions is permitted in the AGC document unless expressly prohibited (par. 14.5), while in the AIA document, consolidation is prohibited unless consent by all parties to a multiparty arbitration has been given (par. 9.1).

Except for the differences outlined above, AIA Document B801 and AGC Document 8d reveal a similarity of approach to that of professional construction management in which the construction manager only manages the construction process and does not guarantee costs or results, nor take responsibility for the work of the multiple prime contractors on the project. Even though the construction manager's involvement with the project during construction is much greater than traditional construction contract administration responsibilities undertaken by the A/E, the nature of the construction manager's responsibility (or lack thereof) is no different. On this point, there is clear agreement that responsibility and liability follow control and that it is inappropriate to place ultimate responsibilities for the cost of the project and performance of the contractors on the independent, professional construction manager, who is not in direct control of the construction process. Owners who wish to have single-point responsibility in this approach to construction management must seek it outside of the context of the standard AIA or AGC forms.

INSTRUCTION SHEET *AIA DOCUMENT B801a*

FOR AIA DOCUMENT B801, STANDARD FORM OF AGREEMENT BETWEEN OWNER AND CONSTRUCTION MANAGER—JUNE 1980 EDITION

A. GENERAL INFORMATION

1. Purpose

This document is intended as the basis for the agreement between the Owner and the Construction Manager for Pre-construction Phase and Construction Phase Services. It establishes the Construction Manager as the Owner's agent. Substantial modifications are required if the Construction Manager is to perform construction work, or if a Guaranteed Maximum Cost is to be established.

2. Related Documents

This document is intended to be used in conjunction with the following AIA Documents:
a) A101/CM, Owner-Contractor Agreement, Stipulated Sum, Construction Management Edition
b) A201/CM, General Conditions of the Contract for Construction, Construction Management Edition
c) B141/CM, Owner-Architect Agreement, Construction Management Edition

3. Use of Non-AIA Forms

If a series of non-AIA documents or a mixture of AIA Documents plus non-AIA documents are to be used, particular care must be taken to achieve consistency of language and intent. Certain owners require the use of owner-architect agreements and other contract forms prepared by them. Such forms should be carefully compared to the standard AIA forms for which they are being substituted before executing an agreement. If there are any significant omissions, additions or variances from the terms of the related standard AIA forms, both legal and insurance counsel should be consulted. Of particular concern is the need for consistency between the Owner-Construction Manager Agreement and the anticipated General Conditions of the Contract for Construction in the delineation of the Construction Manager's Construction Phase services and responsibilities.

4. Letter Forms of Agreement

Letter forms of agreement are generally discouraged by the AIA, as is the performance of a part or the whole of professional services based on oral agreements or understandings. The standard AIA Agreement forms have been developed through more than sixty years of experience and have been tested repeatedly in the courts. In addition, the standard forms have been carefully coordinated with other AIA Documents, including the Owner-Architect Agreement forms, the Owner-Contractor Agreements and the General Conditions of the Contract for Construction. The necessity for specific and complete correlation between these documents and any Owner-Construction Manager Agreement used is of paramount importance.

5. Use of Current Documents

Prior to using any AIA Document, the user should consult the AIA or an AIA component to determine the current edition of each Document.

6. Reproduction

AIA Document B801 is a copyrighted document, and may not be reproduced or excerpted from in substantial part without the express written permission of the AIA. Purchasers of B801 are hereby entitled to reproduce a maximum of ten copies of the completed or executed document for use only in connection with the particular Project. AIA will not permit the reproduction of this Document in blank, or the use of substantial portions of, or language from, this Document, except upon written request and after receipt of written permission from AIA.

B. CHANGES FROM THE PREVIOUS EDITION

1. Format Changes

New terminology, especially as to the "Preconstruction Phase", more accurately describes the role of the Construction Manager. The compensation provisions have been moved to a position near the end of the Document rather than preceding the Terms and Conditions. Also, language dealing with alternative methods of computing compensation has been deleted from the Agreement form and is now provided as part of this Instruction Sheet for incorporation by the user.

2. Content Changes

Numerous changes in content have been made in response to experience with the previous edition, the evolution of construction management, recognition of concerns of Owners and the recommendations of AIA members, committees, legal and insurance counsel. Special emphasis has been placed on a more complete description of the unique role played by the Construction Manager in this form of project delivery.

C. COMPLETING THE B801 FORM

1. Modifications

As with all AIA Documents, users are encouraged to consult an attorney with respect to completing the form. Generally, any necessary modifications can be accomplished by describing in Article 16 changes to the printed text. Legal counsel should also be sought concerning the effect of state and local law on the terms of the Agreement, particularly with respect to registration laws, duties imposed by building codes, interest charges and arbitration.

2. Cover Page

Date: The date represents the date as of which the Agreement becomes effective. It may be the date that an oral agreement was reached, the date the Agreement was originally submitted to the Owner, the date the Owner authorizes the Construction Manager to commence services or the date of actual execution. Professional services should not be performed prior to the effective date of the Agreement.

Identification of Parties: Parties to this Agreement should be identified using the full legal name under which the Agreement is to be executed, including a designation of the legal status of each party: sole proprietorship, partnership, joint venture, unincorporated association, limited partnership or corporation (general, close or professional), etc.

Project Description: The proposed Project should be described in sufficient detail to identify: (1) the official name or title of the Project (2) the location of the site, if known, (3) the proposed building type or usage, and (4) the size, capacity or scope of the Project, if known.

Identification of the Architect: The Architect should be identified by its official name, and if any individuals required to serve the Project are identified in the Owner-Architect Agreement, this information should be included. (Identification of the Architect becomes even more significant if construction management services and architectural services are being provided by the same firm. Identification here helps eliminate possibilities of the Owner not being informed of the dual role being played.)

3. Paragraph 1.1—Preconstruction Phase

The Owner and the Construction Manager should review Preconstruction Phase services, especially as to timing the Construction Manager's entry into the Project. If the Owner-Construction Manager Agreement is effective as of a date when the design has been partially completed, the Owner and the Construction Manager should agree on the extent of design evaluation required of the Construction Manager. Consider whether or not this would include the possible recommendation of changes in designs already completed. Any provisions which would require specific approval by the Owner as a condition of having the Construction Manager continue from the Preconstruction into the Construction Phase should be stated in Article 16.

4. Paragraph 1.3—Additional Services

The Owner and the Construction Manager should carefully review the Additional Services listed. Any that are to be done under Basic Services for the Project should be so noted. Others may be added to, or deleted from, the list.

5. Article 2—The Owner's Responsibilities

If the Owner's responsibilities are to be increased or diminished from those shown in the printed form, appropriate language should be added to Article 16.

6. Article 3—Construction Cost

If a fixed limit of Construction Cost has been agreed to as a condition of the Owner-Construction Manager Agreement, this limit should be included in Article 16.

7. Article 4—Construction Support Activities

The Owner and the Construction Manager should agree whether or not the Construction Manager will provide any temporary construction support activities, and if so, whether materials and services will be purchased through either the Owner's or the Construction Manager's Purchase Order, or some other contracting method independent of this Agreement. If, under this Agreement, the Construction Manager will be authorized to purchase or perform work on behalf of the Owner, appropriate language must be added to Article 16.

AIA DOCUMENT B801a • INSTRUCTION SHEET FOR OWNER-CONSTRUCTION MANAGER AGREEMENT • 1980 EDITION
AIA® • THE AMERICAN INSTITUTE OF ARCHITECTS, 1735 NEW YORK AVENUE, N.W., WASHINGTON, D.C. 20006

8. **Article 5—Direct Personnel Expense**

   If the Owner and the Construction Manager prefer to use Direct Salary Expense, appropriate language must be added in Article 16.

9. **Article 6—Reimbursable Costs**

   These should be carefully reviewed by the Owner and the Construction Manager.

10. **Article 15—Basis of Compensation**

    Paragraph 15.1: Insert the amount of the retainer, if any, and indicate whether it will be credited to the first, last, or proportionately to all of the payments on the Owner's account.

    Paragraph 15.2: Spaces have been provided to insert language covering one form of Basic Compensation for Preconstruction Phase Services and a different form of Basic Compensation for Construction Phase Services, as agreed to by the parties. If preferred, the two categories of services may be combined under a single form of compensation. Sample language covering several forms of compensation, which should be inserted in Subparagraph 15.2.1, is provided below:

    Multiple of Direct Personnel Expense

    "Compensation for services rendered by Principals and employees shall be based on a Multiple of Direct Personnel Expense, in the same manner as described in Subparagraph 15.3.1."

    The Construction Manager may wish to include a method of computing Direct Personnel Expense, such as by using a percentage of employee salaries.

    Professional Fee Plus Expenses

    "Compensation shall be based on a Professional Fee of _____ dollars ($) plus compensation for services rendered by Principals and employees in the same manner as described in Subparagraph 15.3.1."

    Stipulated Sum

    "Compensation shall be a Stipulated Sum of _____ dollars ($)."

    Percentage of Construction Cost

    "Compensation shall be based on _____ percent (%) of the Construction Cost, as defined in Article 3."

    If compensation for the Preconstruction Phase is to be separate from, or on a different basis than, the Construction Phase, both must be clearly described. If compensation for the Preconstruction Phase and the Construction Phase are to be covered by a single form of compensation, and that form is either based on a Stipulated Sum or a Percentage of Construction Cost, the following subparagraph should be completed and inserted:

    "15.2.2 Payments for Basic Services shall be made as provided in Subparagraph 7.1.2 so that Basic Compensation for each Phase shall equal the following percentages of the total Basic Compensation payable:

Preconstruction Phase	percent (%)
Construction Phase	percent (%)"

    In Paragraph 15.3, insert provisions for compensation for Additional Services. For example, if hourly rates are used, the provision might read:
    "Principals' time at the fixed rate of _____ dollars ($) per hour. For the purposes of this Article, the Principals are:

    These billing rates shall be adjusted annually (semi-annually) in accordance with the Construction Manager's adjustments in compensation for Principals and employees." (Add agreed upon limitations.)

    If a Multiple of Direct Personnel Expense is used, insert:
    "Principals' and employees' time at a multiple of _____ ( ) times their Direct Personnel Expense as defined in Article 5."

    If a multiple of direct salaries is used, the term "direct salaries" should be substituted for Direct Personnel Expense above, and this should be noted in Article 16.

    Paragraph 15.5: Establish a due date for payment and insert the percentage rate and basis (monthly or annual) and the time (such as a number of days) after the due date for payment on which interest charges will begin to run. This should be carefully checked against state usury laws which may set a limit on the rate of interest which may legally be charged. In addition, federal truth in lending and similar state and local consumer protection laws may require setting forth the annual percentage rate and other disclosures or waivers for certain types of transactions or with certain types of clients. Advice of legal counsel should be sought on such matters.

Subparagraph 15.6.2: Insert the amount of time after which compensation shall be subject to renegotiation and adjustment.

11. **Article 16—Other Conditions or Services**

Include here any further information required to describe agreements reached between the Owner and the Construction Manager which might include such provisions as:
a) Additional phases
b) Phased construction requirements
c) Additional value management services
d) Use of computers and arrangements for billing computer time
e) Work by the Construction Manager
f) Fixed limit of Construction Cost
g) Guaranteed Maximum Cost, with or without shared savings
h) Additional Basic Services beyond those shown in the standard form
i) Additional Services beyond those shown in the standard form
j) Identification of consultants, if applicable
k) Prequalification of bidders, restrictions on bidding due to the use of public money, etc.
l) Fixed time of performance
m) Modifications to any services or conditions
n) Other additional conditions

Note that any changes in the duties of the Construction Manager during the Construction Phase must be considered with extreme care and correlated with the terms of A201/CM, General Conditions of the Contract for Construction, Construction Management Edition.

D. **EXECUTION OF AGREEMENT**

Each person executing the Agreement should indicate the capacity in which they are acting (i.e., president, secretary, partner, etc.) and the authority under which they are executing the Agreement. Where appropriate, a copy of the resolution authorizing the individual to act on behalf of the firm or entity should be attached.

THE AMERICAN INSTITUTE OF ARCHITECTS

AIA Document B801

Standard Form of Agreement Between Owner and Construction Manager

1980 EDITION

THIS DOCUMENT HAS IMPORTANT LEGAL CONSEQUENCES; CONSULTATION WITH AN ATTORNEY IS ENCOURAGED.

This document is intended to be used in conjunction with
AIA Documents A101/CM, 1980; B141/CM, 1980; and A201/CM, 1980.

AGREEMENT

made as of the   day of   in the year of Nineteen
Hundred and

BETWEEN the Owner:

and the Construction Manager:

For the following Project:
(Include detailed description of Project location and scope.)

the Architect:

The Owner and the Construction Manager agree as set forth below.

TERMS AND CONDITIONS OF AGREEMENT BETWEEN
OWNER AND CONSTRUCTION MANAGER

ARTICLE 1
CONSTRUCTION MANAGER'S SERVICES AND RESPONSIBILITIES

The Construction Manager covenants with the Owner to further the interests of the Owner by furnishing the Construction Manager's skill and judgment in cooperation with, and in reliance upon, the services of an architect. The Construction Manager agrees to furnish business administration and management services and to perform in an expeditious and economical manner consistent with the interests of the Owner.

BASIC SERVICES

The Construction Manager's Basic Services consist of the two Phases described below and any other services included in Article 16 as Basic Services.

1.1   PRECONSTRUCTION PHASE

1.1.1 Provide preliminary evaluation of the program and Project budget requirements, each in terms of the other. With the Architect's assistance, prepare preliminary estimates of Construction Cost for early schematic designs based on area, volume or other standards. Assist the Owner and the Architect in achieving mutually agreed upon program and Project budget requirements and other design parameters. Provide cost evaluations of alternative materials and systems.

1.1.2 Review designs during their development. Advise on site use and improvements, selection of materials, building systems and equipment and methods of Project delivery. Provide recommendations on relative feasibility of construction methods, availability of materials and labor, time requirements for procurement, installation and construction, and factors related to cost including, but not limited to, costs of alternative designs or materials, preliminary budgets and possible economies.

1.1.3 Provide for the Architect's and the Owner's review and acceptance, and periodically update, a Project Schedule that coordinates and integrates the Construction Manager's services, the Architect's services and the Owner's responsibilities with anticipated construction schedules.

1.1.4 Prepare for the Owner's approval a more detailed estimate of Construction Cost, as defined in Article 3, developed by using estimating techniques which anticipate the various elements of the Project, and based on Schematic Design Documents prepared by the Architect. Update and refine this estimate periodically as the Architect prepares Design Development and Construction Documents. Advise the Owner and the Architect if it appears that the Construction Cost may exceed the Project budget. Make recommendations for corrective action.

1.1.5 Coordinate Contract Documents by consulting with the Owner and the Architect regarding Drawings and Specifications as they are being prepared, and recommending alternative solutions whenever design details affect construction feasibility, cost or schedules.

1.1.5.1 Provide recommendations and information to the Owner and the Architect regarding the assignment of responsibilities for safety precautions and programs; temporary Project facilities; and equipment, materials and services for common use of Contractors. Verify that the requirements and assignment of responsibilities are included in the proposed Contract Documents.

1.1.5.2 Advise on the separation of the Project into Contracts for various categories of Work. Advise on the method to be used for selecting Contractors and awarding Contracts. If separate Contracts are to be awarded, review the Drawings and Specifications and make recommendations as required to provide that (1) the Work of the separate Contractors is coordinated, (2) all requirements for the Project have been assigned to the appropriate separate Contract, (3) the likelihood of jurisdictional disputes has been minimized, and (4) proper coordination has been provided for phased construction.

1.1.5.3 Develop a Project Construction Schedule providing for all major elements such as phasing of construction and times of commencement and completion required of each separate Contractor. Provide the Project Construction Schedule for each set of Bidding Documents.

1.1.5.4 Investigate and recommend a schedule for the Owner's purchase of materials and equipment requiring long lead time procurement, and coordinate the schedule with the early preparation of portions of the Contract Documents by the Architect. Expedite and coordinate delivery of these purchases.

1.1.6 Provide an analysis of the types and quantities of labor required for the Project and review the availability of appropriate categories of labor required for critical Phases. Make recommendations for actions designed to minimize adverse effects of labor shortages.

1.1.6.1 Identify or verify applicable requirements for equal employment opportunity programs for inclusion in the proposed Contract Documents.

1.1.7 Make recommendations for pre-qualification criteria for Bidders and develop Bidders' interest in the Project. Establish bidding schedules. Assist the Architect in issuing Bidding Documents to Bidders. Conduct pre-bid conferences to familiarize Bidders with the Bidding Documents and management techniques and with any special systems, materials or methods. Assist the Architect with the receipt of questions from Bidders, and with the issuance of Addenda.

1.1.7.1 With the Architect's assistance, receive Bids, prepare bid analyses and make recommendations to the Owner for award of Contracts or rejection of Bids.

1.1.8 With the Architect's assistance, conduct pre-award conferences with successful Bidders. Assist the Owner in preparing Construction Contracts and advise the Owner on the acceptability of Subcontractors and material suppliers proposed by Contractors.

1.2   CONSTRUCTION PHASE

The Construction Phase will commence with the award of the initial Construction Contract or purchase order and, together with the Construction Manager's obligation to provide Basic Services under this Agreement, will end 30 days after final payment to all Contractors is due.

1.2.1 Unless otherwise provided in this Agreement and incorporated in the Contract Documents, the Construction Manager, in cooperation with the Architect, shall provide administration of the Contracts for Construction as set forth below and in the 1980 Edition of AIA Document A201/CM, General Conditions of the Contract for Construction, Construction Management Edition.

1.2.2 Provide administrative, management and related services as required to coordinate Work of the Contractors with each other and with the activities and responsibilities of the Construction Manager, the Owner and the Architect to complete the Project in accordance with the Owner's objectives for cost, time and quality. Provide sufficient organization, personnel and management to carry out the requirements of this Agreement.

1.2.2.1 Schedule and conduct pre-construction, construction and progress meetings to discuss such matters as procedures, progress, problems and scheduling. Prepare and promptly distribute minutes.

1.2.2.2 Consistent with the Project Construction Schedule issued with the Bidding Documents, and utilizing the Contractors' Construction Schedules provided by the separate Contractors, update the Project Construction Schedule incorporating the activities of Contractors on the Project, including activity sequences and durations, allocation of labor and materials, processing of Shop Drawings, Product Data and Samples, and delivery of products requiring long lead time procurement. Include the Owner's occupancy requirements showing portions of the Project having occupancy priority. Update and reissue the Project Construction Schedule as required to show current conditions and revisions required by actual experience.

1.2.2.3 Endeavor to achieve satisfactory performance from each of the Contractors. Recommend courses of action to the Owner when requirements of a Contract are not being fulfilled, and the nonperforming party will not take satisfactory corrective action.

1.2.3 Revise and refine the approved estimate of Construction Cost, incorporate approved changes as they occur, and develop cash flow reports and forecasts as needed.

1.2.3.1 Provide regular monitoring of the approved estimate of Construction Cost, showing actual costs for activities in progress and estimates for uncompleted tasks. Identify variances between actual and budgeted or estimated costs, and advise the Owner and the Architect whenever projected costs exceed budgets or estimates.

1.2.3.2 Maintain cost accounting records on authorized Work performed under unit costs, additional Work performed on the basis of actual costs of labor and materials, or other Work requiring accounting records.

1.2.3.3 Recommend neccessary or desirable changes to the Architect and the Owner, review requests for changes, assist in negotiating Contractors' proposals, submit recommendations to the Architect and the Owner, and if they are accepted, prepare and sign Change Orders for the Architect's signature and the Owner's authorization.

1.2.3.4 Develop and implement procedures for the review and processing of Applications by Contractors for progress and final payments. Make recommendations to the Architect for certification to the Owner for payment.

1.2.4 Review the safety programs developed by each of the Contractors as required by their Contract Documents and coordinate the safety programs for the Project.

1.2.5 Assist in obtaining building permits and special permits for permanent improvements, excluding permits required to be obtained directly by the various Contractors. Verify that the Owner has paid applicable fees and assessments. Assist in obtaining approvals from authorities having jurisdiction over the Project.

1.2.6 If required, assist the Owner in selecting and retaining the professional services of surveyors, special consultants and testing laboratories. Coordinate their services.

1.2.7 Determine in general that the Work of each Contractor is being performed in accordance with the requirements of the Contract Documents. Endeavor to guard the Owner against defects and deficiencies in the Work. As appropriate, require special inspection or testing, or make recommendations to the Architect regarding special inspection or testing, of Work not in accordance with the provisions of the Contract Documents whether or not such Work be then fabricated, installed or completed. Subject to review by the Architect, reject Work which does not conform to the requirements of the Contract Documents.

1.2.7.1 The Construction Manager shall not be responsible for construction means, methods, techniques, sequences and procedures employed by Contractors in the performance of their Contracts, and shall not be responsible for the failure of any Contractor to carry out Work in accordance with the Contract Documents.

1.2.8 Consult with the Architect and the Owner if any Contractor requests interpretations of the meaning and intent of the Drawings and Specifications, and assist in the resolution of questions which may arise.

1.2.9 Receive Certificates of Insurance from the Contractors, and forward them to the Owner with a copy to the Architect.

1.2.10 Receive from the Contractors and review all Shop Drawings, Product Data, Samples and other submittals. Coordinate them with information contained in related documents and transmit to the Architect those recommended for approval. In collaboration with the Architect, establish and implement procedures for expediting the processing and approval of Shop Drawings, Product Data, Samples and other submittals.

1.2.11 Record the progress of the Project. Submit written progress reports to the Owner and the Architect including information on each Contractor and each Contractor's Work, as well as the entire Project, showing percentages of completion and the number and amounts of Change Orders. Keep a daily log containing a record of weather, Contractors' Work on the site, number of workers, Work accomplished, problems encountered, and other similar relevant data as the Owner may require. Make the log available to the Owner and the Architect.

1.2.11.1 Maintain at the Project site, on a current basis: a record copy of all Contracts, Drawings, Specifications, Addenda, Change Orders and other Modifications, in good order and marked to record all changes made during construction; Shop Drawings; Product Data; Samples; submittals; purchases; materials; equipment; applicable handbooks; maintenance and operating manuals and instruc-

AIA DOCUMENT B801 • OWNER-CONSTRUCTION MANAGER AGREEMENT • JUNE 1980 EDITION • AIA®
©1980 • THE AMERICAN INSTITUTE OF ARCHITECTS, 1735 NEW YORK AVE., N.W., WASHINGTON, D.C 20006

tions; other related documents and revisions which arise out of the Contracts or Work. Maintain records, in duplicate, of principal building layout lines, elevations of the bottom of footings, floor levels and key site elevations certified by a qualified surveyor or professional engineer. Make all records available to the Owner and the Architect. At the completion of the Project, deliver all such records to the Architect for the Owner.

1.2.12 Arrange for delivery and storage, protection and security for Owner-purchased materials, systems and equipment which are a part of the Project, until such items are incorporated into the Project.

1.2.13 With the Architect and the Owner's maintenance personnel, observe the Contractors' checkout of utilities, operational systems and equipment for readiness and assist in their initial start-up and testing.

1.2.14 When the Construction Manager considers each Contractor's Work or a designated portion thereof substantially complete, the Construction Manager shall prepare for the Architect a list of incomplete or unsatisfactory items and a schedule for their completion. The Construction Manager shall assist the Architect in conducting inspections. After the Architect certifies the Date of Substantial Completion of the Work, the Construction Manager shall coordinate the correction and completion of the Work.

1.2.15 Assist the Architect in determining when the Project or a designated portion thereof is substantially complete. Prepare for the Architect a summary of the status of the Work of each Contractor, listing changes in the previously issued Certificates of Substantial Completion of the Work and recommending the times within which Contractors shall complete uncompleted items on their Certificate of Substantial Completion of the Work.

1.2.16 Following the Architect's issuance of a Certificate of Substantial Completion of the Project or designated portion thereof, evaluate the completion of the Work of the Contractors and make recommendations to the Architect when Work is ready for final inspection. Assist the Architect in conducting final inspections. Secure and transmit to the Owner required guarantees, affidavits, releases, bonds and waivers. Deliver all keys, manuals, record drawings and maintenance stocks to the Owner.

1.2.17 The extent of the duties, responsibilities and limitations of authority of the Construction Manager as a representative of the Owner during construction shall not be modified or extended without the written consent of the Owner, the Contractors, the Architect and the Construcion Manager, which consent shall not be unreasonably withheld.

1.3 ADDITIONAL SERVICES

The following Additional Services shall be performed upon authorization in writing from the Owner and shall be paid for as provided in this Agreement.

1.3.1 Services related to investigations, appraisals or evaluations of existing conditions, facilities or equipment, or verification of the accuracy of existing drawings or other information furnished by the Owner.

1.3.2 Services related to Owner-furnished furniture, furnishings and equipment which are not a part of the Project.

1.3.3 Services for tenant or rental spaces.

1.3.4 Consultation on replacement of Work damaged by fire or other cause during construction, and furnishing services in conjunction with the replacement of such Work.

1.3.5 Services made necessary by the default of a Contractor.

1.3.6 Preparing to serve or serving as a witness in connection with any public hearing, arbitration proceeding or legal proceeding.

1.3.7 Recruiting or training maintenance personnel.

1.3.8 Inspections of, and services related to, the Project after the end of the Construction Phase.

1.3.9 Providing any other services not otherwise included in this Agreement.

1.4 TIME

1.4.1 The Construction Manager shall perform Basic and Additional Services as expeditiously as is consistent with reasonable skill and care and the orderly progress of the Project.

ARTICLE 2
THE OWNER'S RESPONSIBILITIES

2.1 The Owner shall provide full information regarding the requirements of the Project, including a program, which shall set forth the Owner's objectives, constraints and criteria, including space requirements and relationships, flexibility and expandability requirements, special equipment and systems and site requirements.

2.2 The Owner shall provide a budget for the Project, based on consultation with the Construction Manager and the Architect, which shall include contingencies for bidding, changes during construction and other costs which are the responsibility of the Owner. The Owner shall, at the request of the Construction Manager, provide a statement of funds available for the Project and their source.

2.3 The Owner shall designate a representative authorized to act in the Owner's behalf with respect to the Project. The Owner, or such authorized representative, shall examine documents submitted by the Construction Manager and shall render decisions pertaining thereto promptly to avoid unreasonable delay in the progress of the Construction Manager's services.

2.4 The Owner shall retain an architect whose services, duties and responsibilities are described in the agreement between the Owner and the Architect, AIA Document B141/CM, 1980 Edition. The Terms and Conditions of the Owner-Architect Agreement will be furnished to the Construction Manager, and will not be modified without written consent of the Construction Manager, which consent shall not be unreasonably withheld. Actions taken by the Architect as agent of the Owner shall be the acts of the Owner and the Construction Manager shall not be responsible for them.

2.5 The Owner shall furnish structural, mechanical, chemical and other laboratory tests, inspections and reports as required by law or the Contract Documents.

2.6 The Owner shall furnish such legal, accounting and insurance counseling services as may be necessary for the Project, including such auditing services as the Owner may require to verify the Project Applications for Payment or to ascertain how or for what purposes the Contractors have used the monies paid by or on behalf of the Owner.

2.7 The Owner shall furnish the Construction Manager a sufficient quantity of construction documents.

2.8 The services, information and reports required by Paragraphs 2.1 through 2.7, inclusive, shall be furnished at the Owner's expense, and the Construction Manager shall be entitled to rely upon their accuracy and completeness.

2.9 If the Owner observes or otherwise becomes aware of any fault or defect in the Project, or nonconformance with the Contract Documents, prompt written notice thereof shall be given by the Owner to the Construction Manager and the Architect.

2.10 The Owner reserves the right to perform work related to the Project with the Owner's own forces, and to award contracts in connection with the Project which are not part of the Construction Manager's responsibilities under this Agreement. The Construction Manager shall notify the Owner if any such independent action will in any way compromise the Construction Manager's ability to meet the Construction Manager's responsibilities under this Agreement.

2.11 The Owner shall furnish the required information and services and shall render approvals and decisions as expeditiously as necessary for the orderly progress of the Construction Manager's services and the Work of the Contractors.

ARTICLE 3
CONSTRUCTION COST

3.1 Construction Cost shall be the total of the final Contract Sums of all of the separate Contracts, actual Reimbursable Costs relating to the Construction Phase as defined in Article 6, and the Construction Manager's compensation.

3.2 Construction Cost does not include the compensation of the Architect and the Architect's consultants, the cost of the land, rights-of-way or other costs which are the responsibility of the Owner as provided in Paragraphs 2.3 through 2.7, inclusive.

3.3 Evaluations of the Owner's Project budget and cost estimates prepared by the Construction Manager represent the Construction Manager's best judgment as a professional familiar with the construction industry. It is recognized, however, that neither the Construction Manager nor the Owner has control over the cost of labor, materials or equipment, over Contractors' methods of determining Bid prices or other competitive bidding or negotiating conditions. Accordingly, the Construction Manager cannot and does not warrant or represent that Bids or negotiated prices will not vary from the Project budget proposed, established or approved by the Owner, or from any cost estimate or evaluation prepared by the Construction Manager.

3.4 No fixed limit of Construction Cost shall be established as a condition of this Agreement by the furnishing, proposal or establishment of a Project budget under Subparagraph 1.1.1 or Paragraph 2.2, or otherwise, unless such fixed limit has been agreed upon in writing and signed by the parties to this Agreement. If such a fixed limit has been established, the Construction Manager shall include contingencies for design, bidding and price escalation, and

shall consult with the Architect to determine what materials, equipment, component systems and types of construction are to be included in the Contract Documents, to suggest reasonable adjustments in the scope of the Project, and to suggest alternate Bids in the Construction Documents to adjust the Construction Cost to the fixed limit. Any such fixed limit shall be increased in the amount of any increase in the Contract Sums occurring after the execution of the Contracts for Construction.

3.4.1 If Bids are not received within the time scheduled at the time the fixed limit of Construction Cost was established, due to causes beyond the Construction Manager's control, any fixed limit of Construction Cost established as a condition of this Agreement shall be adjusted to reflect any change in the general level of prices in the construction industry occurring between the originally scheduled date and the date on which Bids are received.

3.4.2 If a fixed limit of Construction Cost (adjusted as provided in Subparagraph 3.4.1) is exceeded by the sum of the lowest figures from bona fide Bids or negotiated proposals plus the Construction Manager's estimate of other elements of Construction Cost for the Project, the Owner shall (1) give written approval of an increase in such fixed limit, (2) authorize rebidding or renegotiation of the Project or portions of the Project within a reasonable time, (3) if the Project is abandoned, terminate in accordance with Paragraph 10.2, or (4) cooperate in revising the scope and quality of the Work as required to reduce the Construction Cost. In the case of item (4), the Construction Manager, without additional compensation, shall cooperate with the Architect as necessary to bring the Construction Cost within the fixed limit.

ARTICLE 4
CONSTRUCTION SUPPORT ACTIVITIES

4.1 Construction support activities, if provided by the Construction Manager, shall be governed by separate contractual arrangements unless otherwise provided in Article 16.

ARTICLE 5
DIRECT PERSONNEL EXPENSE

5.1 Direct Personnel Expense is defined as the direct salaries of all of the Construction Manager's personnel engaged on the Project, excluding those whose compensation is included in the fee, and the portion of the cost of their mandatory and customary contributions and benefits related thereto such as employment taxes and other statutory employee benefits, insurance, sick leave, holidays, vacations, pensions, and similar contributions and benefits.

ARTICLE 6
REIMBURSABLE COSTS

6.1 The term Reimbursable Costs shall mean costs necessarily incurred in the proper performance of services and paid by the Construction Manager. Such costs shall be at rates not higher than the standard paid in the locality of the Project, except with prior consent of the Owner. Reimbursable Costs and costs not to be reimbursed shall be listed in Article 16.

6.2 Trade discounts, rebates and refunds, and returns from sale of surplus materials and equipment shall accrue to the Owner, and the Construction Manager shall make provisions so that they can be secured.

ARTICLE 7
PAYMENTS TO THE CONSTRUCTION MANAGER

7.1   PAYMENTS ON ACCOUNT OF BASIC SERVICES

7.1.1 An initial payment as set forth in Paragraph 15.1 is the minimum payment under this Agreement.

7.1.2 Subsequent payments for Basic Services shall be made monthly and shall be in proportion to services performed within each Phase of Services, on the basis set forth in Article 15.

7.1.3 If and to the extent that the time initially established for the Construction Phase of the Project is exceeded or extended through no fault of the Construction Manager, compensation for Basic Services required for such extended period of Administration of the Construction Contract shall be computed as set forth in Paragraph 15.3 for Additional Services.

7.1.4 When compensation is based on a percentage of the total of the Contract Sums of all the separate Contracts, and any portions of the Project are deleted or otherwise not constructed, compensation for such portions of the Project shall be payable to the extent services are performed on such portions, in accordance with the schedule set forth in Subparagraph 15.2.1, based on (1) the lowest figures from bona fide Bids or negotiated proposals, or (2) if no such Bids or proposals are received, the most recent estimate of the total of the Contract Sums of all the separate Contracts for such portions of the Project.

7.2   PAYMENTS ON ACCOUNT OF ADDITIONAL SERVICES AND REIMBURSABLE COSTS

7.2.1 Payments on account of the Construction Manager's Additional Services, as defined in Paragraph 1.3, and for Reimbursable Costs, as defined in Article 16, shall be made monthly upon presentation of the Construction Manager's statement of services rendered or costs incurred.

7.3   PAYMENTS WITHHELD

7.3.1 No deductions shall be made from the Construction Manager's compensation on account of penalty, liquidated damages or other sums withheld from payments to Contractors, or on account of the cost of changes in Work other than those for which the Construction Manager is held legally liable.

7.4   PROJECT SUSPENSION OR ABANDONMENT

7.4.1 If the Project is suspended or abandoned in whole or in part for more than three months, the Construction Manager shall be compensated for all services performed prior to receipt of written notice from the Owner of such suspension or abandonment, together with Reimbursable Costs then due and all Termination Expenses as defined in Paragraph 10.4. If the Project is resumed after being suspended for more than three months, the Construction Manager's compensation shall be equitably adjusted.

7.4.2 If construction of the Project has started and is stopped by reason of circumstances not the fault of the Construction Manager, the Owner shall reimburse the Construction Manager for the costs of the Construction

Manager's Project-site staff as provided for by this Agreement. The Construction Manager shall reduce the size of the Project-site staff after 30 days' delay, or sooner if feasible, for the remainder of the delay period as directed by the Owner and, during that period, the Owner shall reimburse the Construction Manager for the costs of such staff prior to reduction plus any relocation or employment termination costs. Upon the termination of the stoppage, the Construction Manager shall provide the necessary Project-site staff as soon as practicable.

ARTICLE 8
CONSTRUCTION MANAGER'S ACCOUNTING RECORDS

8.1 Records of Reimbursable Costs and costs pertaining to services performed on the basis of a Multiple of Direct Personnel Expense shall be kept on the basis of generally accepted accounting principles and shall be available to the Owner or the Owner's authorized representative at mutually convenient times.

ARTICLE 9
ARBITRATION

9.1 All claims, disputes and other matters in question between the parties to this Agreement arising out of or relating to this Agreement or the breach thereof, shall be decided by arbitration in accordance with the Construction Industry Arbitration Rules of the American Arbitration Association then obtaining unless the parties mutually agree otherwise. No arbitration arising out of or relating to this Agreement shall include, by consolidation, joinder or in any other manner, any additional person not a party to this Agreement except by written consent containing a specific reference to this Agreement and signed by the Construction Manager, the Owner, and any other person sought to be joined. Any consent to arbitration involving an additional person or persons shall not constitute consent to arbitration of any dispute not described therein or with any person not named or described therein. This agreement to arbitrate and any agreement to arbitrate with an additional person or persons duly consented to by the parties to this Agreement shall be specifically enforceable under the prevailing arbitration law.

9.2 Notice of demand for arbitration shall be filed in writing with the other party to this Agreement and with the American Arbitration Association, and a copy shall also be filed with the Architect. The demand shall be made within a reasonable time after the claim, dispute or other matter in question has arisen. In no event shall the demand for arbitration be made after the date when institution of legal or equitable proceedings based on such claim, dispute or other matter in question would be barred by the applicable statute of limitations.

9.3 The award rendered by the arbitrators shall be final, and judgment may be entered upon it in accordance with applicable law in any court having jurisdiction thereof.

ARTICLE 10
TERMINATION OF AGREEMENT

10.1 This Agreement may be terminated by either party upon seven days' written notice should the other party fail substantially to perform in accordance with its terms through no fault of the party initiating the termination.

10.2 This Agreement may be terminated by the Owner upon at least fourteen days' written notice to the Construction Manager in the event that the Project is permanently abandoned.

10.3 In the event of termination not the fault of the Construction Manager, the Construction Manager shall be compensated for all services performed to the termination date together with Reimbursable Costs then due and all Termination Expenses.

10.4 Termination Expenses are defined as Reimbursable Costs directly attributable to termination for which the Construction Manager is not otherwise compensated.

ARTICLE 11
MISCELLANEOUS PROVISIONS

11.1 Unless otherwise specified, this Agreement shall be governed by the law in effect at the location of the Project.

11.2 Terms in this Agreement shall have the same meaning as those in the 1980 Edition of AIA Document A201/CM, General Conditions of the Contract for Construction, Construction Management Edition.

11.3 As between the parties to this Agreement: as to all acts or failures to act by either party to this Agreement, any applicable statute of limitations shall commence to run, and any alleged cause of action shall be deemed to have accrued, in any and all events not later than the relevant Date of Substantial Completion of the Project, and as to any acts or failures to act occurring after the relevant Date of Substantial Completion of the Project, not later than the date of issuance of the final Project Certificate for Payment.

11.4 The Owner and the Construction Manager waive all rights against each other, and against the contractors, consultants, agents and employees of the other, for damages covered by any property insurance during construction, as set forth in the 1980 Edition of AIA Document A201/CM, General Conditions of the Contract for Construction, Construction Management Edition. The Owner and the Construction Manager shall each require appropriate similar waivers from their contractors, consultants and agents.

ARTICLE 12
SUCCESSORS AND ASSIGNS

12.1 The Owner and the Construction Manager, respectively, bind themselves, their partners, successors, assigns and legal representatives to the other party to this Agreement, and to the partners, successors, assigns and legal representatives of such other party with respect to all covenants of this Agreement. Neither the Owner nor the Construction Manager shall assign, sublet or transfer any interest in this Agreement without the written consent of the other.

ARTICLE 13
EXTENT OF AGREEMENT

13.1 This Agreement represents the entire and integrated agreement between the Owner and the Construction Manager and supersedes all prior negotiations, representations or agreements, either written or oral. This Agreement may be amended only by written instrument signed by both the Owner and the Construction Manager.

13.2 Nothing contained herein shall be deemed to create any contractual relationship between the Construction Manager and the Architect or any of the Contractors, Subcontractors or material suppliers on the Project; nor shall anything contained in this Agreement be deemed to give any third party any claim or right of action against the Owner or the Construction Manager which does not otherwise exist without regard to this Agreement.

ARTICLE 14
INSURANCE

14.1 The Construction Manager shall purchase and maintain insurance for protection from claims under workers' or workmen's compensation acts; claims for damages because of bodily injury, including personal injury, sickness, disease or death of any of the Construction Manager's employees or of any person; from claims for damages because of injury to or destruction of tangible property including loss of use resulting therefrom; and from claims arising out of the performance of this Agreement and caused by negligent acts for which the Construction Manager is legally liable.

AIA DOCUMENT B801 • OWNER-CONSTRUCTION MANAGER AGREEMENT • JUNE 1980 EDITION • AIA®
©1980 • THE AMERICAN INSTITUTE OF ARCHITECTS, 1735 NEW YORK AVE., N.W., WASHINGTON, D.C. 20006

ARTICLE 15
BASIS OF COMPENSATION

The Owner shall compensate the Construction Manager for the Scope of Services provided, in accordance with Article 7; Payments to the Construction Manager, and the other Terms and Conditions of this Agreement, as follows:

15.1   AN INITIAL PAYMENT of   dollars ($) shall be made upon execution of this Agreement and credited to the Owner's account as follows:

15.2   **BASIC COMPENSATION**

15.2.1   FOR BASIC SERVICES, as described in Paragraphs 1.1 and 1.2, and any other services included in Article 16 as part of Basic Services, Basic Compensation shall be computed as follows:

For Preconstruction Phase Services, compensation shall be:
(Here insert basis of compensation, including fixed amounts, multiples or percentages.)

For Construction Phase Services, compensation shall be:
(Here insert basis of compensation, including fixed amounts, multiples or percentages.)

15.3   **COMPENSATION FOR ADDITIONAL SERVICES**

15.3.1   FOR ADDITIONAL SERVICES OF THE CONSTRUCTION MANAGER, as described in Paragraph 1.3, and any other services included in Article 16 as Additional Services, compensation shall be computed as follows:
(Here insert basis of compensation, including fixed amounts, multiples or percentages.)

15.4   FOR REIMBURSABLE COSTS, as described in Article 6 and Article 16, the actual costs incurred by the Construction Manager in the interest of the Project.

15.5   Payments due the Construction Manager and unpaid under this Agreement shall bear interest from the date payment is due at the rate entered below, or in the absence thereof, at the legal rate prevailing at the principal place of business of the Construction Manager.
(Here insert any rate of interest agreed upon.)

(Usury laws and requirements under the Federal Truth in Lending Act, similar state and local consumer credit laws, and other regulations at the Owner's and Construction Manager's principal places of business, the location of the Project and elsewhere may affect the validity of this provision. Specific legal advice should be obtained with respect to deletion, modification or other requirements such as written disclosures or waivers.)

15.6   The Owner and the Construction Manager agree in accordance with the Terms and Conditions of this Agreement that:

15.6.1   IF THE SCOPE of the Project or the Construction Manager's Services is changed materially, the amounts of compensation shall be equitably adjusted.

15.6.2   IF THE SERVICES covered by this Agreement have not been completed within (   ) months of the date hereof, through no fault of the Construction Manager, the amounts of compensation, rates and multiples set forth herein shall be equitably adjusted.

AIA DOCUMENT B801 • OWNER-CONSTRUCTION MANAGER AGREEMENT • JUNE 1980 EDITION • AIA®
©1980 • THE AMERICAN INSTITUTE OF ARCHITECTS, 1735 NEW YORK AVE., N.W., WASHINGTON, D.C. 20006   **B801 — 1980 8**

ARTICLE 16
OTHER CONDITIONS OR SERVICES

(List Reimbursable Costs and costs not to be reimbursed.)

This Agreement entered into as of the day and year first written above.

OWNER   CONSTRUCTION MANAGER

_____   _____
_____   _____
_____   _____

AIA DOCUMENT B801 • OWNER-CONSTRUCTION MANAGER AGREEMENT • JUNE 1980 EDITION • AIA®
©1980 • THE AMERICAN INSTITUTE OF ARCHITECTS, 1735 NEW YORK AVE., N.W., WASHINGTON, D.C. 20006   **B801 — 1980 9**

THE ASSOCIATED GENERAL CONTRACTORS

STANDARD FORM OF AGREEMENT BETWEEN OWNER AND CONSTRUCTION MANAGER

(OWNER AWARDS ALL TRADE CONTRACTS)

This Document has important legal and insurance consequences; consultation with an attorney is encouraged with respect to its completion or modification.

AGREEMENT

Made this day   of   in the year Nineteen Hundred and

BETWEEN

the Owner, and

the Construction Manager.

For services in connection with the following described Project: (Include complete Project location and scope)

The Architect/Engineer for the Project is

The Owner and the Construction Manager agree as set forth below:

Certain provisions of this document have been derived, with modifications, from the following documents published by The American Institute of Architects. AIA Document A111, Owner-Contractor Agreement, ©1974, AIA Document A201, General Conditions, ©1976, AIA Document B801, Owner-Construction Manager Agreement, ©1973, by The American Institute of Architects. Usage made of AIA language, with the permission of AIA, does not imply AIA endorsement or approval of this document. Further reproduction of copyrighted AIA materials without separate written permission from AIA is prohibited.

AGC Document No. 8d   Owner-Construction Manager Agreement (Owner Awards Contracts)   June 1979
©1979 Associated General Contractors of America

TABLE OF CONTENTS

ARTICLE 1

The Construction Team and Extent of Agreement

The CONSTRUCTION MANAGER accepts the relationship of trust and confidence established between him and the Owner by this Agreement. He covenants with the Owner to furnish his best skill and judgment and to cooperate with the Architect/Engineer in furthering the interests of the Owner. He agrees to furnish efficient business administration and superintendence and to use his best efforts to complete the Project in the best and soundest way and in the most expeditious and economical manner consistent with the interest of the Owner.

1.1 *The Construction Team:* The Construction Manager, the Owner, and the Architect/Engineer called the "Construction Team" shall work from the beginning of design through construction completion. The Construction Manager shall provide leadership to the Construction Team on all matters relating to construction.

1.2 *Extent of Agreement:* This Agreement represents the entire agreement between the Owner and the Construction Manager and supersedes all prior negotiations, representations or agreements. When Drawings and Specifications are complete, they shall be identified by amendment to this Agreement. This Agreement shall not be superseded by any provisions of the documents for construction and may be amended only by written instrument signed by both Owner and Construction Manager.

1.3 Definitions

1.3.1 The Project is the total construction to be managed under this Agreement.

1.3.2 The Work is that part of the construction that a particular Trade Contractor is to perform and such General Condition Items that the Construction Manager performs with his own forces.

1.3.3 General Condition Items as used herein shall be deemed to mean provision of facilities or performance of work by the Construction Manager for items which do not lend themselves readily to inclusion in one of the separate trade contracts. General condition items may include (but are not limited to) the following: watchmen; scaffolding; hoists; signs; safety barricades; water boys; cleaning; dirt chutes; cranes; shanties; preparation for ceremonies including minor construction activity in connection therewith; temporary toilets; fencing; sidewalk bridges; first aid station; trucking; temporary elevator; special equipment; winter protection; temporary heat, water, and electricity; temporary protective enclosures; field office and related costs thereof such as equipment furnishings and office supplies; progress, final and miscellaneous photographs; messengers; installation of Owner furnished items; post and planking; general maintenance; subsoil exploration; refuse disposal; field and laboratory tests of concrete, steel and soils; surveys; bench marks and monuments; storage on-site and off-site of long lead procurement items; and miscellaneous minor construction work when it is not feasible for the Owner to secure competitive bids or proposals thereon.

1.3.4 The term day shall mean calendar day unless otherwise specifically designated.

ARTICLE 2

Construction Manager's Services

The Construction Manager will perform the following services under this Agreement in each of the two phases described below.

2.1 Design Phase

2.1.1 *Consultation During Project Development:* Advise the Owner on site selection and participate in development of project feasibility studies. Schedule and attend regular meetings with the Architect/Engineer during the development of conceptual and preliminary design to advise on site use and improvements, selection of materials, building systems and equipment. Provide recommendations on construction feasibility, availability of materials and labor, time requirements for installation and construction, and factors related to cost including costs of alternative designs or materials, preliminary budgets, and possible economies.

2.1.2 *Scheduling:* Develop a Project Schedule that coordinates and integrates the Architect/Engineer's design efforts and the Owner's activities with construction schedules. Update the Project Schedule incorporating a detailed schedule for all activities of the project, including realistic activity sequences and durations, allocation of labor and materials, processing of shop drawings and samples, and delivery of products requiring long lead-time procurement. Include the Owner's occupancy requirements showing portions of the Project having occupancy priority.

2.1.3 *Project Construction Budget:* Prepare a Project budget as soon as major Project requirements have been identified, and update periodically for the Owner's approval. Prepare an estimate based on a quantity survey of Drawings and Specifications at the end of the schematic design phase for approval by the Owner as the Project Construction Budget. Update and refine this estimate for Owner's approval as the development of the Drawings and Specificas proceeds, and advise the Owner and the Architect/Engineer if it appears that the Project Construction Budget will not be met and make recommendations for corrective action.

2.1.4 *Coordination of Contract Documents:* Review the Drawings and Specifications as they are being prepared, recommending alternative solutions whenever design details affect construction feasibility or schedules without, however, assuming any of the Architect/Engineer's responsibilities for design.

2.1.5 *Construction Planning:* Recommend for purchase by the Owner and expedite the procurement of long-lead items to ensure their delivery by the required dates.

2.1.5.1 Make recommendations to the Owner and the Architect/Engineer regarding the division of Work in the Drawings and Specifications to facilitate the bidding and awarding of Trade Contracts, allowing for phased construction taking into consideration such factors as time of performance, availability of labor, overlapping trade jurisdictions, and provisions for temporary facilities.

2.1.5.2 Review the Drawings and Specifications with the Architect/Engineer to eliminate areas of conflict and overlapping in the Work to be performed by the various Trade Contractors and prepare prequalification criteria for bidders.

2.1.5.3 Develop Trade Contractor interest in the Project and as working Drawings and Specifications are completed, assist the Owner in taking competitive bids on the Work of the various Trade Contractors. After analyzing the bids, recommend to the Owner that such contracts be awarded.

2.1.6 *Equal Employment Opportunity:* Determine applicable requirements for equal employment opportunity programs for inclusion in Project bidding documents.

2.2 Construction Phase

2.2.1 *Project Control:* Monitor the Work of the Trade Contractors and coordinate the Work with the activities and responsibilities of the Owner, Architect/Engineer and Construction Manager to complete the Project in accordance with the Owner's objectives of cost, time and quality.

2.2.1.1 Maintain a competent full-time staff at the Project site to coordinate and provide general direction of the Work and progress of the Trade Contractors on the Project.

2.2.1.2 Establish on-site organization and lines of authority in order to carry out the overall plans of the Construction Team.

2.2.1.3 Establish procedures for coordination among the Owner, Architect/Engineer, Trade Contractors and Construction Manager with respect to all aspects of the Project and implement such procedures.

2.2.1.4 Schedule and conduct progress meetings at which Trade Contractors, Owner, Architect/Engineer and Construction Manager can discuss jointly such matters as procedures, progress, problems and scheduling.

2.2.1.5 Provide regular monitoring of the Project Schedule as construction progresses. Identify potential variances between scheduled and probable completion dates. Review schedule for Work not started or incomplete and recommend to the Owner and Trade Contractors adjustments in the schedule to meet the probable complete date. Provide summary reports of each monitoring and document all changes in schedule.

2.2.1.6 Determine the adequacy of the Trade Contractors' personnel and equipment and the availability of materials and supplies to meet the schedule. Recommend courses of action to the Owner when requirements of a Trade Contract are not being met.

2.2.2 *Provision of Facilities:* Provide all supervision, labor, materials, construction equipment, tools and subcontract items which are necessary for the completion of the General Condition Items which are not provided by either the Trade Contractors or the Owner.

2.2.3 *Cost Control:* Develop and monitor an effective system of Project cost control. Revise and refine the initially approved Project Construction Budget, incorporate approved changes as they occur, and develop cash flow reports and forecasts as needed. Identify variances between actual and budgeted or estimated costs and advise the Owner and Architect/Engineer whenever projected cost exceeds budgets or estimates.

2.2.3.1 Maintain cost accounting records on authorized Work performed under unit costs, actual costs for labor and materials, or other basis requiring accounting records. Afford the Owner access to these records and preserve them for a period of three [3] years after final payment.

2.2.4 *Change Order:* Develop and implement a system for the preparation, review and processing of Change Orders. Recommend necessary or desirable changes to the Owner and the Architect/Engineer, review requests for changes, submit recommendations to the Owner and the Architect/Engineer, and assist in negotiating Change Orders.

2.2.5 *Payments to Trade Contractors:* Develop and implement a procedure for the review, approval, processing and payment of applications by Trade Contractors for progress and final payments.

2.2.6 *Permits and Fees:* Assist the Owner and Architect/Engineer in obtaining all building permits and special permits for permanent improvements, excluding any permits required to be obtained by the various Trade Contractors, such as permits for inspection, temporary facilities, etc. Assist in obtaining approvals from all the authorities having jurisdiction.

2.2.7 *Owner's Consultants:* If required, assist the Owner in selecting and retaining professional services of a surveyor, testing laboratories and special consultants, and coordinate these services.

2.2.8 *Review of Work and Safety:* Review the Work of Trade Contractors for defects and deficiencies in the Work without assuming any of the Architect/Engineer's legal responsibilities for design and inspection. Review the safety programs of each of the Trade Contractors and make appropriate recommendations. In making such reviews, he shall not be required to make exhaustive or continuous inspections to check quality of work, safety precautions and programs in connection with the Project. The performance of such services by the Construction Manager shall not relieve the Trade Contractors of their responsibilities for performance of the work and for the safety of persons and property, and for compliance with all federal, state and local statutes, rules, regulations and orders applicable to the conduct of the Work.

2.2.9 *Document Interpretation:* Refer all questions for interpretation of the documents prepared by the Architect/Engineer to the Architect/Engineer.

2.2.10 *Shop Drawings and Samples:* In collaboration with the Architect/Engineer, establish and implement procedures for expediting the processing and approval of shop drawings and samples.

2.2.11 *Reports and Project Site Documents:* Record the progress of the Project. Submit written progress reports to the Owner and the Architect/Engineer including information on the Trade Contractors' Work, and the percentage of completion. Keep a daily log available to the Owner and the Architect/Engineer.

2.2.11.1 Maintain at the Project site, on a current basis: records of all necessary Contracts, Drawings, samples, purchases, materials, equipment, maintenance and operating manuals and instructions, and other construction related documents, including all revisions. Obtain data from Trade Contractors and maintain a current set of record Drawings, Specifications and operating manuals. At the completion of the Project, deliver all such records to the Owner.

2.2.12 *Substantial Completion:* Determine Substantial Completion of Work or designated portions thereof and prepare for the Architect/Engineer a list of incomplete or unsatisfactory items and a schedule for their completion.

2.2.13 *Start-Up:* With the Owner's maintenance personnel, direct the checkout of utilities, operations systems and equipment for readiness and assist in their initial start-up and testing by the Trade Contractors.

2.2.14 *Final Completion:* Determine final completion and provide written notice to the Owner and Architect/Engineer that the Work is ready for final inspection. Secure and transmit to the Architect/Engineer required guarantees, affidavits, releases, bonds and waivers. Turn over to the Owner all keys, manuals, record drawings and maintenance stocks.

2.2.15 *Warranty:* The Construction Manager shall collect and deliver to the Owner any specific written warranties or guarantees given by others, including all required Trade Contractor guarantees and warranties.

2.3 Additional Services

At the request of the Owner, the Construction Manager will provide the following additional services upon written agreement between the Owner and Construction Manager defining the extent of such additional services and the amount and manner in which the Construction Manager will be compensated for such additional services.

2.3.1 Services related to investigation, appraisals or valuations of existing conditions, facilities or equipment, or verifying the accuracy of existing drawings or other Owner-furnished information.

2.3.2 Services related to Owner-furnished equipment, furniture and furnishings which are not a part of the Work.

2.3.3 Services for tenant or rental spaces.

2.3.4 Services related to construction performed by the Owner.

2.3.5 Consultation on replacement of Work damaged by fire or other cause during construction, and furnishing services for the replacement of such Work.

2.3.6 Services made necessary by the default of a Trade Contractor.

2.3.7 Preparing to serve or serving as an expert witness in connection with any public hearing, arbitration proceeding,   legal proceeding.

2.3.8 Finding housing for construction labor, and defining requirements for establishment and maintenance of base camps.

2.3.9 Obtaining or training maintenance personnel or negotiating maintenance service contracts.

2.3.10 Services related to Work performed by the Construction Manager other than General Conditions items.

2.3.11 Inspections of and services related to the Project after completion of the services under this Agreement.

2.3.12 Providing any other service not otherwise included in this Agreement.

ARTICLE 3

Owner's Responsibilities

3.1 The Owner shall provide full information regarding his requirements for the Project.

3.2 The Owner shall designate a representative who shall be fully acquainted with the Project and has authority to approve Project Construction Budgets, Changes in the Project, render decisions promptly consistent with Project Schedule and furnish information expeditiously.

3.3 The Owner shall retain an Architect/Engineer for design and to prepare construction documents for the Project. The Architect/Engineer's services, duties and responsibilities are described in the Agreement between the Owner and the Archtiect/Engineer, a copy of which will be furnished to the Construction Manager. The Agreement between the Owner and the Architect/Engineer shall not be modified without written notification to the Construction Manager.

3.4 The Owner shall furnish for the site of the Project all necessary surveys describing the physical characteristics, soil reports and subsurface investigations, legal limitations, utility locations, and a legal description.

3.5 The Owner shall secure and pay for necessary approvals, easements, assessments and charges required for the construction, use or occupancy of permanent structures or for permanent changes in existing facilities.

3.6 The Owner shall furnish such legal services as may be necessary for providing the items set forth in Paragraph 3.5, and such auditing services as he may require.

3.7 The Construction Manager will be furnished without charge all copies of Drawings and Specifications reasonably necessary for the execution of the Work.

3.8 The Owner shall provide the insurance for the Project as provided in Paragraph 10.4, and shall bear the cost of any bonds required.

3.9 The services, information, surveys and reports required by the above paragraphs shall be furnished with reasonable promptness at the Owner's expense, and the Construction Manager shall be entitled to rely upon the accuracy and completeness thereof.

3.10 If the Owner becomes aware of any fault or defect in the Project or non-conformance with the Drawings and Specifications, he shall give prompt written notice thereof to the Construction Manager.

3.11 The Owner shall furnish reasonable evidence satisfactory to the Construction Manager that sufficient funds are available and committed for the entire cost of the Project. Unless such reasonable evidence is furnished, the Construction Manager is not required to commence any Work, or may, if such evidence is not presented within a reasonable time, stop the Project upon 15 days notice to the Owner.

3.12 The Owner shall communicate with the Trade Contractors only through the Construction Manager.

ARTICLE 4

Trade Contracts

4.1 All portions of the Project other than General Condition Items shall be performed under Trade Contracts including contracts for general construction work with several construction trades. The Construction Manager shall assist the Owner in requesting and receiving proposals from Trade Contractors and Trade Contracts will be awarded by the Owner after the proposals are reviewed by the Architect/Engineer, Construction Manager and Owner.

4.2 Trade Contracts will be between the Owner and the Trade Contractors. The form of the Trade Contracts including the General and Supplementary Conditions shall be satisfactory to the Construction Manager.

AGC DOCUMENT NO. 8d • OWNER CONSTRUCTION MANAGER AGREEMENT • JUNE 1979

ARTICLE 5

Project Schedule

5.1   The services to be provided under this Contract shall be in general accordance with the following schedule:

5.2   At the time work commences, a Date of Substantial Completion of the project shall also be established.

5.3   The Date of Substantial Completion of the Project or a designated portion thereof is the date when construction is sufficiently complete in accordance with the Drawings and Specifications so the Owner can occupy or utilize the Project or designated portion thereof for the use for which it is intended. Warranties called for by this Agreement or by the Drawings and Specifications shall commence on the Date of Substantial Completion of the Project or designated portion thereof.

5.4   If the Construction Manager is delayed at any time in the progress of the Project by any act or neglect of the Owner or the Architect/Engineer or by any employee of either, or by any separate contractor employed by the Owner, or by changes ordered in the Project, or by labor disputes, fire, unusual delay in transportation, adverse weather conditions not reasonably anticipatable, unavoidable casualties or any causes beyond the Construction Manager's control, or by delay authorized by the Owner pending arbitration, the Construction Completion Date shall be extended by Change Order for a reasonable length of time.

ARTICLE 6

Construction Manager's Fee

6.1   In consideration of the performance of the Contract, the Owner agrees to pay the Construction Manager in current funds as compensation for his services a Construction Manager's Fee as set forth in Subparagraphs 6.1.1 and 6.1.2.

6.1.1   For the performance of the Design Phase services, a fee of

which shall be paid monthly, in equal proportions, based on the scheduled Design Phase time.

6.1.2   For work or services performed during the Construction Phase, a fee of

which shall be paid proportionately to the ratio the monthly payment for the Cost of the Project bears to the estimated cost. Any balance of this fee shall be paid at the time of final payment.

6.2   Adjustments in Fee shall be made as follows:

6.2.1   For changes in the Project as provided in Article 8, the Construction Manager's Fee shall be adjusted by   % of the additional cost of the work for increases in the scope of the Project.

6.2.2   For delays in the Project not the responsibility of the Construction Manager, there will be an equitable adjustment in the fee to compensate the Construction Manager for his increased expenses, overhead and profit.

6.2.3   The Construction Manager shall be paid an additional fee in the same proportion as set forth in 6.2.1 if the Construction Manager performs services in connection with the reconstruction of any insured or uninsured loss.

6.3   Including in the Construction Manager's Fee are the following:

6.3.1   Salaries or other compensation of the Construction Manager's employees at the principal office and branch offices, except employees listed in Subparagraph 7.2.2.

6.3.2   General operating expenses of the Construction Manager's principal and branch offices other than the field office.

6.3.3   Any part of the Construction Manager's capital expenses, including interest on the Construction Manager's capital employed for the Project.

6.3.4   Overhead or general expenses of any kind, except as may be expressly included in Article 7.

ARTICLE 7

Reimbursable Costs

7.1   The Owner agrees to pay the Construction Manager for the reimbursable costs he incurs as defined in Paragraph 7.2. Such payment shall be in addition to the Construction Manager's Fee stipulated in Article 6.

7.2   Reimbursable Cost Items

7.2.1   Wages paid for labor in the direct employ of the Construction Manager in the performance of General Condition Items under applicable collective bargaining agreements, or under a salary or wage schedule agreed upon by the Owner and Construction Manager, and including such welfare or other benefits, if any, as may be payable with respect thereto.

7.2.2   Salaries of the Construction Manager's employees when stationed at the field office, in whatever capacity employed, employees engaged on the road in expediting the production or transportation of materials and equipment, and employees in the main or branch office performing the functions listed below:

7.2.3   Cost of all employee benefits and taxes for such items as unemployment compensation and social security, insofar as such cost is based on wages, salaries, or other remuneration paid to employees of the Construction Manager referred to in Subparagraphs 7.2.1 and 7.2.2.

7.2.4   The proportion of reasonable transportation, traveling, moving, and hotel expenses of the Construction Manager or of his officers or employees incurred in discharge of duties connected with the Project.

7.2.5   Cost including rental of all materials, supplies and equipment required to perform General Condition Items, including costs of transportation, storage and maintenance thereof.

7.2.6   Cost of the premiums for all insurance which the Construction Manager is required to procure by this Agreement or is deemed necessary by the Construction Manager.

7.2.7   Sales, use, gross receipts or similar taxes related to the Project imposed by any governmental authority, and for which the Construction Manager is liable.

7.2.8   Permit fees, licenses, tests, royalties, damages for infringement of patents and costs of defending suits therefor, and deposits lost for causes other than the Construction Manager's negligence. If royalties or losses and damages, including costs of defense, are incurred which arise from a particular design, process, or the product of a particular manufacturer or manufacturers specified by the Owner or Architect/Engineer, and the ConstructionManager has no reason to believe there will be infringement of patent rights, such royalties, losses and damages shall be paid by the Owner.

7.2.9   Losses, expenses or damages to the extent not compensated by insurance or otherwise (including settlement made with the written approval of the Owner.)

7.2.10   Minor expenses such as telegrams, long-distance telephone calls, telephone service at the site, expressage, and similar petty cash items in connection with the Project.

7.2.11   Cost of removal of all debris.

7.2.12   Cost incurred due to an emergency affecting the safety of persons and property.

7.2.13   Cost of data processing services required in the performance of the services outlined in Article 2.

7.2.14   Legal costs reasonably and properly resulting from prosecution of the Project for the Owner.

7.2.15   All costs directly incurred in the performance of the Project and not included in the Construction Manager's Fee as set forth in Paragraph 6.3.

AGC DOCUMENT NO. 8d • OWNER-CONSTRUCTION MANAGER AGREEMENT • JUNE 1979

ARTICLE 8

Changes in the Project

8.1 The Owner, without invalidating this Agreement, may order Changes in the Project within the general scope of this Agreement consisting of additions, deletions or other revisions, the Guaranteed Maximum Price, if established, the Construction Manager's Fee and the Construction Completion Date being adjusted accordingly. All such Changes in the Project shall be authorized by Change Order.

8.2 A Change Order is a written order to the Construction Manager signed by the Owner or his authorized agent issued after the execution of this Agreement, authorizing a change in the scope of the Project, services to be provided, and/or the Construction Manager's Fee or the Construction Completion Date. The Construction Manager's Fee shall be adjusted as provided in Paragraph 6.2.

ARTICLE 9

Payments to the Construction Manager

9.1 The Construction Manager shall submit monthly to the Owner a statement, sworn to if required, showing in detail all moneys paid out, costs accumulated or costs incurred on account of the Cost of the Project during the previous month and the amount of the Construction Manager's Fee due as provided in Article 6. Payment by the Owner to the Construction Manager of the statement amount shall be made within ten (10) days after it is submitted.

9.2 Final payment constituting the unpaid balance of the Cost of the Project and the Construction Manager's Fee shall be due and payable when the Project is delivered to the Owner, ready for beneficial occupancy, or when the Owner occupies the Project, whichever event first occurs, provided that the Project be then substantially completed and this Agreement substantially performed.

9.3 If the Owner should fail to pay the Construction Manager within seven (7) days after the time the payment of any amount becomes due, then the Construction Manager may, upon seven (7) additional days' written notice to the Owner and the Architect/Engineer, stop the Project until payment of the amount owing has been received.

9.4 Payments due but unpaid shall bear interest at the rate the Owner is paying on his construction loan or at the legal rate, whichever is higher.

ARTICLE 10

Insurance, Indemnity and Waiver of Subrogation

10.1   Indemnity

10.1.1 The Construction Manager agrees to indemnify and hold the Owner harmless from all claims for bodily injury and property damage (other than the work itself and other property insured under Paragraph 10.4) that may arise from the Construction Manager's operations under this Agreement.

10.1.2 The Owner shall cause all Trade Contractors to agree to indemnify the Owner and the Construction Manager and hold them harmless from all claims for bodily injury and property damage (other than property insured under Paragraph 10.4) that may arise from that contractor's operations. Such provisions shall be in a form satisfactory to the Construction Manager.

10.2   Construction Manager's Liability Insurance

10.2.1 The Construction Manager shall purchase and maintain such insurance as will protect him from the claims set forth below which may arise out of or result from the Construction Manager's operations under this Agreement.

10.2.1.1 Claims under workers' compensation, disability benefit and other similar employee benefit acts which are applicable to the work to be performed.

10.2.1.2 Claims for damages because of bodily injury, occupational sickness or disease, or death of his employees under any applicable employer's liability law.

10.2.1.3 Claims for damages because of bodily injury, or death of any person other than his employees.

10.2.1.4 Claims for damages insured by usual personal injury liability coverage which are sustained (1) by any person as a result of an offense directly or indirectly related to the employment of such person by the Construction Manager or (2) by any other person.

AGC DOCUMENT NO. 8d • OWNER-CONSTRUCTION MANAGER AGREEMENT • JUNE 1979   7

10.2.1.5 Claims for damages, other than to the work itself, because of injury to or destruction of tangible property, including loss of use therefrom.

10.2.1.6 Claims for damages because of bodily injury or death of any person or property damage arising out of the ownership, maintenance or use of any motor vehicle.

10.2.2 The Construction Manager's Comprehensive General Liability Insurance shall include premises — operations (including explosion, collapse and underground coverage) elevators, independent contractors, completed operations, and blanket contractual liability on all written contracts, all including broad form property damage coverage.

10.2.3 The Construction Manager's Comprehensive General and Automobile Liability Insurance, as required by Subparagraphs 10.2.1 and 10.2.2 shall be written for not less than limits of liability as follows:

a. Comprehensive General Liability
1. Personal Injury

$_____ Each Occurrence

$_____ Aggregate

(Completed Operations)

2. Property Damage

$_____ Each Occurrence

$_____ Aggregate

b. Comprehensive Automobile Liability
1. Bodily Injury

$_____ Each Person

$_____ Each Occurrence

2. Property Damage

$_____ Each Occurrence

10.2.4 Comprehensive General Liability Insurance may be arranged under a single policy for the full limits required or by a combination of underlying policies with the balance provided by an Excess or Umbrella Liability policy.

10.2.5 The foregoing policies shall contain a provision that coverages afforded under the policies will not be cancelled or nor renewed until at least sixty (60) days' prior written notice has been given to the Owner. Certificates of Insurance showing such coverages to be in Force shall be filed with the Owner prior to commencement of the Work.

10.3 Owner's Liability Insurance

10.3.1 The Owner shall be responsible for purchasing and maintaining his own liability insurance and, at his option, may purchase and maintain such insurance as will protect him against claims which may arise from operations under this Agreement.

10.4 Insurance to Protect Project

10.4.1 The Owner shall purchase and maintain property insurance in a form acceptable to the Construction Manager upon the entire Project for the full cost of replacement as of the time of any loss. This insurance shall include as named insureds the Owner, the Construction Manager, Trade Contractors and their Trade Subcontractors and shall insure against loss from the perils of Fire, Extended Coverage, and shall include "All Risk" insurance for physical loss or damage including without duplication of coverage at least theft, vandalism, malicious mischief, transit, collapse, flood, earthquake, testing, and damage resulting from defective design, workmanship or material. The Owner will increase limits of coverage, if necessary, to reflect estimated replacement cost. The Owner will be responsible for any co-insurance penalties or deductibles. If the Project covers an addition to or is adjacent to an existing building, the Construction Manager, Trade Contractors and their Trade Subcontractors shall be named as additional insureds under the Owner's Property Insurance covering such building and its contents.

10.4.1.1 If the Owner finds it necessary to occupy or use a portion or portions of the Project prior to Substantial Completion thereof, such occupancy shall not commence prior to a time mutually agreed to by the Owner and Construction Manager and to which the insurance company or companies providing the property insurance have consented by endorsement to the policy or policies. This insurance shall not be cancelled or lapsed on account of such partial occupancy. Consent of the Construction Manager and of the insurance company or companies to such occupancy or use shall not be unreasonably withheld.

10.4.2 The Owner shall purchase and maintain such boiler and machinery insurance as may be required or necessary. This insurance shall include the interests of the Owner, the Construction Manager, Trade Contractors and their Trade Subcontractors in the Work.

10.4.3 The Owner shall purchase and maintain such insurance as will protect the Owner and Construction Manager against loss of use of Owner's property due to those perils insured pursuant to Subparagraph 10.4.1. Such policy will provide coverage for expediting expenses of materials, continuing overhead of the Owner and Construction Manager, necessary labor expense including overtime, loss of income by the Owner and other determined exposures. Exposures of the Owner and the Construction Manager shall be determined by mutual agreement and separate limits of coverage fixed for each item.

10.4.4.   The Owner shall file a copy of all policies with the Construction Manager before an exposure to loss may occur. Copies of any subsequent endorsements will be furnished to the Construction Manager. The Construction Manager will be given sixty (60) days notice of cancellation, non-renewal, or any endorsements restricting or reducing coverage. If the Owner does not intend to purchase such insurance, he shall inform the Construction Manager in writing prior to the commencement of the Work. The Construction Manager may then effect insurance which will protect the interest of himself, the Trade Contractors and their Trade Subcontractors in the Project, the cost of which shall be reimbursable pursuant to Article 7. If the Construction Manager is damaged by failure of the Owner to purchase or maintain such insurance or to so notify the Construction Manager, the Owner shall bear all reasonable costs properly attributable thereto.

10.5   Property Insurance Loss Adjustment

10.5.1   Any insured loss shall be adjusted with the Owner and the Construction Manager and made payable to the Owner and Construction Manager as trustees for the insureds, as their interests may appear, subject to any applicable mortgagee clause.

10.5.2   Upon the occurrence of an insured loss, monies received will be deposited in a separate account and the trustees shall make distribution in accordance with the agreement of the parties in interest, or in the absence of such agreement, in accordance with an arbitration award pursuant to Article 14. If the trustees are unable to agree on the settlement of the loss, such dispute shall also be submitted to arbitration pursuant to Article 14.

10.6   Waiver of Subrogation

10.6.1   The Owner and Construction Manager waive all rights against each other, the Architect/Engineer, Trade Contractors, and their Trade Subcontractors for damages caused by perils covered by insurance provided under Paragraph 10.4, except such rights as they may have to the proceeds of such insurance held by the Owner and Construction Manager as trustees. The Owner shall require similar waivers from all Trade Contractors and their Trade Subcontractors.

10.6.2   The Owner and Construction Manager waive all rights against each other and the Architect/Engineer, Trade Contractors and their Trade Subcontractors for loss or damage to any equipment used in connection with the Project and covered by any property insurance. The Owner shall require similar waivers from all Trade Contractors and their Trade Subcontractors.

10.6.3   The Owner waives subrogation against the Construction Manager, Architect/Engineer, Trade Contractors, and their Trade Subcontractors on all property and consequential loss policies carried by the Owner on adjacent properties and under property and consequential loss policies purchased for the Project after its completion.

10.6.4   If the policies of insurance referred to in this Paragraph require an endorsement to provide for continued coverage where there is a waiver of subrogation, the owners of such policies will cause them to be so endorsed.

ARTICLE 11

**Termination of the Agreement and Owner's
Right to Perform Construction Manager's Obligations**

11.1   Termination by the Construction Manager

11.1.1   If the Project is stopped for a period of thirty days under an order of any court or other public authority having jurisdiction, or as a result of an act of government, such as a declaration of a national emergency making materials unavailable, through no act or fault of the Construction Manager, or if the Project should be stopped for a period of thirty days by the Construction Manager for the Owner's failure to make payment thereon, then the Construction Manager may, upon seven days' written notice to the Owner and the Architect/Engineer, terminate this Agreement and recover from the Owner payment for all work executed, the Construction Manager's Fee earned to date, and for any proven loss sustained upon any materials, equipment, tools, construction equipment and machinery, including reasonable profit and damages.

11.2   Owner's Right to Perform Construction Manager's Obligations and Termination by the Owner for Cause

11.2.1   If the Construction Manager fails to perform any of his obligations under this Agreement, the Owner may, after seven days' written notice during which period the Construction Manager fails to perform such obligation, make good such deficiencies.

11.2.2   If the Construction Manager is adjudged a bankrupt, or if he makes a general assignment for the benefit of his creditors, or if a receiver is appointed on account of his insolvency, or persistently disregards laws, ordinances, rules, regulations or orders of any public authority having jurisdiction, or otherwise is guilty of a substantial violation of a provision of the Agreement, then the Owner may, without prejudice to any right or remedy and after giving the Construction Manager and his surety, if any, seven days' written notice, during which period Construction Manager fails to cure the violation, terminate the employment of the Construction Manager and take possession of the site and of all materials, equipment, tools, construction equipment and machinery thereon owned by the Construction

Manager and may finish the Project by whatever method he may deem expedient. In such case, the Construction Manager shall not be entitled to receive any further payment until the Project is finished nor shall he be relieved from his obligations assumed under Article 6.

11.3 Termination by Owner Without Cause

11.3.1 If the Owner terminates this Agreement other than pursuant to Subparagraph 11.2.2 or Subparagraph 11.3.2, he shall reimburse the Construction Manager for any unpaid Reimbursable Cost of the Project due him under Article 7, plus (1) the unpaid balance of the Fee computed upon the Cost of the Project to the date of termination at the rate of the percentage named in Subparagraph 6.2.1 or if the Construction Manager's Fee be stated as a fixed sum, such an amount as will increase the payment on account of his fee to a sum which bears the same ratio to the said fixed sum as the Cost of the Project at the time of termination bears to a reasonable estimated Cost of the Project when completed. The Owner shall also pay to the Construction Manager fair compensation, either by purchase or rental at the election of the Owner, for any equipment retained. In case of such termination of the Agreement the Owner shall further assume and become liable for obligations, commitments and unsettled claims that the Construction Manager has previously undertaken or incurred in good faith in connection with said Project. The Construction Manager shall, as a condition of receiving the payments referred to in this Article 11, execute and deliver all such papers and take all such steps, including the legal assignment of his contractual rights, as the Owner may require for the purpose of fully vesting in him the rights and benefits of the Construction Manager under such obligations or commitments.

11.3.2 After the completion of the Design Phase, if the final cost estimates make the Project no longer feasible from the standpoint of the Owner, the Owner may terminate this Agreement and pay the Construction Manager his Fee in accordance with Subparagraph 6.1.1 plus any costs incurred pursuant to Article 7.

ARTICLE 12

Assignment and Governing Law

12.1 Neither the Owner nor the Construction Manager shall assign his interest in this Agreement without the written consent of the other except as to the assignment of proceeds.

12.2 This Agreement shall be governed by the law of the place where the Project is located.

ARTICLE 13

Miscellaneous Provisions

AGC DOCUMENT NO. 8d • OWNER-CONSTRUCTION MANAGER AGREEMENT • JUNE 1979

ARTICLE 14

Arbitration

14.1   All claims, disputes and other matters in question arising out of, or relating to, this Agreement or the breach thereof, except with respect to the Architect/Engineer's decision on matters relating to artistic effect, and except for claims which have been waived by the making or acceptance of final payment shall be decided by arbitration in accordance with the Construction Industry Arbitration Rules of the American Arbitration Association then obtaining unless the parties mutually agree otherwise. This Agreement to arbitrate shall be specifically enforceable under the prevailing arbitration law.

14.2   Notice of the demand for arbitration shall be filed in writing with the other party to this Agreement and with the American Arbitration Association. The demand for arbitration shall be made within a reasonable time after the claim, dispute or other matter in question has arisen, and in no event shall it be made after the date when institution of legal or equitable proceedings based on such claim, dispute or other matter in question would be barred by the applicable statute of limitations.

14.3   The award rendered by the arbitrators shall be final and judgment may be entered upon it in accordance with applicable law in any court having jurisdiction thereof.

14.4   Unless otherwise agreed in writing, the Construction Manager shall continue to carry out his responsibilities under this Agreement and maintain the Contract Completion Date during any arbitration proceedings, and the Owner shall continue to make payments in accordance with this Agreement.

14.5   All claims which are related to or dependent upon each other, shall be heard by the same arbitrator or arbitrators even though the parties are not the same unless a specific contract prohibits such consolidation. •

This Agreement executed the day and year first written above.

ATTEST: OWNER:

By_____

ATTEST: CONSTRUCTION MANAGER:

By_____

Contractor Construction Management Approach

AGC Document 8 is the only form document published which contemplates that the construction manager will commit to a date of completion, guarantee a maximum cost for the project, and warrant its own work and that the trade contractors performing the greater part of the construction.

Whereas the NSPE/ACEC construction management agreement might be described as a traditional owner-engineer agreement with the inclusion of some additional services, and AGC construction management agreement may fairly be described as a cost-plus construction contract in which the contractor undertakes to provide some consultation during the design phase, then acts as a broker during the construction phase. With minor exceptions, the services performed by the construction manager under AGC Document 8 are identical to those provided under AGC Document 8d, described above. Some of the terms and conditions have been changed to account for the changed role of the construction manager as follows:

In paragraph 2.1.5.3, the construction manager is authorized to take competitive bids and to award trade contracts for portions of the work. In paragraph 2.2.2, the construction manager agrees to provide all construction work not provided by the owner or by the "Trade Contractors," and to perform such work in accordance with the "Plans and Specifications" and "the procedure applicable to the Project." In paragraph 2.2.8, the construction manager "inspects" (as opposed to "reviews" in Document 8d) the work of the trade contractors. In paragraph 2.2.8.1, the disclaimer of any responsibility for exhaustive or continuous inspections "to check quality of work" found in Document 8d is omitted, but such an underlying responsibility to inspect the work is not expressly undertaken. Finally, the construction manager warrants its own work and that of the trade contractors whom it has engaged and undertakes an obligation to correct defective work for a one-year period of time, in paragraph 2.2.15. A limited number of additional services are available under AGC Document 8.

In article 4, the AGC owner–construction manager agreement specifies that the general rule will be that "Trade Contracts" will be with the construction manager, not the owner as in AGC Document 8d; and article 4 reiterates that the construction manager is responsible to the owner for the acts and omissions of the trade contractors. Article 5 provides for the establishment of a date of substantial completion at the time a guaranteed maximum price is offered.

Article 6 sets forth the construction manager's obligation to give a guaranteed maximum price to the owner, in which case the owner effectively loses the option of hiring separate contractors. The GMP is accomplished by execution of AGC Document 8a, *Amendment to Owner–Construction Manager Contract.* The guaranteed maximum price is subject to adjustment because of changes in the work ordered by the owner (par. 9.1), changes in the quantities of work performed on a unit price basis (par. 9.1.4), concealed subsurface conditions or concealed or unknown conditions in an existing structure (par. 9.1.5), work on account of emergencies (par. 9.4.1), costs resulting from delays caused by the owner or the A/E (par. 6.1), increases in taxes (par. 6.3), and insurance provided by the construction manager when the owner does not procure it (par. 12.4.4). In exchange for obtaining a guaranteed maximum price, the owner essentially gives the construction manager a cost-plus contract to perform the work (art. 8), subject to the GMP. Although the GMP sets a limit on the owner's financial risk, the costs of the project for which the construction manager is reimbursed are liberally defined, including, for instance, the cost of corrective work (par. 8.2.13). Costs that may be charged to the owner are limited to those appearing in the definition of the construction manager's fee in article 7. There is also a general cost "catch-all" in paragraph 8.2.19.

Article 10 provides the owner with the benefit of cash discounts, provided the owner makes the funds available to the construction manager to take advantage of such discounts. Otherwise, discounts accrue to the construction manager and are not deducted from the GMP or the cost of the work.

In article 11, the construction manager is paid its fee and for the work performed during the previous month, regardless of whether the construction manager has actually paid for such labor and materials. Paragraph 11.3 makes it clear that the construction manager in fact has no obligation to pay trade contractors until it has been paid by the owner. Final payment to the construction manager is due upon substantial completion, less a retainage of 150 percent of the estimated cost of correction or completing any work that remains to be done. The construction manager offers an "unconditional promise to complete said items within a reasonable time."

The remainder of the provisions in AGC Document 8 are essentially the same as in AGC Document 8d.

Owners contemplating use of the contractor construction management approach should be aware of several aspects of the contractual arrangement provided under AGC Document 8, as follows:

1. *Absence of General Conditions*. In a normal cost-plus construction contract, the contractor is subject to all the terms and conditions found in AIA Document A201, *General Conditions of the Contract for Construction*, which has been endorsed and approved by the AGC, and portions of which have been included in Document 8. There are no detailed general conditions which apply to the construction manager acting in the role of the general contractor, other than the provisions which have been directly incorporated in Document 8 itself.

2. *Absence of Oversight by A/E*. The AGC document eliminates a role for the architect or engineer as the owner's representative, to judge the performance of the work of the construction manager, approve payment for work completed, render decisions on disputes, or certify substantial or final completion of the work. Rather, the AGC Document 8 places the construction manager in the position of being responsible for overseeing its own work, approving payments to itself and its subcontractors ("trade contractors"), and determining that its own work is complete, without the benefit of oversight by the A/E. If its work must be corrected, the construction manager is reimbursed for the cost of correction (although the guaranteed maximum price is not affected).

3. *Risk Avoidance by the Construction Manager.* The contractor–construction manager attempts to shift most of the responsibilities that are normally those of the general contractor to the individual trade contractors, who will be operating under AGC Document 8b, *General Conditions for Trade Contractors under Construction Management Agreements,* as general conditions applicable to their work. For example, when work is covered up and must be uncovered in order to permit its inspection, the contractor normally must bear the expense of uncovering the work if it is found to be defective, whether or not it can recover this cost from the subcontractor. Although the construction manager is in the position of general contractor under this agreement, it bears no such risk; however, the AGC takes pains in its general conditions, Document 8b, to ensure that the requirement applies to the trade contractor. In similar fashion, the responsibility for construction safety is disclaimed by the construction manager in the owner–construction manager agreement and placed on the individual trade contractors in the AGC general conditions.

In light of the conditions outlined above, the AGC owner–construction manager agreement provides the construction manager with a most favorable relationship with the owner, in which it has all the power and authority it could desire, takes little financial risk or responsibility, and avoids having an architect or engineer interfering with its business relationship with the owner. Under the AGC general conditions, the construction manager's role replaces that of the owner, architect engineer, or general contractor, depending on which is most favorable to the construction manager in a particular situation. The potential for abuse in the AGC documents is clear; the documents actually legitimize practices such as the failure or refusal of the construction manager to pay subcontractors and the withholding of money on subcontracts, despite the construction manager's having been paid in full by the owner (no provision is made for the owner to withhold from the construction manager under the AGC documents, although the construction manager has all the authority to withhold payment from the trade contractors that normally belongs to the owner). It would not be an understatement to say that under the AGC owner–construction manager agreement the construction manager is operating under clear and irreconcilable conflicts of interest when it is responsible for reviewing and approving its own work and that of its subcontractors.

The only alternatives for the owner in the use of the AGC document are to develop a manuscript agreement or to amend or modify the AGC document. Because AGC Document 8 was developed largely by assembling selected excerpts from the AIA owner–construction manager agreement (B801), the owner-contractor agreement for cost-plus work (A111) and the general conditions (A201), a similar approach can be taken with far more satisfactory results from the owner's standpoint. This approach would be to enter into a two-phased agreement with the contractor–construction manager for a fix fee, in which the first (or design) phase requires the contractor–construction manager to perform the design phase services listed in either the AIA or AGC owner–construction manager agreement (the differences between the services in the two being minimal), and in which the second phase is a traditional cost-plus construction contract using AIA Documents A111 and A201 (both of which, incidentally, have been approved and endorsed by the AGC). Conditions enabling the owner to elect not to proceed with the construction phase (if, for instance, it is not satisfied with the guaranteed maximum price offered by the construction manager) should be included in the first-phase agreement; alternatively, the second-phase agreement (the cost-plus construction contract) can be limited to a letter of intent. Such an approach clearly separates the roles of the construction manager when acting first as the owner's consultant and adviser and then as a broker-contractor with potentially adverse interests.

The second alternative—modifying or amending the AGC document— may be more difficult and time-consuming than it is worth. The first step in making such modifications is to analyze both the AGC owner–construction manager agreement, Document 8, and the AGC general conditions, Document 8b, in light of the same AIA documents mentioned above. This comparison will reveal the terms and conditions designed to protect the owner's interests which have been omitted in the AGC documents and which the careful owner will want to have amended back into the AGC owner–construction manager agreement. At the least, the work performed by or under contract with the construction manager should be subject to a form of general conditions which will thoroughly cover the construction manager's responsibilities to the owner.

THE ASSOCIATED GENERAL CONTRACTORS

STANDARD FORM OF AGREEMENT BETWEEN OWNER AND CONSTRUCTION MANAGER

(GUARANTEED MAXIMUM PRICE OPTION)

(See AGC Document No. 8a for Establishing the
Guaranteed Maximum Price)

This Document has important legal and insurance consequences; consultation with an attorney is encouraged with respect to its completion or modification.

AGREEMENT

Made this   day of   in the year of Nineteen Hundred and

BETWEEN   the Owner, and

the Construction Manager.

For services in connection with the following described Project: (Include complete Project location and scope)

The Architect/Engineer for the Project is

The Owner and the Construction Manager agree as set forth below:

TABLE OF CONTENTS

ARTICLE 1

The Construction Team and Extent of Agreement

The CONSTRUCTION MANAGER accepts the relationship of trust and confidence established between him and the Owner by this Agreement. He covenants with the Owner to furnish his best skill and judgment and to cooperate with the Architect/Engineer in furthering the interests of the Owner. He agrees to furnish efficient business administration and superintendence and to use his best efforts to complete the Project in an expeditious and economical manner consistent with the interest of the Owner.

1.1 *The Construction Team:* The Construction Manager, the Owner, and the Architect/Engineer called the "Construction Team" shall work from the beginning of design through construction completion. The Construction Manager shall provide leadership to the Construction Team on all matters relating to construction.

1.2 *Extent of Agreement:* This Agreement represents the entire agreement between the Owner and the Construction Manager and supersedes all prior negotiations, representations or agreements. When Drawings and Specifications are complete, they shall be identified by amendment to this Agreement. This Agreement shall not be superseded by any provisions of the documents for construction and may be amended only by written instrument signed by both the Owner and the Construction Manager.

1.3 *Definitions:* The Project is the total construction to be performed under this Agreement. The Work is that part of the construction that the Construction Manager is to perform with his own forces or that part of the construction that a particular Trade Contractor is to perform. The term day shall mean calendar day unless otherwise specifically designated.

ARTICLE 2

Construction Manager's Services

The Construction Manager will perform the following services under this Agreement in each of the two phases described below.

2.1 Design Phase

2.1.1 *Consultation During Project Development:* Schedule and attend regular meetings with the Architect/Engineer during the development of conceptual and preliminary design to advise on site use and improvements, selection of materials, building systems and equipment. Provide recommendations on construction feasibility, availability of materials and labor, time requirements for installation and construction, and factors related to cost including costs of alternative designs or materials, preliminary budgets, and possible economies.

2.1.2 *Scheduling:* Develop a Project Time Schedule that coordinates and integrates the Architect/Engineer's design efforts with construction schedules. Update the Project Time Schedule incorporating a detailed schedule for the construction operations of the Project, including realistic activity sequences and durations, allocation of labor and materials, processing of shop drawings and samples, and delivery of products requiring long lead-time procurement. Include the Owner's occupancy requirements showing portions of the Project having occupancy priority.

2.1.3 *Project Construction Budget:* Prepare a Project budget as soon as major Project requirements have been identified, and update periodically for the Owner's approval. Prepare an estimate based on a quantity survey of Drawings and Specifications at the end of the schematic design phase for approval by the Owner as the Project Construction Budget. Update and refine this estimate for the Owner's approval as the development of the Drawings and Specifications proceeds, and advise the Owner and the Architect/Engineer if it appears that the Project Construction Budget will not be met and make recommendations for corrective action.

2.1.4 *Coordination of Contract Documents:* Review the Drawings and Specifications as they are being prepared, recommending alternative solutions whenever design details affect construction feasibility or schedules without, however, assuming any of the Architect/Engineer's responsibilities for design.

2.1.5 *Construction Planning:* Recommend for purchase and expedite the procurement of long-lead items to ensure their delivery by the required dates.

2.1.5.1 Make recommendations to the Owner and the Architect/Engineer regarding the division of Work in the Drawings and Specifications to facilitate the bidding and awarding of Trade Contracts, allowing for phased construction taking into consideration such factors as time of performance, availability of labor, overlapping trade jurisdictions, and provisions for temporary facilities.

2.1.5.2 Review the Drawings and Specifications with the Architect/Engineer to eliminate areas of conflict and overlapping in the Work to be performed by the various Trade Contractors and prepare prequalification criteria for bidders.

2.1.5.3 Develop Trade Contractor interest in the Project and as working Drawings and Specifications are completed, take competitive bids on the Work of the various Trade Contractors. After analyzing the bids, either award contracts or recommend to the Owner that such contracts be awarded.

2.1.6 *Equal Employment Opportunity:* Determine applicable requirements for equal emloyment opportunity programs for inclusion in Project bidding documents.

2.2 Construction Phase

2.2.1 *Project Control:* Monitor the Work of the Trade Contractors and coordinate the Work with the activities and responsibilities of the Owner, Architect/Engineer and Construction Manager to complete the Project in accordance with the Owner's objectives of cost, time and quality.

2.2.1.1 Maintain a competent full-time staff at the Project site to coordinate and provide general direction of the Work and progress of the Trade Contractors on the Project.

2.2.1.2 Establish on-site organization and lines of authority in order to carry out the overall plans of the Construction Team.

2.2.1.3 Establish procedures for coordination among the Owner, Architect/Engineer, Trade Contractors and Construction Manager with respect to all aspects of the Project and implement such procedures.

2.2.1.4 Schedule and conduct progress meetings at which Trade Contractors, Owner, Architect/Engineer and Construction Manager can discuss jointly such matters as procedures, progress, problems and scheduling.

2.2.1.5 Provide regular monitoring of the schedule as construction progresses. Identify potential variances between scheduled and probable completion dates. Review schedule for Work not started or incomplete and recommend to the Owner and Trade Contractors adjustments in the schedule to meet the probable completion date. Provide summary reports of each monitoring and document all changes in schedule.

2.2.1.6 Determine the adequacy of the Trade Contractors' personnel and equipment and the availability of materials and supplies to meet the schedule. Recommend courses of action to the Owner when requirements of a Trade Contract are not being met.

2.2.2 *Physical Construction:* Provide all supervision, labor, materials, construction equipment, tools and subcontract items which are necessary for the completion of the Project which are not provided by either the Trade Contractors or the Owner. To the extent that the Construction Manager performs any Work with his own forces, he shall, with respect to such Work, perform in accordance with the Plans and Specifications and in accordance with the procedure applicable to the Project.

2.2.3 *Cost Control:* Develop and monitor an effective system of Project cost control. Revise and refine the initially approved Project Construction Budget, incorporate approved changes as they occur, and develop cash flow reports and forecasts as needed. Identify variances between actual and budgeted or estimated costs and advise Owner and Architect/Engineer whenever projected cost exceeds budgets or estimates.

2.2.3.1 Maintain cost accounting records on authorized Work performed under unit costs, actual costs for labor and material, or other bases requiring accounting records. Afford the Owner access to these records and preserve them for a period of three (3) years after final payment.

2.2.4 *Change Orders:* Develop and implement a system for the preparation, review and processing of Change Orders. Recommend necessary or desirable change to the Owner and the Architect/Engineer, review requests for changes, submit recommendations to the Owner and the Architect/Engineer, and assist in negotiating Change Orders.

2.2.5 *Payments to Trade Contractors:* Develop and implement a procedure for the review, processing and payment of applications by Trade Contractors for progress and final payments.

2.2.6 *Permits and Fees:* Assist the Owner and Architect/Engineer in obtaining all building permits and special permits for permanent improvements, excluding permits for inspection or temporary facilities required to be obtained directly by the various Trade Contractors. Assist in obtaining approvals from all the authorities having jurisdiction.

2.2.7 *Owner's Consultants:* If required, assist the Owner in selecting and retaining professional services of a surveyor, testing laboratories and special consultants, and coordinate these services, without assuming any responsibility or liability of or for these consultants.

2.2.8 *Inspection:* Inspect the Work of Trade Contractors for defects and deficiencies in the Work without assuming any of the Architect/Engineer's responsibilities for inspection.

2.2.8.1 Review the safety programs of each of the Trade Contractors and make appropriate recommendations. In making such recommendations and carrying out such reviews, he shall not be required to make exhaustive or continuous inspections to check safety precautions and programs in connection with the Project. The performance of such services by the Construction Manager shall not relieve the Trade Contractors of their responsibilities for the safety of persons and property, and for compliance with all federal, state and local statutes, rules, regulations and orders applicable to the conduct of the Work.

2.2.9 *Document Interpretation:* Refer all questions for interpretation of the documents prepared by the Architect/Engineer to the Architect/Engineer.

2.2.10 *Shop Drawings and Samples:* In collaboration with the Architect/Engineer, establish and implement procedures for expediting the processing and approval of shop drawings and samples.

2.2.11 *Reports and Project Site Documents:* Record the progress of the Project. Submit written progress reports to the Owner and the Architect/Engineer including information on the Trade Contractors' Work, and the percentage of completion. Keep a daily log available to the Owner and the Architect/Engineer.

2.2.11.1 Maintain at the Project site, on a current basis: records of all necessary Contracts, Drawings, samples, purchases, materials, equipment, maintenance and operating manuals and instructions, and other construction related documents, including all revisions. Obtain data from Trade Contractors and maintain a current set of record Drawings, Specifications and operating manuals. At the completion of the Project, deliver all such records to the Owner.

2.2.12 *Substantial Completion:* Determine Substantial Completion of the Work or designated portions thereof and prepare for the Architect/Engineer a list of incomplete or unsatisfactory items and a schedule for their completion.

2.2.13 *Start-Up:* With the Owner's maintenance personnel, direct the checkout of utilities, operations systems and equipment for readiness and assist in their initial start-up and testing by the Trade Contractors.

2.2.14 *Final Completion:* Determine final completion and provide written notice to the Owner and Architect/Engineer that the Work is ready for final inspection. Secure and transmit to the Architect/Engineer required guarantees, affidavits, releases, bonds and waivers. Turn over to the Owner all keys, manuals, record drawings and maintenance stocks.

2.2.15 *Warranty:* Where any Work is performed by the Construction Manager's own forces or by Trade Contractors under contract with the Construction Manager, the Construction Manager shall warrant that all materials and equipment included in such Work will be new, unless otherwise specified, and that such Work will be of good quality, free from improper workmanship and defective materials and in conformance with the Drawings and Specifications. With respect to the same Work, the

Construction Manager further agrees to correct all Work defective in material and workmanship for a period of one year from the Date of Substantial Completion or for such longer periods of time as may be set forth with respect to specific warranties contained in the trade sections of the Specifications. The Construction Manager shall collect and deliver to the Owner any specific written warranties given by others.

2.3 Additional Services

2.3.1 At the request of the Owner the Construction Manager will provide the following additional services upon written agreement between the Owner and Construction Manager defining the extent of such additional services and the amount and manner in which the Construction Manager will be compensated for such additional services.

2.3.2 Services related to investigation, appraisals or valuations of existing conditions, facilities or equipment, or verifying the accuracy of existing drawings or other Owner-furnished information.

2.3.3 Services related to Owner-furnished equipment, furniture and furnishings which are not a part of this Agreement.

2.3.4 Services for tenant or rental spaces not a part of this Agreement.

2.3.5 Obtaining or training maintenance personnel or negotiating maintenance service contracts.

ARTICLE 3

Owner's Responsibilities

3.1 The Owner shall provide full information regarding his requirements for the Project.

3.2 The Owner shall designate a representative who shall be fully acquainted with the Project and has authority to issue and approve Project Construction Budgets, issue Change Orders, render decisions promptly and furnish information expeditiously.

3.3 The Owner shall retain an Architect/Engineer for design and to prepare construction documents for the Project. The Architect/Engineer's services, duties and responsibilities are described in the Agreement between the Owner and the Architect/Engineer, a copy of which will be furnished to the Construction Manager. The Agreement between the Owner and the Architect/Engineer shall not be modified without written notification to the Construction Manager.

3.4 The Owner shall furnish for the site of the Project all necessary surveys describing the physical characteristics, soil reports and subsurface investigations, legal limitations, utility locations, and a legal description.

3.5 The Owner shall secure and pay for necessary approvals, easements, assessments and charges required for the construction, use or occupancy of permanent structures or for permanent changes in existing facilities.

3.6 The Owner shall furnish such legal services as may be necessary for providing the items set forth in Paragraph 3.5, and such auditing services as he may require.

3.7 The Construction Manager will be furnished without charge all copies of Drawings and Specifications reasonably necessary for the execution of the Work.

3.8 The Owner shall provide the insurance for the Project as provided in Paragraph 12.4, and shall bear the cost of any bonds required.

3.9 The services, information, surveys and reports required by the above paragraphs or otherwise to be furnished by other consultants employed by the Owner, shall be furnished with reasonable promptness at the Owner's expense and the Construction Manager shall be entitled to rely upon the accuracy and completeness thereof.

3.10 If the Owner becomes aware of any fault or defect in the Project or non-conformance with the Drawings and Specifications, he shall give prompt written notice thereof to the Construction Manager.

AGC DOCUMENT NO. 8 • OWNER-CONSTRUCTION MANAGER AGREEMENT   JULY 1980

3.11   The Owner shall furnish, prior to commencing work and at such future times as may be requested, reasonable evidence satisfactory to the Construction Manager that sufficient funds are available and committed for the entire cost of the Project. Unless such reasonable evidence is furnished, the Construction Manager is not required to commence or continue any Work, or may, if such evidence is not presented within a reasonable time, stop the Project upon 15 days notice to the Owner. The failure of the Construction Manager to insist upon the providing of this evidence at any one time shall not be a waiver of the Owner's obligation to make payments pursuant to this Agreement nor shall it be a waiver of the Construction Manager's right to request or insist that such evidence be provided at a later date.

3.12   The Owner shall communicate with the Trade Contractors only through the Construction Manager.

ARTICLE 4

Trade Contracts

4.1   All portions of the Project that the Construction Manager does not perform with his own forces shall be performed under Trade Contracts. The Construction Manager shall request and receive proposals from Trade Contractors and Trade Contracts will be awarded after the proposals are reviewed by the Architect/Engineer, Construction Manager and Owner.

4.2   If the Owner refuses to accept a Trade Contractor recommended by the Construction Manager, the Construction Manager shall recommend an acceptable substitute and the Guaranteed Maximum Price if applicable shall be increased or decreased by the difference in cost occasioned by such substitution and an appropriate Change Order shall be issued.

4.3   Unless otherwise directed by the Owner, Trade Contracts will be between the Construction Manager and the Trade Contractors. Whether the Trade Contracts are with the Construction Manager or the Owner, the form of the Trade Contracts including the General and Supplementary Conditions shall be satisfactory to the Construction Manager.

4.4   The Construction Manager shall be responsible to the Owner for the acts and omissions of his agents and employees, Trade Contractors performing Work under a contract with the Construction Manager, and such Trade Contractors' agents and employees.

ARTICLE 5

Schedule

5.1   The services to be provided under this Contract shall be in general accordance with the following schedule:

5.2   At the time a Guaranteed Maximum Price is established, as provided for in Article 6, a Date of Substantial Completion of the project shall also be established.

5.3   The Date of Substantial Completion of the Project or a designated portion thereof is the date when construction is sufficiently complete in accordance with the Drawings and Specifications so the Owner can occupy or utilize the Project or designated portion thereof for the use for which it is intended. Warranties called for by this Agreement or by the Drawings and Specifications shall commence on the Date of Substantial Completion of the Project or designated portion thereof.

5.4   If the Construction Manager is delayed at any time in the progress of the Project by any act or neglect of the Owner or the Architect/Engineer or by any employee of either, or by any separate contractor employed by the Owner, or by changes ordered in the Project, or by labor disputes, fire, unusual delay in transportation, adverse weather conditions not reasonably anticipatable, unavoidable casualties or any causes beyond the Construction Manager's control, or by delay authorized by the Owner pending arbitration, the Construction Completion Date shall be extended by Change Order for a reasonable length of time.

ARTICLE 6

Guaranteed Maximum Price

6.1 When the design, Drawings and Specifications are sufficiently complete, the Construction Manager will, if desired by the Owner, establish a Guaranteed Maximum Price, guaranteeing the maximum price to the Owner for the Cost of the Project and the Construction Manager's Fee. Such Guaranteed Maximum Price will be subject to modification for Changes in the Project as provided in Article 9, and for additional costs arising from delays caused by the Owner or the Architect/Engineer.

6.2 When the Construction Manager provides a Guaranteed Maximum Price, the Trade Contracts will either be with the Construction Manager or will contain the necessary provisions to allow the Construction Manager to control the performance of the Work. The Owner will also authorize the Construction Manager to take all steps necessary in the name of the Owner, including arbitration or litigation, to assure that the Trade Contractors perform their contracts in accordance with their terms.

6.3 The Guaranteed Maximum Price will only include those taxes in the Cost of the Project which are legally enacted at the time the Guaranteed Maximum Price is established.

ARTICLE 7

Construction Manager's Fee

7.1 In consideration of the performance of the Contract, the Owner agrees to pay the Construction Manager in current funds as compensation for his services a Construction Manager's Fee as set forth in subparagraphs 7.1.1 and 7.1.2.

7.1.1 For the performance of the Design Phase services, a fee of which shall be paid monthly, in equal proportions, based on the scheduled Design Phase time.

7.1.2 For work or services performed during the Construction Phase, a fee of which shall be paid proportionately to the ratio the monthly payment for the Cost of the Project bears to the estimated cost. Any balance of this fee shall be paid at the time of final payment.

7.2 Adjustments in Fee shall be made as follows:

7.2.1 For Changes in the Project as provided in Article 9, the Construction Manager's Fee shall be adjusted as follows:

7.2.2 For delays in the Project not the responsibility of the Construction Manager, there will be an equitable adjustment in the fee to compensate the Constructon Manager for his increased expenses.

7.2.3 The Construction Manager shall be paid an additional fee in the same proportion as set forth in 7.2.1 if the Construction Manager is placed in charge of the reconstruction of any insured or uninsured loss.

7.3 Included in the Construction Manager's Fee are the following:

7.3.1 Salaries or other compensation of the Construction Manager's employees at the principal office and branch offices, except employees listed in Subparagraph 8.2.2.

---

6   **AGC DOCUMENT NO. 8 • OWNER-CONSTRUCTION MANAGER AGREEMENT   JULY 1980**

7.3.2 General operating expenses of the Construction Manager's principal and branch offices other than the field office.

7.3.3 Any part of the Construction Manager's capital expenses, including interest on the Construction Manager's capital employed for the project.

7.3.4 Overhead or general expenses of any kind, except as may be expressly included in Article 8.

7.3.5 Costs in excess of the Guaranteed Maximum Price.

ARTICLE 8

Cost of the Project

8.1 The term Cost of the Project shall mean costs necessarily incurred in the Project during either the Design or Construction Phase, and paid by the Construction Manager, or by the Owner if the Owner is directly paying Trade Contractors upon the Construction Manager's approval and direction. Such costs shall include the items set forth below in this Article.

8.1.1 The Owner agrees to pay the Construction Manager for the Cost of the Project as defined in Article 8. Such payment shall be in addition to the Construction Manager's Fee stipulated in Article 7.

8.2 Cost Items

8.2.1 Wages paid for labor in the direct employ of the Construction Manager in the performance of his Work under applicable collective bargaining agreements, or under a salary or wage schedule agreed upon by the Owner and Construction Manager, and including such welfare or other benefits, if any, as may be payable with respect thereto.

8.2.2 Salaries of the Construction Manager's employees when stationed at the field office, in whatever capacity employed, employees engaged on the road in expediting the production or transportation of materials and equipment, and employees in the main or branch office performing the functions listed below:

8.2.3 Cost of all employee benefits and taxes for such items as unemployment compensation and social security, insofar as such cost is based on wages, salaries, or other remuneration paid to employees of the Construction Manager and included in the Cost of the Project under Subparagraphs 8.2.1 and 8.2.2.

8.2.4 Reasonable transportation, traveling, moving, and hotel expenses of the Construction Manager or of his officers or employees incurred in discharge of duties connected with the Project.

8.2.5 Cost of all materials, supplies and equipment incorporated in the Project, including costs of transportation and storage thereof.

8.2.6 Payments made by the Construction Manager or Owner to Trade Contractors for their Work performed pursuant to contract under this Agreement.

8.2.7 Cost, including transportation and maintenance, of all materials, supplies, equipment, temporary facilities and hand tools not owned by the workmen, which are employed or consumed in the performance of the Work, and cost less salvage value on such items used but not consumed which remain the property of the Construciton Manager.

8.2.8 Rental charges of all necessary machinery and equipment, exclusive of hand tools, used at the site of the Project, whether rented from the Construction Manager or other, including installation, repairs and replacements, dismantling, removal, costs of lubrication, transportation and delivery costs thereof, at rental charges consistent with those prevailing in the area.

8.2.9   Cost of the premiums for all insurance which the Construction Manager is required to procure by this Agreement or is deemed necessary by the Construction Manager.

8.2.10   Sales, use, gross receipts or similar taxes related to the Project imposed by any governmental authority, and for which the Construction Manager is liable.

8.2.11   Permit fees, licenses, tests, royalties, damages for infringement of patents and costs of defending suits therefor, and deposits lost for causes other than the Construction Manager's negligence. If royalties or losses and damages, including costs of defense, are incurred which arise from a particular design, process, or the product of a particular manufacturer or manufacturers specified by the Owner or Architect/Engineer, and the Construction Manager has no reason to believe there will be infringement of patent rights, such royalties, losses and damages shall be paid by the Owner and not considered as within the Guaranteed Maximum Price.

8.2.12   Losses, expenses or damages to the extent not compensated by insurance or otherwise (including settlement made with the written approval of the Owner).

8.2.13   The cost of corrective work subject, however, to the Guaranteed Maximum Price.

8.2.14   Minor expenses such as telegrams, long-distance telephone calls, telephone service at the site, expressage, and similar petty cash items in connection with the Project.

8.2.15   Cost of removal of all debris.

8.2.16   Cost incurred due to an emergency affecting the safety of persons and property.

8.2.17   Cost of data processing services required in the performance of the services outlined in Article 2.

8.2.18   Legal costs reasonably and properly resulting from prosecution of the Project for the Owner.

8.2.19   All costs directly incurred in the performance of the Project and not included in the Construction Manager's Fee as set forth in Paragraph 7.3.

ARTICLE 9

Changes in the Project

9.1   The Owner, without invalidating this Agreement, may order Changes in the Project within the general scope of this Agreement consisting of additions, deletions or other revisions, the Guaranteed Maximum Price, if established, the Construction Manager's Fee and the Construction Completion Date being adjusted accordingly. All such Changes in the Project shall be authorized by Change Order.

9.1.1   A Change Order is a written order to the Construction Manager signed by the Owner or his authorized agent issued after the execution of this Agreement, authorizing a Change in the Project or the method or manner of performance and/or an adjustment in the Guaranteed Maximum Price, the Construction Manager's Fee, or the Construction Completion Date. Each adjustment in the Guaranteed Maximum Price resulting from a Change Order shall clearly separate the amount attributable to the Cost of the Project and the Construction Manager's Fee.

9.1.2   The increase or decrease in the Guaranteed Maximum Price resulting from a Chance in the Project shall be determined in one or more of the following ways:

.1   by mutual acceptance of a lump sum properly itemized and supported by sufficient substantiating data to permit evaluation;

.2   by unit prices stated in the Agreement or subsequently agreed upon;

.3   by cost as defined in Article 8 and a mutually acceptable fixed or percentage fee; or

.4   by the method provided in Subparagraph 9.1.3.

9.1.3 If none of the methods set forth in Clauses 9.1.2.1 through 9.1.2.3 is agreed upon, the Construction Manager, provided he receives a written order signed by the Owner, shall promptly proceed with the Work involved. The cost of such Work shall then be determined on the basis of the reasonable expenditures and savings of those performing the Work attributed to the change, including, in the case of an increase in the Guaranteed Maximum Price, a reasonable increase in the Construction Manager's Fee. In such case, and also under Clauses 9.1.2.3 and 9.1.2.4 above, the Construction Manager shall keep and present, in such form as the Owner may prescribe, an itemized accounting together with appropriate supporting data of the increase in the Cost of the Project as outlined in Article 8. The amount of decrease in the Guaranteed Maximum Price to be allowed by the Construction Manager to the Owner for any deletion or change which results in a net decrease in cost will be the amount of the actual net decrease. When both additions and credits are involved in any one change, the increase in Fee shall be figured on the basis of net increase, if any.

9.1.4 If unit prices are stated in the Agreement or subsequently agreed upon, and if the quantities originally contemplated are so changed in a proposed Change Order or as a result of several Change Orders that application of the agreed unit prices to the quantities of Work proposed will cause substantial inequity to the Owner or the Construction Manager, the applicable unit prices and Guaranteed Maximum Price shall be equitably adjusted.

9.1.5 Should concealed conditions encountered in the performance of the Work below the surface of the ground or should concealed or unknown conditions in an existing structure be at variance with the conditions indicated by the Drawings, Specifications, or Owner-furnished information or should unknown physical conditions below the surface of the ground or should concealed or unknown conditions in an existing structure of an unusual nature, differing materially from those ordinarily encountered and generally recognized as inherent in work of the character provided for in this Agreement, be encountered, the Guaranteed Maximum Price and the Construction Completion Date shall be equitably adjusted by Change Order upon claim by either party made within a reasonable time after the first observance of the conditions.

9.2 Claims for Additional Cost or Time

9.2.1 If the Construction Manager wishes to make a claim for an increase in the Guaranteed Maximum Price, an increase in his fee, or an extension in the Construction Completion Date, he shall give the Owner written notice thereof within a reasonable time after the occurrence of the event giving rise to such claim. This notice shall be given by the Construction Manager before proceeding to execute any Work, except in an emergency endangering life or property in which case the Construction Manager shall act, at his discretion, to prevent threatened damage, injury or loss. Claims arising from delay shall be made within a reasonable time after the delay. No such claim shall be valid unless so made. If the Owner and the Construction Manager cannot agree on the amount of the adjustment in the Guaranteed Maximum Price, Construction Manager's Fee or Construction Completion Date, it shall be determined pursuant to the provisions of Article 16. Any change in the Guaranteed Maximum Price, Construction Manager's Fee or Construction Completion Date resulting from such claim shall be authorized by Change Order.

9.3. Minor Changes in the Project

9.3.1 The Architect/Engineer will have authority to order minor Changes in the Project not involving an adjustment in the Guaranteed Maximum Price or an extension of the Construction Completion Date and not inconsistent with the intent of the Drawings and Specifications. Such Changes may be effected by written order and shall be binding on the Owner and the Construction Manager.

9.4 Emergencies

9.4.1 In any emergency affecting the safety of persons or property, the Construction Manager shall act, at his discretion, to prevent threatened damage, injury or loss. Any increase in the Guaranteed Maximum Price or extension of time claimed by the Construction Manager on account of emergency work shall be determined as provided in this Article.

ARTICLE 10

Discounts

All discounts for prompt payment shall accrue to the Owner to the extent the Cost of the Project is paid directly by the

Owner or from a fund made available by the Owner to the Construction Manager for such payments. To the extent the Cost of the Project is paid with funds of the Construction Manager, all cash discounts shall accrue to the Construction Manager. All trade discounts, rebates and refunds, and all returns from sale of surplus materials and equipment, shall be credited to the Cost of the Project.

ARTICLE 11

Payments to the Construction Manager

11.1 The Construction Manager shall submit monthly to the Owner a statement, sworn to if required, showing in detail all moneys paid out, costs accumulated or costs incurred on account of the Cost of the Project during the previous month and the amount of the Construction Manager's Fee due as provided in Article 7. Payment by the Owner to the Construction Manager of the statement amount shall be made within ten (10) days after it is submitted.

11.2 Final payment constituting the unpaid balance of the Cost of the Project and the Construction Manager's Fee shall be due and payable when the Project is delivered to the Owner, ready for beneficial occupancy, or when the Owner occupies the Project, whichever event first occurs, provided that the Project be then substantially completed and this Agreement substantially performed. If there should remain minor items to be completed, the Construction Manager and Architect/Engineer shall list such items and the Construction Manager shall deliver, in writing, his unconditional promise to complete said items within a reasonable time thereafter. The Owner may retain a sum equal to 150% of the estimated cost of completing any unfinished items, provided that said unfinished items are listed separately and the estimated cost of completing any unfinished items likewise listed separately. Thereafter, Owner shall pay to Construction Manager, monthly, the amount retained for incomplete items as each of said items is completed.

11.3 The Construction Manager shall promptly pay all the amounts due Trade Contractors or other persons with whom he has a contract upon receipt of any payment from the Owner, the application for which includes amounts due such Trade Contractor or other persons. Before issuance of final payment, the Construction Manager shall submit satisfactory evidence that all payrolls, materials bills and other indebtedness connected with the Project have been paid or otherwise satisfied.

11.4 If the Owner should fail to pay the Construction Manager within seven (7) days after the time the payment of any amount becomes due, then the Construction Manager may, upon seven (7) additional days' written notice to the Owner and the Architect/Engineer, stop the Project until payment of the amount owing has been received.

11.5 Payments due but unpaid shall bear interest at the rate the Owner is paying on his construction loan or at the legal rate, whichever is higher.

ARTICLE 12

Insurance, Indemnity and Waiver of Subrogation

12.1 Indemnity

12.1.1 The Construction Manager agrees to indemnify and hold the Owner harmless from all claims for bodily injury and property damage (other than the Work itself and other property insured under Paragraph 12.4) that may arise from the Construction Manager's operations under this Agreement.

12.1.2 The Owner shall cause any other contractor who may have a contract with the Owner to perform construction or installation work in the areas where Work will be performed under this Agreement, to agree to indemnify the Owner and the Construction Manager and hold them harmless from all claims for bodily injury and property damage (other than property insured under Paragraph 12.4) that may arise from that contractor's operations. Such provisions shall be in a form satisfactory to the Construction Manager.

12.2 Construction Manager's Liability Insurance

12.2.1 The Construction Manager shall purchase and maintain such insurance as will protect him from the claims set forth below which may arise out of or result from the Construction Manager's operations under this Agreement whether such operations be by himself or by any Trade Contractor or by anyone directly or indirectly employed by any of them, or by anyone for whose acts any of them may be liable:

12.2.1.1 Claims under workers' compensation, disability benefit and other similar employee benefit acts which are applicable to the Work to be performed.

12.2.1.2 Claims for damages because of bodily injury, occupational sickness or disease, or death of his employees under any applicable employer's liability law.

12.2.1.3 Claims for damages because of bodily injury, death of any person other than his employees.

12.2.1.4 Claims for damages insured by usual personal injury liability coverage which are sustained (1) by any person as a result of an offense directly or indirectly related to the employment of such person by the Construction Manager or (2) by any other person.

12.2.1.5 Claims for damages, other than to the Work itself, because of injury to or destruction of tangible property, including loss of use therefrom.

12.2.1.6 Claims for damages because of bodily injury or death of any person or property damage arising out of the ownership, maintenance or use of any motor vehicle.

12.2.2 The Construction Manager's Comprehensive General Liability Insurance shall include premises — operations (including explosion, collapse and underground coverage) elevators, independent contractors, completed operations, and blanket contractual liability on all written contracts and including broad form property damage coverage.

12.2.3 The Construction Manager's Comprehensive General and Automobile Liability Insurance, as required by Subparagraphs 12.2.1 and 12.2.2 shall be written for not less than limits of liability as follows:

a. Comprehensive General Liability
 1. Personal Injury

$_____ Each Occurrence

$_____ Aggregate
(Completed Operations)

 2. Property Damage

$_____ Each Occurrence

$_____ Aggregate

b. Comprehensive Automobile Liability
 1. Bodily Injury

$_____ Each Person

$_____ Each Occurrence

 2. Property Damage

$_____ Each Occurrence

12.2.4 Comprehensive General Liability Insurance may be arranged under a single policy for the full limits required or by a combination of underlying policies with the balance provided by an Excess or Umbrella Liability policy.

12.2.5 The foregoing policies shall contain a provision that coverages afforded under the policies will not be cancelled or not renewed until at least sixty (60) days' prior written notice has been given to the Owner. Certificates of Insurance showing such coverages to be in force shall be filed with the Owner prior to commencement of the Work.

12.3 Owner's Liability Insurance

12.3.1 The Owner shall be responsible for purchasing and maintaining his own liability insurance and, at his option, may purchase and maintain such insurance as will protect him against claims which may arise from operations under this Agreement.

12.4 Insurance to Protect Project

12.4.1 The Owner shall purchase and maintain property insurance in a form acceptable to the Construction Manager upon the entire Project for the full cost of replacement as of the time of any loss. This insurance shall include as named insureds the Owner, the Construction Manager, Trade Contractors and their Trade Subcontractors and shall insure against loss from the perils of Fire, Extended Coverage, and shall include "All Risk" insurance for physical loss or damage including, without duplication of coverage, at least theft, vandalism, malicious mischief, transit, collapse, flood, earthquake, testing, and damage resulting from defective design, workmanship or material. The Owner will increase limits of coverage, if necessary, to reflect estimated replacement cost. The Owner will be responsible for any co-insurance penalties or deductibles. If the Project covers an addition to or is adjacent to an existing building, the Construction Manager, Trade Contractors and their Trade Subcontractors shall be named as additional insureds under the Owner's Property Insurance covering such building and its contents.

12.4.1.1 If the Owner finds it necessary to occupy or use a portion or portions of the Project prior to Substantial Completion thereof, such occupancy shall not commence prior to a time mutually agreed to by the Owner and Construction Manager and to which the insurance company or companies providing the property insurance have consented by endorsement to the policy or policies. This insurance shall not be cancelled or lapsed on account of such partial occupancy. Consent of the Construction Manager and of the insurance company or companies to such occupancy or use shall not be unreasonably withheld.

12.4.2 The Owner shall purchase and maintain such boiler and machinery insurance as may be required or necessary. This insurance shall include the interests of the Owner, the Construction Manager, Trade Contractors and their Trade Subcontractors in the Work.

12.4.3 The Owner shall purchase and maintain such insurance as will protect the Owner and Construction Manager against loss of use of Owner's property due to those perils insured pursuant to Subparagraph 12.4.1. Such policy will provide coverage for expediting expenses of materials, continuing overhead of the Owner and Construction Manager, necessary labor expense including overtime, loss of income by the Owner and other determined exposures. Exposures of the Owner and the Construction Manager shall be determined by mutual agreement and separate limits of coverage fixed for each item.

12.4.4 The Owner shall file a copy of all policies with the Construction Manager before an exposure to loss may occur. Copies of any subsequent endorsements will be furnished to the Construction Manager. The Construction Manager will be given sixty (60) days notice of cancellation, non-renewal, or any endorsements restricting or reducing coverage. If the Owner does not intend to purchase such insurance, he shall inform the Construction Manager in writing prior to the commencement of the Work. The Construction Manager may then effect insurance which will protect the interest of himself, the Trade Contractors and their Trade Subcontractors in the Project, the cost of which shall be a Cost of the Project pursuant to Article 8, and the Guaranteed Maximum Price shall be increased by Change Order. If the Construction Manager is damaged by failure of the Owner to purchase or maintain such insurance or to so notify the Construction Manager, the Owner shall bear all reasonable costs properly attributable thereto.

12.5 Property Insurance Loss Adjustment

12.5.1 Any insured loss shall be adjusted with the Owner and the Construction Manager and made payable to the Owner and Construction Manager as trustees for the insureds, as their interests may appear, subject to any applicable mortgagee clause.

12.5.2 Upon the occurrence of an insured loss, monies received will be deposited in a separate account and the trustees shall make distribution in accordance with the agreement of the parties in interest, or in the absence of such agreement, in accordance with an arbitration award pursuant to Article 16. If the trustees are unable to agree on the settlement of the loss, such dispute shall also be submitted to arbitration pursuant to Article 16.

12.6 Waiver of Subrogation

12.6.1 The Owner and Construction Manager waive all rights against each other, the Architect/Engineer, Trade Contractors, and their Trade Subcontractors for damages caused by perils covered by insurance provided under Paragraph 12.4, except such rights as they may have to the proceeds of such insurance held by the Owner and Construction Manager as trustees. The Construction Manager shall require similar waivers from all Trade Contractors and their Trade Subcontractors.

12.6.2 The Owner and Construction Manager waive all rights against each other and the Architect/Engineer, Trade Contractors and their Trade Subcontractors for loss or damage to any equipment used in connection with the Project and covered by any property insurance. The Construction Manager shall require similar waivers from all Trade Contractors and their Trade Subcontractors.

12.6.3 The Owner waives subrogation against the Construction Manager, Architect/Engineer, Trade Contractors, and their Trade Subcontractors on all property and consequential loss policies carried by the Owner on adjacent properties and under property and consequential loss policies purchased for the Project after its completion.

12.6.4 If the policies of insurance referred to in this Paragraph require an endorsement to provide for continued coverage where there is a waiver of subrogation, the owners of such policies will cause them to be so endorsed.

ARTICLE 13

**Termination of the Agreement and Owner's
Right to Perform Construction Manager's Obligations**

13.1 Termination by the Construction Manager

13.1.1 If the Project, in whole or substantial part, is stopped for a period of thirty days under an order of any court or other public authority having jurisdiction, or as a result of an act of government, such as a declaration of a national emergency making materials unavailable, through no act or fault of the Construction Manager, or if the Project should be stopped for a period of thirty days by the Construction Manager for the Owner's failure to make payment thereon, then the Construction Manager may, upon seven days' written notice to the Owner and the Architect/Engineer, terminate this Agreement and recover from the Owner payment for all work executed, the Construction Manager's Fee earned to date, and for any proven loss sustained upon any materials, equipment, tools, construction equipment and machinery, cancellation charges on existing obligations of the Construction Manager, and a reasonable profit.

13.2 Owner's Right to Perform Construction Manager's Obligations and Termination by the Owner for Cause

13.2.1 If the Construction Manager fails to perform any of his obligations under this Agreement including any obligation he assumes to perform Work with his own forces, the Owner may, after seven days' written notice during which period the Construction Manager fails to perform such obligation, make good such deficiencies. The Guaranteed Maximum Price, if any, shall be reduced by the cost to the Owner of making good such deficiencies.

13.2.2 If the Construction Manager is adjudged a bankrupt, or if he makes a general assignment for the benefit of his creditors, or if a receiver is appointed on account of his insolvency, or if he persistently or repeatedly refuses or fails, except in cases for which extension of time is provided, to supply enough properly skilled workmen or proper materials, or if he fails to make proper payment to Trade Contractors or for materials or labor, or persistently disregards laws, ordinances, rules, regulations or orders of any public authority having jurisdiction, or otherwise is guilty of a substantial violation of a provision of the Agreement, then the Owner may, without prejudice to any right or remedy and after giving the Construction Manager and his surety, if any, seven days' written notice, during which period the Construction Manager fails to cure the violation, terminate the employment of the Construction Manager and take possession of the site and of all materials, equipment, tools, construction equipment and machinery thereon owned by the Construction Manager and may finish the Project by whatever reasonable method he may deem expedient. In such case, the Construction Manager shall not be entitled to receive any further payment until the Project is finished nor shall he be relieved from his obligations assumed under Article 6.

13.3 Termination by Owner Without Cause

13.3.1 If the Owner terminates this Agreement other than pursuant to Subparagraph 13.2.2 or Subparagraph 13.3.2, he shall reimburse the Construction Manager for any unpaid Cost of the Project due him under Article 8, plus (1) the unpaid balance of the Fee computed upon the Cost of the Project to the date of termination at the rate of the percentage named in Subparagraph 7.2.1 or if the Construction Manager's Fee be stated as a fixed sum, such an amount as will increase the payment on account of his fee to a sum which bears the same ratio to the said fixed sum as the Cost of the Project at the time of termination bears to the adjusted Guaranteed Maximum Price, if any, otherwise to a reasonable estimated Cost of the Project when completed. The Owner shall also pay to the Construction Manager fair compensation, either by purchase or rental at the election of the Owner, for any equipment retained. In case of such termination of the Agreement the Owner shall further assume and become liable for obligations, commitments and unsettled claims that the Construction Manager has previously undertaken or incurred in good faith in connection with said Project. The Construction Manager shall, as a condition of receiving the payments referred to in this Article 13, execute and deliver all such papers and take all such steps, including the legal assignment of his contractual rights, as the Owner may require for the purpose of fully vesting in him the rights and benefits of the Construction Manager under such obligations or commitments.

13.3.2   After the completion of the Design Phase, if the final cost estimates make the Project no longer feasible from the standpoint of the Owner, the Owner may terminate this Agreement and pay the Construction Manager his Fee in accordance with Subparagraph 7.1.1 plus any costs incurred pursuant to Article 9.

ARTICLE 14

Assignment and Governing Law

14.1   Neither the Owner nor the Construction Manager shall assign his interest in this Agreement without the written consent of the other except as to the assignment of proceeds.

14.2   This Agreement shall be governed by the law of the place where the Project is located.

ARTICLE 15

Miscellaneous Provisions

15.1   It is expressly understood that the Owner shall not directly retaining the services of an Architect/Engineer.

ARTICLE 16

Arbitration

16.1   All claims, disputes and other matters in questions arising out of, or relating to, this Agreement or the breach thereof, except with respect to the Architect/Engineer's decision on matters relating to artistic effect, and except for claims which have been waived by the making or acceptance of final payment shall be decided by arbitration in accordance with the Construction Industry Arbitration Rules of the American Arbitration Association then obtaining unless the parties mutually agree otherwise. This Agreement to arbitrate shall be specifically enforceable under the prevailing arbitration law.

16.2   Notice of the demand for arbitration shall be filed in writing with the other party to this Agreement and with the American Arbitration Association. The demand for arbitration shall be made within a reasonable time after the claim, dispute or other matter in question has arisen, and in no event shall it be made after the date when institution of legal or equitable proceedings based on such claim, dispute or other matter in question would be barred by the applicable statute of limitations.

16.3   The award rendered by the arbitrators shall be final and judgment may be entered upon it in accordance with applicable law in any court having jurisdiction thereof.

16.4   Unless otherwise agreed in writing, the Construction Manager shall carry on the Work and maintain the Contract Completion Date during any arbitration proceedings, and the Owner shall continue to make payments in accordance with this Agreement.

16.5   All claims which are related to or dependent upon each other, shall be heard by the same arbitrator or arbitrators even though the parties are not the same unless a specific contract prohibits such consolidation.

This Agreement executed the day and year first written above.

ATTEST: OWNER:

ATTEST: CONSTRUCTION MANAGER:

SAMPLE

INSTRUCTION SHEET *AIA DOCUMENT A111a*

FOR AIA DOCUMENT A111, STANDARD FORM OF AGREEMENT BETWEEN OWNER AND CONTRACTOR where the basis of payment is the COST OF THE WORK PLUS A FEE— 1978 EDITION

A. GENERAL INFORMATION:

AIA Document A111, Standard Form of Agreement Between Owner and Contractor, is for use where the basis of payment to the Contractor is the cost of the Work plus a fixed or percentage fee. The 1978 Edition has been prepared for use in conjunction with the 1976 Edition of AIA Document A201, General Conditions of the Contract for Construction and contains provisions for stipulating a Guaranteed Maximum Cost. Although the Cost Plus Fee Arrangement lacks the financial certainty of a lump sum agreement, it may be desirable when fixed prices on portions of the Work cannot be obtained, or when construction must be started before Drawings and Specifications are completed, as well as under other circumstances.

B. CHANGES FROM THE PREVIOUS EDITION:

Provisions which have been revised in or added to the 1978 Edition of the Cost Plus Fee Owner-Contractor Agreement are listed below:

1. ARTICLE 4—TIME OF COMMENCEMENT AND SUBSTANTIAL COMPLETION:

The word "SUBSTANTIAL" has been added to the article name. The General Conditions, AIA Document A201, 1976 Edition, make it clear that the Contract Time runs until the Date of Substantial Completion; the Owner should be aware that an additional period of time will be required to reach final completion.

Paragraph 4.1: Revised to clarify that the Contract Time is subject to Owner-authorized adjustments by Change Order.

2. ARTICLE 7—CHANGES IN THE WORK:

Paragraph 7.1: Revised to reference the Contract Documents rather than just the General Conditions, in determining the procedure for and the amount of Change Orders.

3. ARTICLE 8—COSTS TO BE REIMBURSED:

Subparagraph 8.1.16: A parenthetical statement has been added calling attention to the space for desired modifications.

4. ARTICLE 9—COSTS NOT TO BE REIMBURSED:

Subparagraph 9.1.4: New subparagraph. Rental costs not specifically provided for in Subparagraph 8.1.8 or in modifications thereto are not to be reimbursed.

5. ARTICLE 11—SUBCONTRACTS AND OTHER AGREEMENTS:

Article modified to clarify that agreements other than subcontracts are also subject to Article 11.

6. ARTICLE 13—APPLICATIONS FOR PAYMENT:

Paragraph 13.1 Revised to clarify that the Owner and Architect may require such supporting data as necessary to certify the Contractor's right to payment.

7. ARTICLE 14—PAYMENTS TO THE CONTRACTOR:

Paragraph 4.1: Revised pursuant to the 1976 Edition of A201 to clarify that the Architect will take appropriate action in response to the Contractor's Application for Payment. The Owner and Contractor should be aware that the Architect will recommend payment only after the Contractor's right to payment has been substantiated.

Subparagraph 14.1.1: A new provision added to clarify that the Architect must and does rely on the information provided by the Contractor in recommending payment. The Architect is not expected to confirm that the Contractor has actually paid all expenses incurred.

Paragraph 14.3: A new provision, conforming with the 1976 Edition of A201, to allow prior agreement on a rate of interest for overdue payments. A parenthetical statement, calling attention to Federal and state laws applicable to interest provisions, has also been added.

8. ARTICLE 16—MISCELLANEOUS PROVISIONS:

Paragraph 16.1: Revised to reference the Contract Documents rather than just the General Conditions.

C. COMPLETING THE FORM:

(NOTE: Prospective bidders should be made aware of any additional provisions which may be included in A111, such as liquidated damages, retainage, or payment for stored materials, by an appropriate notice in the Bidding Documents.)

1. Cover Page:

The names of the Owner and the Architect should be shown in the same form as in the other Contract Documents; include the full legal or corporate names under which the Owner and Contractor are entering the Agreement.

2. ARTICLE 2—THE WORK:

The general scope of the Work should be carefully defined here since changes by Change Order, under Paragraph 12.1 of A201, must be within the general scope of the Work contemplated by the Contract. This Article should be used to describe the portions of the Project for which the Contractor is responsible, if separate contracts are used.

3. ARTICLE 4—TIME OF COMMENCEMENT AND SUBSTANTIAL COMPLETION:

The following items should be included as appropriate:

a. Date of commencement of the Work: This should not be earlier than the date of execution of the Contract. When time of performance is to be strictly enforced, the statement of starting time should be carefully considered. At the end of the first line, enter either the specific date of commencement of the Work, or if a notice to proceed is to be used, enter the words, "on the date stipulated in the notice to proceed."

b. Substantial Completion of the Work: Substantial Completion of the Work may be expressed as a number of days (preferably calendar days) or as a specified date.

c. Provision for liquidated damages: If liquidated damages are to be assessed because delayed construction will result in the Owner actually suffering loss, the entire provision for liquidated damages should be entered in the instructions to bidders as well as the Agreement. This provision should be drafted by the Owner's attorney. Liquidated damages are not a penalty to be inflicted on the Contractor, but must bear an actual and reasonably estimated relationship to the loss of the Owner if the building is not completed on time; for example, the cost per day of renting space to house students if a dormitory cannot be occupied when needed, additional financing costs, loss of profits, etc.

4. ARTICLE 5—COST OF THE WORK AND GUARANTEED MAXIMUM COST:

Any incentive provisions for distribution of savings under a Guaranteed Maximum Cost should be included in Article 5. Delete Paragraph 5.2 if no maximum cost is established.

5. ARTICLE 6—CONTRACTOR'S FEE:

The Contractor's fee may be stated as a stipulated sum or as a percentage of Cost of the Work. Fee adjustments adding to and subtracting from the Contractor's fee may be related to the cost of the Change or some other means.

6. ARTICLE 8—COSTS TO BE REIMBURSED and ARTICLE 9—COSTS NOT TO BE REIMBURSED:

All costs which are established as reimbursable or expressly not reimbursable should be stated in Articles 8 and 9.

An appropriate provision should be included in Article 8 if the Contractor's overhead costs associated with corrective work are to be reimbursed, when such overhead costs are incurred after final payment.

7. ARTICLE 14—PAYMENTS TO THE CONTRACTOR:

Due dates for payments should be established in consideration of the time required for the Contractor to prepare an Application for Payment, for the Architect to take appropriate action, and for the Owner to make payment, within the time limits set in Article 9 of A201 and in this Article of A111. Note that the Architect does not "certify" payment under this contract form, and if the Architect uses AIA Document G702, Application and Certificate for Payment, the certification provisions should be deleted, not only from G702, but also from the Owner-Architect Agreement.

8. ARTICLE 16—MISCELLANEOUS PROVISIONS:

An accurate, detailed enumeration of all Contract Documents must be made in this Article.

9. Signatures:

Subparagraph 1.2.1 of AIA Document A201 states that the Contract Documents shall be executed in not less than triplicate by the Owner and the Contractor. The Agreement should be executed by the parties in their capacities as individuals, partners, officers, etc., as appropriate.

D. REPRODUCTION:

AIA Document A111 is a copyrighted document, and may not be reproduced or excerpted from in substantial part without the express written permission of AIA. Purchasers of A111 are hereby entitled to reproduce a maximum of ten copies of the completed or executed document for use only in connection with the particular Project. AIA will not permit the reproduction of this document in blank, or the use of substantial portions of, or language from, this Document, except upon written request and after receipt of written permission from AIA.

THE AMERICAN INSTITUTE OF ARCHITECTS

AIA Document A111

Standard Form of Agreement Between Owner and Contractor

where the basis of payment is the

COST OF THE WORK PLUS A FEE

1978 EDITION

*THIS DOCUMENT HAS IMPORTANT LEGAL CONSEQUENCES; CONSULTATION WITH
AN ATTORNEY IS ENCOURAGED WITH RESPECT TO ITS COMPLETION OR MODIFICATION*

Use only with the 1976 Edition of AIA Document A201, General Conditions of the Contract for Construction.

This document has been approved and endorsed by The Associated General Contactors of America

AGREEMENT

made as of the   day of   in the year of Nineteen
Hundred and

BETWEEN the Owner:

and the Contractor:

the Project:

the Architect:

The Owner and the Contractor agree as set forth below.

AIA DOCUMENT A111 • COST-PLUS OWNER-CONTRACTOR AGREEMENT • NINTH EDITION • APRIL 1978 • AIA®
© 1978 • THE AMERICAN INSTITUTE OF ARCHITECTS, 1735 NEW YORK AVE., N.W., WASHINGTON, D.C. 20006   **A111-1978 1**

ARTICLE 1

THE CONTRACT DOCUMENTS

1.1   The Contract Documents consist of this Agreement, the Conditions of the Contract (General, Supplementary and other Conditions), the Drawings, the Specifications, all Addenda issued prior to and all Modifications issued after execution of this Agreement. These form the Contract, and all are as fully a part of the Contract as if attached to this Agreement or repeated herein. An enumeration of the Contract Documents appears in Article 16. If anything in the Contract Documents is inconsistent with this Agreement, the Agreement shall govern.

ARTICLE 2

THE WORK

2.1   The Contractor shall perform all the Work required by the Contract Documents for

(Here insert the caption descriptive of the Work as used on other Contract Documents.)

ARTICLE 3

THE CONTRACTOR'S DUTIES AND STATUS

3.1   The Contractor accepts the relationship of trust and confidence established between him and the Owner by this Agreement. He covenants with the Owner to furnish his best skill and judgment and to cooperate with the Architect in furthering the interests of the Owner. He agrees to furnish efficient business administration and superintendence and to use his best efforts to furnish at all times an adequate supply of workmen and materials, and to perform the Work in the best way and in the most expeditious and economical manner consistent with the interests of the Owner.

ARTICLE 4

TIME OF COMMENCEMENT AND SUBSTANTIAL COMPLETION

4.1   The Work to be performed under this Contract shall be commenced

and, subject to authorized adjustments,

Substantial Completion shall be achieved not later than

(Here insert any special provisions for liquidated damages relating to failure to complete on time.)

AIA DOCUMENT A111 • COST-PLUS OWNER-CONTRACTOR AGREEMENT • NINTH EDITION • APRIL 1978 • AIA®
© 1978 • THE AMERICAN INSTITUTE OF ARCHITECTS, 1735 NEW YORK AVE., N.W., WASHINGTON, D.C. 20006   **A111-1978**   **2**

ARTICLE 5

COST OF THE WORK AND GUARANTEED MAXIMUM COST

5.1   The Owner agrees to reimburse the Contractor for the Cost of the Work as defined in Article 8. Such reimbursement shall be in addition to the Contractor's Fee stipulated in Article 6.

5.2   The maximum cost to the Owner, including the Cost of the Work and the Contractor's Fee, is guaranteed not to exceed the sum of   dollars ($); such Guaranteed Maximum Cost shall be increased or decreased for Changes in the Work as provided in Article 7.

(Here insert any provision for distribution of any savings Delete Paragraph 5 2 if there is no Guaranteed Maximum Cost)

ARTICLE 6

CONTRACTOR'S FEE

6.1   In consideration of the performance of the Contract, the Owner agrees to pay the Contractor in current funds as compensation for his services a Contractor's Fee as follows:

6.2   For Changes in the Work, the Contractor's Fee shall be adjusted as follows:

6.3   The Contractor shall be paid   percent (%) of the proportional amount of his Fee with each progress payment, and the balance of his Fee shall be paid at the time of final payment.

ARTICLE 7

CHANGES IN THE WORK

7.1   The Owner may make Changes in the Work as provided in the Contract Documents. The Contractor shall be reimbursed for Changes in the Work on the basis of Cost of the Work as defined in Article 8.

7.2   The Contractor's Fee for Changes in the Work shall be as set forth in Paragraph 6.2, or in the absence of specific provisions therein, shall be adjusted by negotiation on the basis of the Fee established for the original Work.

ARTICLE 8

COSTS TO BE REIMBURSED

8.1   The term Cost of the Work shall mean costs necessarily incurred in the proper performance of the Work and paid by the Contractor. Such costs shall be at rates not higher than the standard paid in the locality of the Work except with prior consent of the Owner, and shall include the items set forth below in this Article 8.

8.1.1   Wages paid for labor in the direct employ of the Contractor in the performance of the Work under applicable collective bargaining agreements, or under a salary or wage schedule agreed upon by the Owner and Contractor, and including such welfare or other benefits, if any, as may be payable with respect thereto.

8.1.2   Salaries of Contractor's personnel when stationed at the field office, in whatever capacity employed. Personnel engaged, at shops or on the road, in expediting the production or transportation of materials or equipment, shall be considered as stationed at the field office and their salaries paid for that portion of their time spent on this Work.

8.1.3   Cost of contributions, assessments or taxes incurred during the performance of the Work for such items as unemployment compensation and social security, insofar as such cost is based on wages, salaries, or other remuneration paid to employees of the Contractor and included in the Cost of the Work under Subparagraphs 8.1.1 and 8.1.2.

8.1.4   The portion of reasonable travel and subsistence expenses of the Contractor or of his officers or employees incurred while traveling in discharge of duties connected with the Work.

8.1.5   Cost of all materials, supplies and equipment incorporated in the Work, including costs of transportation thereof.

8.1.6   Payments made by the Contractor to Subcontractors for Work performed pursuant to Subcontracts under this Agreement.

8.1.7   Cost, including transportation and maintenance, of all materials, supplies, equipment, temporary facilities and hand tools not owned by the workers, which are consumed in the performance of the Work, and cost less salvage value on such items used but not consumed which remain the property of the Contractor.

8.1.8   Rental charges of all necessary machinery and equipment, exclusive of hand tools, used at the site of the Work, whether rented from the Contractor or others, including installation, minor repairs and replacements, dismantling, removal, transportation and delivery costs thereof, at rental charges consistent with those prevailing in the area.

8.1.9   Cost of premiums for all bonds and insurance which the Contractor is required by the Contract Documents to purchase and maintain.

8.1.10   Sales, use or similar taxes related to the Work and for which the Contractor is liable imposed by any governmental authority.

8.1.11   Permit fees, royalties, damages for infringement of patents and costs of defending suits therefor, and deposits lost for causes other than the Contractor's negligence.

8.1.12   Losses and expenses, not compensated by insurance or otherwise, sustained by the Contractor in connection with the Work, provided they have resulted from causes other than the fault or neglect of the Contractor. Such losses shall include settlements made with the written consent and approval of the Owner. No such losses and expenses shall be included in the Cost of the Work for the purpose of determining the Contractor's Fee. If, however, such loss requires reconstruction and the Contractor is placed in charge thereof, he shall be paid for his services a Fee proportionate to that stated in Paragraph 6.1.

8.1.13   Minor expenses such as telegrams, long distance telephone calls, telephone service at the site, expressage, and similar petty cash items in connection with the Work.

8.1.14   Cost of removal of all debris.

8.1.15 Costs incurred due to an emergency affecting the safety of persons and property.

8.1.16 Other costs incurred in the performance of the Work if and to the extent approved in advance in writing by the Owner.

(Here insert modifications or limitations to any of the above Subparagraphs, such as equipment rental charges and small tool charges applicable to the Work.)

ARTICLE 9

COSTS NOT TO BE REIMBURSED

9.1 The term Cost of the Work shall not include any of the items set forth below in this Article 9.

9.1.1 Salaries or other compensation of the Contractor's personnel at the Contractor's principal office and branch offices.

9.1.2 Expenses of the Contractor's principal and branch offices other than the field office.

9.1.3 Any part of the Contractor's capital expenses, including interest on the Contractor's capital employed for the Work.

9.1.4 Except as specifically provided for in Subparagraph 8.1.8 or in modifications thereto, rental costs of machinery and equipment.

9.1.5 Overhead or general expenses of any kind, except as may be expressly included in Article 8.

9.1.6 Costs due to the negligence of the Contractor, any Subcontractor, anyone directly or indirectly employed by any of them, or for whose acts any of them may be liable, including but not limited to the correction of defective or nonconforming Work, disposal of materials and equipment wrongly supplied, or making good any damage to property.

9.1.7 The cost of any item not specifically and expressly included in the items described in Article 8.

9.1.8 Costs in excess of the Guaranteed Maximum Cost, if any, as set forth in Article 5 and adjusted pursuant to Article 7.

ARTICLE 10

DISCOUNTS, REBATES AND REFUNDS

10.1   All cash discounts shall accrue to the Contractor unless the Owner deposits funds with the Contractor with which to make payments, in which case the cash discounts shall accrue to the Owner. All trade discounts, rebates and refunds, and all returns from sale of surplus materials and equipment shall accrue to the Owner, and the Contractor shall make provisions so that they can be secured.

(Here insert any provisions relating to deposits by the Owner to permit the Contractor to obtain cash discounts.)

ARTICLE 11

SUBCONTRACTS AND OTHER AGREEMENTS

11.1   All portions of the Work that the Contractor's organization does not perform shall be performed under Subcontracts or by other appropriate agreement with the Contractor. The Contractor shall request bids from Subcontractors and shall deliver such bids to the Architect. The Owner will then determine, with the advice of the Contractor and subject to the reasonable objection of the Architect, which bids will be accepted.

11.2   All Subcontracts shall conform to the requirements of the Contract Documents. Subcontracts awarded on the basis of the cost of such work plus a fee shall also be subject to the provisions of this Agreement insofar as applicable.

ARTICLE 12

ACCOUNTING RECORDS

12.1   The Contractor shall check all materials, equipment and labor entering into the Work and shall keep such full and detailed accounts as may be necessary for proper financial management under this Agreement, and the system shall be satisfactory to the Owner. The Owner shall be afforded access to all the Contractor's records, books, correspondence, instructions, drawings, receipts, vouchers, memoranda and similar data relating to this Contract, and the Contractor shall preserve all such records for a period of three years, or for such longer period as may be required by law, after the final payment.

ARTICLE 13

APPLICATIONS FOR PAYMENT

13.1   The Contractor shall, at least ten days before each payment falls due, deliver to the Architect an itemized statement, notarized if required, showing in complete detail all moneys paid out or costs incurred by him on account of the Cost of the Work during the previous month for which he is to be reimbursed under Article 5 and the amount of the Contractor's Fee due as provided in Article 6, together with payrolls for all labor and such other data supporting the Contractor's right to payment for Subcontracts or materials as the Owner or the Architect may require.

ARTICLE 14

PAYMENTS TO THE CONTRACTOR

14.1 The Architect will review the Contractor's Applications for Payment and will promptly take appropriate action thereon as provided in the Contract Documents. Such amount as he may recommend for payment shall be payable by the Owner not later than the _____ day of the month.

14.1.1 In taking action on the Contractor's Applications for Payment, the Architect shall be entitled to rely on the accuracy and completeness of the information furnished by the Contractor and shall not be deemed to represent that he has made audits of the supporting data, exhaustive or continuous on-site inspections or that he has made any examination to ascertain how or for what purposes the Contractor has used the moneys previously paid on account of the Contract.

14.2 Final payment, constituting the entire unpaid balance of the Cost of the Work and of the Contractor's Fee, shall be paid by the Owner to the Contractor _____ days after Substantial Completion of the Work unless otherwise stipulated in the Certificate of Substantial Completion, provided the Work has been completed, the Contract fully performed, and final payment has been recommended by the Architect.

14.3 Payments due and unpaid under the Contract Documents shall bear interest from the date payment is due at the rate entered below, or in the absence thereof, at the legal rate prevailing at the place of the Project.

(Here insert any rate of interest agreed upon.)

(Usury laws and requirements under the Federal Truth in Lending Act, similar state and local consumer credit laws and other regulations at the Owner's and Contractor's principal places of business, the location of the Project and elsewhere may affect the validity of this provision. Specific legal advice should be obtained with respect to deletion, modification, or other requirements such as written disclosures or waivers.)

ARTICLE 15

TERMINATION OF CONTRACT

15.1 The Contract may be terminated by the Contractor as provided in the Contract Documents.

15.2 If the Owner terminates the Contract as provided in the Contract Documents, he shall reimburse the Contractor for any unpaid Cost of the Work due him under Article 5, plus (1) the unpaid balance of the Fee computed upon the Cost of the Work to the date of termination at the rate of the percentage named in Article 6, or (2) if the Contractor's Fee be stated as a fixed sum, such an amount as will increase the payments on account of his Fee to a sum which bears the same ratio to the said fixed sum as the Cost of the Work at the time of termination bears to the adjusted Guaranteed Maximum Cost, if any, otherwise to a reasonable estimated Cost of the Work when completed. The Owner shall also pay to the Contractor fair compensation, either by purchase or rental at the election of the Owner, for any equipment retained. In case of such termination of the Contract the Owner shall further assume and become liable for obligations, commitments and unsettled claims that the Contractor has previously undertaken or incurred in good faith in connection with said Work. The Contractor shall, as a condition of receiving the payments referred to in this Article 15, execute and deliver all such papers and take all such steps, including the legal assignment of his contractual rights, as the Owner may require for the purpose of fully vesting in himself the rights and benefits of the Contractor under such obligations or commitments.

ARTICLE 16

MISCELLANEOUS PROVISIONS

16.1 Terms used in this Agreement which are defined in the Contract Documents shall have the meanings designated in those Contract Documents.

16.2 The Contract Documents, which constitute the entire agreement between the Owner and the Contractor, are listed in Article 1 and, except for Modifications issued after execution of this Agreement, are enumerated as follows:

(List below the Agreement, the Conditions of the Contract, [General, Supplementary, and other Conditions], the Drawings, the Specifications, and any Addenda and accepted alternates, showing page or sheet numbers in all cases and dates where applicable.)

This Agreement entered into as of the day and year first written above.

OWNER   CONTRACTOR

_____   _____

_____   _____

_____   _____

Related Forms

The AIA and AGC publish related forms to facilitate the use of the construction management approach. These forms are briefly described below.

AIA Document A101/CM, *Standard Form of Agreement between Owner and Contractor* (CM edition).   This form is similar to AIA's traditional owner–contractor agreement form, AIA Document A101, and is used in conjunction with A201/CM, *General Conditions of the Contract for Construction,* to form the contract between the owner and each of the separate contractors who will be performing construction work. The form establishes the scope of the contractor's work, the fixed price for such work, and a completion date. In article 5, the amount of retainage and any provision for reducing retainage, such as upon substantial completion, can be inserted. In article 7, all the drawings and specifications describing the contractor's work and the form of general conditions (AIA Document A101/CM may be used in conjunction with AGC Document 8b, *General Conditions for Trade Contractors under Construction Management*) should be specifically enumerated. If the contractor is to be provided any special facilities or services or will be subject to special working conditions (such as to allow the owner continued access to and use of a facility being remodeled or added to), these provisions should be included.

INSTRUCTION SHEET
AIA DOCUMENT A101/CMa

FOR AIA DOCUMENT A101/CM, STANDARD FORM OF AGREEMENT BETWEEN OWNER AND CONTRAC-
TOR, CONSTRUCTION MANAGEMENT EDITION — JUNE 1980 EDITION

A. GENERAL INFORMATION:

AIA Document A101/CM, Standard Form of Agreement Between Owner and Contractor, Construction Management Edition, is for use where the basis of payment is a stipulated sum (fixed price). The 1980 Edition has been
prepared for use with the 1980 Edition of AIA Document A201/CM, General Conditions of the Contract for Construction, Construction Management Edition. It is suitable for any arrangement between the Owner and the
Contractor where the cost has been set in advance by either bidding or negotiation. It has been prepared with
multiple prime Contractors in mind, so if a single general Contract is let, AIA Document A101 should be used.

B. CHANGES FROM THE PREVIOUS EDITION:

Provisions which have been revised or added to the current edition of A101/CM include the following:

1. Article 3 — Time of Commencement and Substantial Completion

The word "SUBSTANTIAL" has been added to the article name. A201/CM makes it clear that Contract Time
runs until the Date of Substantial Completion; the Owner should be aware that an additional period of time
will be required to reach final completion.

2. Article 4 — Contract Sum

Revised to reference the Contract Documents for the determination of Change Orders. A separate paragraph
has been added to outline the basis on which the Contract Sum has been determined, as applicable.

3. Article 5 — Progress Payments

The first paragraph stipulates a specific day of the month as the end of the period for which progress payments will be made. The Owner is now required to make progress payments not later than a designated
number of days following the end of that period.

The provision for interest on payments due and unpaid has been expanded to provide for the entry of a specific rate of interest agreed upon in advance. A parenthetical statement has been added drawing attention to
truth in lending and other laws which may govern the use and form of an interest provision under certain
circumstances.

4. Article 6 — Final Payment

Modified to provide that final payment is due when the Work has been completed; the reference to a stipulated number of days after Substantial Completion of the Work has been deleted. The Certificate of Substantial Completion will provide the time period within which the Contractor will bring the Work to final
completion.

C. COMPLETING THE A101/CM FORM:

(NOTE: Prospective bidders should have been made aware of any additional provisions which may be included
in the Agreement, such as liquidated damages, retainage, or payment for stored materials, by an appropriate
notice in the Bidding Documents.)

1. Cover Page

The names of the Owner and the Architect should be shown in the same form as in the other Contract Documents; include the full legal or corporate names under which the Owner and the Contractor are entering the
Agreement.

2. Article 2 — The Work

The general scope of the Work should be carefully defined here, since changes by Change Order must be
within the general scope of the Work contemplated by the Contract. This Article should be used to describe
the portions of the Project for which the Contractor is responsible.

3. Article 3 — Time of Commencement and Substantial Completion

The following information should be included as appropriate:

a. Date of commencement of the Work: This should not be earlier than the date of execution of the Contract. When time of performance is to be strictly enforced, the statement of starting time should be care-

fully considered. At the end of the first line, enter either the specific date of commencement of the Work, or if a notice to proceed is to be used, enter the words, "on the date stipulated in the notice to proceed."

b. Substantial Completion of the Work: Substantial Completion of the Work may be expressed as a number of days (preferably calendar days) or as a specified date. The time requirements will ordinarily have been fulfilled when the Work is Substantially Complete, as defined in the General Conditions, even if a few minor items may remain to be completed or corrected.

c. Coordination with Project Schedule: Include appropriate provisions delineating the Contractor's responsibility for commencing the Work and achieving Substantial Completion in accordance with the Construction Manager's comprehensive Project Schedule. It may also be appropriate to designate the Contractor's responsibilities, including any limitations thereon, for accommodating the Work to subsequent changes in the Project Schedule.

d. Provision for liquidated damages: If liquidated damages are to be assessed because delayed construction will result in the Owner actually suffering loss, the entire provision for liquidated damages should be included in the Instructions to Bidders, as well as in the Agreement. This provision should be drafted by the Owner's attorney. Liquidated damages are not a penalty to be inflicted on the Contractor, but must bear an actual and reasonably estimated relationship to the loss to the Owner if the building is not completed on time; for example, the cost per day of renting space to house students if a dormitory cannot be occupied when needed, additional financing costs, loss of profits, etc.

4. Article 4 — Contract Sum

The Contract Sum should be stated in words and in numerals. The basis on which the Contract Sum has been determined should be fully identified.

If unit prices are included in connection with the Contract Sum, and if not covered elsewhere in the Contract Documents in more detail, the following provision is suggested for Article 4:

"The unit prices listed below shall determine the value of changes in the Work, as applicable. They shall be considered complete including all material and equipment, labor, installation costs, overhead and profit."

Specific allowances for overhead and profit on Change Orders may also be included here if they have been stipulated in the Bidding Documents or agreed to by negotiation between the Owner and the Contractor.

5. Article 5 — Progress Payments

The following items should be included as appropriate: due dates for payments, payment for materials stored off the site, retained percentage and any provision for adjustments thereto, and designation of the rate of interest applying to payments due and unpaid under the Contract Documents.

Note that A201/CM provides that the Contractor must submit Applications for Payment to the Construction Manager at least 15 days before the date of each progress payment. The Construction Manager has 7 days in which to assemble the Application with similar applications from other Contractors on the Project into a combined Project Application for Payment and forward it with recommendations to the Architect. The Architect has 7 days after receipt to issue a Project Certificate for Payment. Consult with the Owner to establish dates that account for these intervals and allow the Owner time to process the Certificate and make payment.

6. Article 7 — Miscellaneous Provisions

The Contract Documents must be enumerated in detail in Paragraph 7.2. If unit prices are incorporated in the Contractor's bid, the bid itself may be incorporated into the Contract in lieu of repeating all of the unit prices in Article 4. Similarly, other Bidding Documents, bonds, etc., may be modified to designate such other items as part of the Contract Documents. For Paragraphs 7.3 and 7.4, attach additional sheets as necessary. Each party should date and initial each sheet.

7. Signatures

AIA Document A201/CM requires execution in not less than quadruplicate. The Agreement should be executed by the parties in their capacities as individuals, partners, officers, etc., as appropriate.

D. REPRODUCTION:

AIA Document A101/CM is a copyrighted document, and may not be reproduced or excerpted from in substantial part without the express written permission of AIA. Purchasers of A101/CM are hereby entitled to reproduce a maximum of ten copies of the completed or executed document for use only in connection with the particular Project. AIA will not permit the reproduction of this Document in blank, or the use of substantial portions of, or language from, this Document, except upon written request and after receipt of written permission from AIA.

AIA DOCUMENT A101/CMa • INSTRUCTION SHEET FOR OWNER-CONTRACTOR AGREEMENT
CONSTRUCTION MANAGEMENT EDITION • 1980 EDITION • AIA® • THE AMERICAN
INSTITUTE OF ARCHITECTS, 1735 NEW YORK AVENUE, N.W., WASHINGTON, D.C. 20006

THE AMERICAN INSTITUTE OF ARCHITECTS

AIA Document A101/CM

CONSTRUCTION MANAGEMENT EDITION

Standard Form of Agreement Between Owner and Contractor

where the basis of payment is a

STIPULATED SUM

1980 EDITION

THIS DOCUMENT HAS IMPORTANT LEGAL CONSEQUENCES; CONSULTATION WITH AN ATTORNEY IS ENCOURAGED.

This document is intended to be used in conjunction with AIA Documents A201/CM, 1980; B141/CM, 1980; and B801, 1980.

AGREEMENT

made as of the   day of   in the year of Nineteen
Hundred and

BETWEEN the Owner:

and the Contractor:

the Project:

the Construction Manager:

the Architect:

The Owner and the Contractor agree as set forth below.

AIA DOCUMENT A101/CM • OWNER-CONTRACTOR AGREEMENT • CONSTRUCTION
MANAGEMENT EDITION • JUNE 1980 EDITION • AIA® • ©1980 • THE AMERICAN
INSTITUTE OF ARCHITECTS, 1735 NEW YORK AVE., N.W., WASHINGTON, D.C. 20006   **A101/CM — 1980   1**

ARTICLE 1
THE CONTRACT DOCUMENTS

The Contract Documents consist of this Agreement, the Conditions of the Contract (General, Supplementary and other Conditions), the Drawings, the Specifications, all Addenda issued prior to and all Modifications issued after execution of this Agreement. These form the Contract, and all are as fully a part of the Contract as if attached to this Agreement or repeated herein. An enumeration of the Contract Documents appears in Article 7.

ARTICLE 2
THE WORK

The Contractor shall perform all the Work required by the Contract Documents for
(Here insert the caption descriptive of the Work as used on other Contract Documents.)

ARTICLE 3
TIME OF COMMENCEMENT AND SUBSTANTIAL COMPLETION

The Work to be performed under this Contract shall be commenced

and, subject to authorized adjustments, Substantial Completion of the Work shall be achieved not later than

(Here insert any special provisions for liquidated damages relating to failure to complete on time.)

ARTICLE 4
CONTRACT SUM

The Owner shall pay the Contractor in current funds for the performance of the Work, subject to additions and deductions by Change Order as provided in the Contract Documents, the Contract Sum of

The Contract Sum is determined as follows:
(State here the base bid or other lump sum amount, accepted alternates and unit prices, as applicable.)

ARTICLE 5
PROGRESS PAYMENTS

Based upon Applications for Payment submitted to the Construction Manager by the Contractor and Project Certificates for Payment issued by the Architect, the Owner shall make progress payments on account of the Contract Sum to the Contractor as provided in the Contract Documents for the period ending the   day of each month as follows:

Not later than   days following the end of the period covered by the Application for Payment,   percent (%) of the portion of the Contract Sum properly allocable to labor, materials and equipment incorporated in the Work and   percent (%) of the portion of the Contract Sum properly allocable to materials and equipment suitably stored at the site or at some other location agreed upon in writing, for the period covered by the Application for Payment, less the aggregate of previous payments made by the Owner; and upon Substantial Completion of the Work, a sum sufficient to increase the total payments to   percent (%) of the Contract Sum, less such amounts as the Architect shall determine for all incomplete Work and unsettled claims as provided in the Contract Documents.

(If not covered elsewhere in the Contract Documents, here insert any provision for limiting or reducing the amount retained after the Work reaches a certain stage of completion.)

Payments due and unpaid under the Contract Documents shall bear interest from the date payment is due at the rate entered below, or in the absence thereof, at the legal rate prevailing at the place of the Project.
(Here insert any rate of interest agreed upon.)

(Usury laws and requirements under the Federal Truth in Lending Act, similar state and local consumer credit laws and other regulations at the Owner's and Contractor's principal places of business, the location of the Project and elsewhere may affect the validity of this provision. Specific legal advice should be obtained with respect to deletion, modification or other requirements such as written disclosures or waivers.)

AIA DOCUMENT A101/CM • OWNER-CONTRACTOR AGREEMENT • CONSTRUCTION MANAGEMENT EDITION • JUNE 1980 EDITION • AIA® • ©1980 • THE AMERICAN INSTITUTE OF ARCHITECTS, 1735 NEW YORK AVE., N.W., WASHINGTON, D.C. 20006

A101/CM — 1980   3

ARTICLE 6
FINAL PAYMENT

Final payment, constituting the entire unpaid balance of the Contract Sum, shall be paid by the Owner to the Contractor when the Work has been completed, the Contract fully performed, and the Architect has issued a Project Certificate for Payment which approves the final payment due the Contractor.

ARTICLE 7
MISCELLANEOUS PROVISIONS

7.1 Terms used in this Agreement which are defined in the Conditions of the Contract shall have the meanings designated in those Conditions.

7.2 The Contract Documents, which constitute the entire agreement between the Owner and the Contractor, are listed in Article 1 and, except for Modifications issued after execution of this Agreement, are enumerated as follows:

(List below the Agreement, the Conditions of the Contract [General, Supplementary and other Conditions], the Drawings, the Specifications, and any Addenda and accepted alternates, showing page or sheet numbers in all cases and dates where applicable.)

7.3   Temporary facilities and services:
(Here insert temporary facilities and services which are different from or in addition to those included elsewhere in the Contract Documents.)

7.4   Working Conditions:
(Here list any special conditions affecting the Contract.)

This Agreement entered into as of the day and year first written above.

OWNER   CONTRACTOR

_____   _____

_____   _____

_____   _____

AIA **DOCUMENT A101/CM** • OWNER-CONTRACTOR AGREEMENT • CONSTRUCTION
MANAGEMENT EDITION • JUNE 1980 EDITION • AIA® • ©1980 • THE AMERICAN
INSTITUTE OF ARCHITECTS, 1735 NEW YORK AVE., N.W., WASHINGTON, D.C. 20006

A101/CM — 1980   **5**

AIA Document A201/CM, *General Conditions of the Contract for Construction* (CM edition). A201/CM is based on AIA Document A201, the form of general conditions used with traditional fixed-price or cost-plus construction contracts. In essence, the form incorporates the concept of the independent construction manager and defines the working relationships between the contractors, the construction manager, and the architect during construction. Individual contractors are not relieved of much, if any, responsibility by the existence of the construction manager. The A201/CM form has a bias in favor of a strong role for the architect during construction, despite the existence of the construction manager: the architect continues to have the final approval of payments to the contractors, the authority to initially decide disputes, the authority to reject defective or nonconforming work (the construction manager may also do so, but is subject to being overruled by the architect), the approval of shop drawings and other submittals, the approval of change orders and the authority to order minor changes in the work, and the final determination of substantial completion and final completion. It is fair to say that in the AIA general conditions, the architect has lost none of its authority to the construction manager and that the construction manager, in general, is playing a supporting role for the architect. The construction manager's authority is never more than concurrent with the architect's and it is often subordinate to that of the architect. AIA Document 201/CM poses the risk that effective control over the project may be lost unless the architect and construction manager can work closely together in a mutually supportive relationship. Where the architect and construction manager disagree, the owner will have to be the final arbiter—a role it may not wish or be prepared to fulfill.

Owners who wish to use the architect essentially as a design service (with some consultation on the design effects of any changes during construction and the interpretation of the drawings) and to use the construction manager as their primary agent and representative during construction, will be required to make extensive amendments to the owner-architect agreement (B141/CM), the owner–construction manager agreement (B801) (which incorporates A201/VM by reference), and the general conditions (A201/CM), as well as minor adjustments to the owner-contractor agreement (A101/CM). Such amendments would be designed to limit the role of the architect during construction and to expand that of the construction manager. Comparison of AIA Document A201/CM with AGC Document 8b, which provides for a strong role on the part of the construction manager and a minimal role for the architect, will aid the owner in defining the role of the construction manager.

THE AMERICAN INSTITUTE OF ARCHITECTS

AIA Document A201/CM

CONSTRUCTION MANAGEMENT EDITION

General Conditions of the Contract for Construction

THIS DOCUMENT HAS IMPORTANT LEGAL CONSEQUENCES; CONSULTATION WITH AN ATTORNEY IS ENCOURAGED.

1980 EDITION

SAMPLE

TABLE OF ARTICLES

AIA DOCUMENT A201/CM • GENERAL CONDITIONS OF THE CONTRACT FOR CONSTRUCTION
CONSTRUCTION MANAGEMENT EDITION • JUNE 1980 EDITION • AIA® • © 1980 • THE
AMERICAN INSTITUTE OF ARCHITECTS, 1735 NEW YORK AVE., N.W., WASHINGTON, D.C. 20006

A201/CM — 1980 **1**

INDEX

AIA DOCUMENT A201/CM • GENERAL CONDITIONS OF THE CONTRACT FOR CONSTRUCTION
CONSTRUCTION MANAGEMENT EDITION • JUNE 1980 EDITION • AIA® • © 1980 • THE
AMERICAN INSTITUTE OF ARCHITECTS, 1735 NEW YORK AVE., N.W., WASHINGTON, D.C. 20006

AIA DOCUMENT A201/CM • GENERAL CONDITIONS OF THE CONTRACT FOR CONSTRUCTION
CONSTRUCTION MANAGEMENT EDITION • JUNE 1980 EDITION • AIA® • © 1980 • THE
AMERICAN INSTITUTE OF ARCHITECTS, 1735 NEW YORK AVE., N.W., WASHINGTON, D.C. 20006

A201/CM—1980   3

AIA DOCUMENT A201/CM • GENERAL CONDITIONS OF THE CONTRACT FOR CONSTRUCTION
CONSTRUCTION MANAGEMENT EDITION • JUNE 1980 EDITION • AIA® • © 1980 • THE
AMERICAN INSTITUTE OF ARCHITECTS, 1735 NEW YORK AVE., N.W., WASHINGTON, D.C. 20006

AIA DOCUMENT A201/CM • GENERAL CONDITIONS OF THE CONTRACT FOR CONSTRUCTION
CONSTRUCTION MANAGEMENT EDITION • JUNE 1980 EDITION • AIA® • © 1980 • THE
AMERICAN INSTITUTE OF ARCHITECTS, 1735 NEW YORK AVE., N.W., WASHINGTON, D.C. 20006

A201/CM—1980   5

GENERAL CONDITIONS OF THE CONTRACT FOR CONSTRUCTION

ARTICLE 1
CONTRACT DOCUMENTS

1.1   DEFINITIONS

1.1.1   THE CONTRACT DOCUMENTS

The Contract Documents consist of the Owner-Contractor Agreement, the Conditions of the Contract (General, Supplementary and other Conditions), the Drawings, the Specifications, and all Addenda issued prior to and all Modifications issued after execution of the Contract. A Modification is (1) a written amendment to the Contract signed by both parties, (2) a Change Order, (3) a written interpretation issued by the Architect pursuant to Subparagraph 2.3.11, or (4) a written order for a minor change in the Work issued by the Architect pursuant to Paragraph 12.4. The Contract Documents do not include Bidding Documents such as the Advertisement or Invitation to Bid, the Instructions to Bidders, sample forms, the Contractor's Bid or portions of Addenda relating to any of these, or any other documents unless specifically enumerated in the Owner-Contractor Agreement.

1.1.2   THE CONTRACT

The Contract Documents form the Contract for Construction. This Contract represents the entire and integrated agreement between the parties hereto and supersedes all prior negotiations, representations or agreements, either written or oral. The Contract may be amended or modified only by a Modification as defined in Subparagraph 1.1.1. The Contract Documents shall not be construed to create any contractual relationship of any kind between the Architect and the Contractor, between the Construction Manager and the Contractor or between the Architect and the Construction Manager, but the Architect and the Construction Manager shall be entitled to performance of the obligations of the Contractor intended for their benefit and to enforcement thereof. Nothing contained in the Contract Documents shall create any contractual relationship between the Owner, the Construction Manager or the Architect and any Subcontractor or Sub-subcontractor.

1.1.3   THE WORK

The Work comprises the completed construction required of the Contractor by the Contract Documents, and includes all labor necessary to produce such construction, and all materials and equipment incorporated or to be incorporated in such construction.

1.1.4   THE PROJECT

The Project, as defined in the Owner-Contractor Agreement, is the total construction of which the Work performed under the Contract Documents is a part.

1.2   EXECUTION, CORRELATION AND INTENT

1.2.1   The Contract Documents shall be signed in not less than quadruplicate by the Owner and the Contractor. If either the Owner or the Contractor or both do not sign the Conditions of the Contract, Drawings, Specifications

or any of the other Contract Documents, the Architect shall identify such Documents.

1.2.2   Execution of the Contract by the Contractor is a representation that the Contractor has visited the site, become familiar with the local conditions under which the Work is to be performed, and has correlated personal observations with the requirements of the Contract Documents.

1.2.3   The intent of the Contract Documents is to include all items necessary for the proper execution and completion of the Work. The Contract Documents are complementary, and what is required by any one shall be as binding as if required by all. Work not covered in the Contract Documents will not be required unless it is consistent therewith and is reasonably inferable therefrom as being necessary to produce the intended results. Words and abbreviations which have well-known technical or trade meanings are used in the Contract Documents in accordance with such recognized meanings.

1.2.4   The organization of the Specifications into divisions, sections and articles, and the arrangement of Drawings shall not control the Contractor in dividing the Work among Subcontractors or in establishing the extent of Work to be performed by any trade.

1.3   OWNERSHIP AND USE OF DOCUMENTS

1.3.1   All Drawings, Specifications and copies thereof furnished by the Architect are and shall remain the property of the Architect. They are to be used only with respect to this Project and are not to be used on any other project. With the exception of one contract set for each party to the Contract, such documents are to be returned or suitably accounted for to the Architect on request at the completion of the Work. Submission or distribution to meet official regulatory requirements or for other purposes in connection with the Project is not to be construed as publication in derogation of the Architect's common law copyright or other reserved rights.

ARTICLE 2
ADMINISTRATION OF THE CONTRACT

2.1   THE ARCHITECT

2.1.1   The Architect is the person lawfully licensed to practice architecture, or an entity lawfully practicing architecture, identified as such in the Owner-Contractor Agreement. The term Architect means the Architect or the Architect's authorized representative.

2.2   THE CONSTRUCTION MANAGER

2.2.1   The Construction Manager is the person or entity identified as such in the Owner-Contractor Agreement. The term Construction Manager means the Construction Manager or the Construction Manager's authorized representative.

2.3   ADMINISTRATION OF THE CONTRACT

2.3.1   The Architect and the Construction Manager will provide administration of the Contract as hereinafter described.

2.3.2 The Architect and the Construction Manager will be the Owner's representatives during construction and until final payment to all contractors is due. The Architect and the Construction Manager will advise and consult with the Owner. All instructions to the Contractor shall be forwarded through the Construction Manager. The Architect and the Construction Manager will have authority to act on behalf of the Owner only to the extent provided in the Contract Documents, unless otherwise modified by written instrument in accordance with Subparagraph 2.3.22.

2.3.3 The Construction Manager will determine in general that the Work of the Contractor is being performed in accordance with the Contract Documents, and will endeavor to guard the Owner against defects and deficiencies in the Work of the Contractor.

2.3.4 The Architect will visit the site at intervals appropriate to the stage of construction to become generally familiar with the progress and quality of the Work and to determine in general if the Work is proceeding in accordance with the Contract Documents. However, the Architect will not be required to make exhaustive or continuous on-site inspections to check the quality or quantity of the Work. On the basis of on-site observations as an architect, the Architect will keep the Owner informed of the progress of the Work, and will endeavor to guard the Owner against defects and deficiencies in the Work of the Contractor.

2.3.5 Neither the Architect nor the Construction Manager will be responsible for or have control or charge of construction means, methods, techniques, sequences or procedures, or for safety precautions and programs in connection with the Work, and neither will be responsible for the Contractor's failure to carry out the Work in accordance with the Contract Documents. Neither the Architect nor the Construction Manager will be responsible for or have control or charge over the acts or omissions of the Contractor, Subcontractors, or any of their agents or employees, or any other persons performing any of the Work.

2.3.6 The Architect and the Construction Manager shall at all times have access to the Work wherever it is in preparation and progress. The Contractor shall provide facilities for such access so that the Architect and the Construction Manager may perform their functions under the Contract Documents.

2.3.7 The Construction Manager will schedule and coordinate the Work of all contractors on the Project including their use of the site. The Construction Manager will keep the Contractor informed of the Project Construction Schedule to enable the Contractor to plan and perform the Work properly.

2.3.8 The Construction Manager will review all Applications for Payment by the Contractor, including final payment, and will assemble them with similar applications from other contractors on the Project into a combined Project Application for Payment. The Construction Manager will then make recommendations to the Architect for certification for payment.

2.3.9 Based on the Architect's observations, the recom-

mendations of the Construction Manager and an evaluation of the Project Application for Payment, the Architect will determine the amount owing to the Contractor and will issue a Project Certificate for Payment incorporating such amount, as provided in Paragraph 9.4.

2.3.10 The Architect will be the interpreter of the requirements of the Contract Documents and the judge of the performance thereunder by both the Owner and the Contractor.

2.3.11 The Architect will render interpretations necessary for the proper execution or progress of the Work, with reasonable promptness and in accordance with agreed upon time limits. Either party to the Contract may make written request to the Architect for such interpretations.

2.3.12 Claims, disputes and other matters in question between the Contractor and the Owner relating to the execution or progress of the Work or the interpretation of the Contract Documents shall be referred initially to the Architect for decision. After consultation with the Construction Manager, the Architect will render a decision in writing within a reasonable time.

2.3.13 All interpretations and decisions of the Architect shall be consistent with the intent of and reasonably inferable from the Contract Documents and will be in writing or in graphic form. In this capacity as interpreter and judge the Architect will endeavor to secure faithful performance by both the Owner and the Contractor, will not show partiality to either, and will not be liable for the result of any interpretation or decision rendered in good faith in such capacity.

2.3.14 The Architect's decisions in matters relating to artistic effect will be final if consistent with the intent of the Contract Documents.

2.3.15 Any claim, dispute or other matter in question between the Contractor and the Owner referred to the Architect through the Construction Manager, except those relating to artistic effect as provided in Subparagraph 2.3.14 and those which have been waived by the making or acceptance of final payment as provided in Subparagraphs 9.9.4 through 9.9.6, inclusive, shall be subject to arbitration upon the written demand of either party. However, no demand for arbitration of any such claim, dispute or other matter may be made until the earlier of (1) the date on which the Architect has rendered a written decision, or (2) the tenth day after the parties have presented their evidence to the Architect or have been given a reasonable opportunity to do so, if the Architect has not rendered a written decision by that date. When such a written decision of the Architect states (1) that the decision is final but subject to appeal, and (2) that any demand for arbitration of a claim, dispute or other matter covered by such decision must be made within thirty days after the date on which the party making the demand receives the written decision, failure to demand arbitration within said thirty day period will result in the Architect's decision becoming final and binding upon the Owner and the Contractor. If the Architect renders a decision after arbitration proceedings have been initiated, such decision may be entered as evidence but will not supersede any arbitration proceedings unless the decision is acceptable to all parties concerned.

AIA DOCUMENT A201/CM • GENERAL CONDITIONS OF THE CONTRACT FOR CONSTRUCTION CONSTRUCTION MANAGEMENT EDITION • JUNE 1980 EDITION • AIA® • © 1980 • THE AMERICAN INSTITUTE OF ARCHITECTS, 1735 NEW YORK AVE., N.W., WASHINGTON, D.C. 20006

7   A201/CM — 1980

2.3.16 The Architect will have authority to reject Work which does not conform to the Contract Documents, and to require special inspection or testing, but will take such action only after consultation with the Construction Manager. Subject to review by the Architect, the Construction Manager will have the authority to reject Work which does not conform to the Contract Documents. Whenever, in the Construction Manager's opinion, it is considered necessary or advisable for the implementation of the intent of the Contract Documents, the Construction Manager will have authority to require special inspection or testing of the Work in accordance with Subparagraph 7.7.2 whether or not such Work be then fabricated, installed or completed. The foregoing authority of the Construction Manager will be subject to the provisions of Subparagraphs 2.3.10 through 2.3.16, inclusive, with respect to interpretations and decisions of the Architect. However, neither the Architect's nor the Construction Manager's authority to act under this Subparagraph 2.3.16, nor any decision made by them in good faith either to exercise or not to exercise such authority shall give rise to any duty or responsibility of the Architect or the Construction Manager to the Contractor, any Subcontractor, any of their agents or employees, or any other person performing any of the Work.

2.3.17 The Construction Manager will receive from the Contractor and review all Shop Drawings, Product Data and Samples, coordinate them with information contained in related documents, and transmit to the Architect those recommended for approval.

2.3.18 The Architect will review and approve or take other appropriate action upon the Contractor's submittals such as Shop Drawings, Product Data and Samples, but only for conformance with the design concept of the Work and the information given in the Contract Documents. Such action shall be taken with reasonable promptness so as to cause no delay. The Architect's approval of a specific item shall not indicate approval of an assembly of which the item is a component.

2.3.19 Following consultation with the Construction Manager, the Architect will take appropriate action on Change Orders in accordance with Article 12, and will have authority to order minor changes in the Work as provided in Subparagraph 12.4.1.

2.3.20 The Construction Manager will maintain at the Project site one record copy of all Contracts, Drawings, Specifications, Addenda, Change Orders and other Modifications pertaining to the Project, in good order and marked currently to record all changes made during construction, and approved Shop Drawings, Product Data and Samples. These shall be available to the Architect and the Contractor, and shall be delivered to the Architect for the Owner upon completion of the Project.

2.3.21 The Construction Manager will assist the Architect in conducting inspections to determine the dates of Substantial and final completion, and will receive and forward to the Owner for the Owner's review written warranties and related documents required by the Contract and assembled by the Contractor. The Architect will issue a final Project Certificate for Payment upon compliance with the requirements of Paragraph 9.9.

2.3.22 The duties, responsibilities and limitations of authority of the Architect and the Construction Manager as the Owner's representatives during construction as set forth in the Contract Documents, will not be modified or extended without written consent of the Owner, the Contractor, the Architect and the Construction Manager, which consent shall not be unreasonably withheld. Failure of the Contractor to respond within ten days to a written request shall constitute consent by the Contractor.

2.3.23 In case of the termination of the employment of the Architect or the Construction Manager, the Owner shall appoint an architect or a construction manager against whom the Contractor makes no reasonable objection and whose status under the Contract Documents shall be that of the former architect or construction manager, respectively. Any dispute in connection with such appointments shall be subject to arbitration.

ARTICLE 3
OWNER

3.1   DEFINITION

3.1.1 The Owner is the person or entity identified as such in the Owner-Contractor Agreement. The term Owner means the Owner or the Owner's authorized representative.

3.2   INFORMATION AND SERVICES REQUIRED OF THE OWNER

3.2.1 The Owner shall, at the request of the Contractor, at the time of execution of the Owner-Contractor Agreement furnish to the Contractor reasonable evidence that the Owner has made financial arrangements to fulfill the Owner's obligations under the Contract. Unless such reasonable evidence is furnished, the Contractor is not required to execute the Owner-Contractor Agreement or to commence the Work.

3.2.2 The Owner shall furnish all surveys describing the physical characteristics, legal limitations and utility locations for the site of the Project, and a legal description of the site.

3.2.3 Except as provided in Subparagraph 4.7.1, the Owner shall secure and pay for necessary approvals, easements, assessments and charges required for the construction, use or occupancy of permanent structures or for permanent changes in existing facilities.

3.2.4 Information or services under the Owner's control shall be furnished by the Owner with reasonable promptness to avoid delay in the orderly progress of the Work.

3.2.5 Unless otherwise provided in the Contract Documents, the Contractor will be furnished, free of charge, all copies of Drawings and Specifications reasonably necessary for the execution of the Work.

3.2.6 The Owner shall forward all instructions to the Contractor through the Construction Manager, with simultaneous notification to the Architect.

3.2.7 The foregoing are in addition to other duties and responsibilities of the Owner enumerated herein and especially those in respect to Work By Owner or By Separate Contractors, Payments and Completion, and Insurance in Articles 6, 9 and 11, respectively.

AIA DOCUMENT A201/CM • GENERAL CONDITIONS OF THE CONTRACT FOR CONSTRUCTION CONSTRUCTION MANAGEMENT EDITION • JUNE 1980 EDITION • AIA® • © 1980 • THE AMERICAN INSTITUTE OF ARCHITECTS, 1735 NEW YORK AVE., N.W., WASHINGTON, D.C. 20006

A201/CM — 1980   8

3.3   OWNER'S RIGHT TO STOP THE WORK

3.3.1   If the Contractor fails to correct defective Work as required by Paragraph 13.2, or persistently fails to carry out the Work in accordance with the Contract Documents, the Owner, by a written order signed personally or by an agent specifically so empowered by the Owner in writing, may order the Contractor to stop the Work, or any portion thereof, until the cause for such order has been eliminated; however, this right of the Owner to stop the Work shall not give rise to any duty on the part of the Owner to exercise this right for the benefit of the Contractor or any other person or entity, except to the extent required by Subparagraph 6.1.3.

3.4   OWNER'S RIGHT TO CARRY OUT THE WORK

3.4.1.   If the Contractor defaults or neglects to carry out the Work in accordance with the Contract Documents, and fails within seven days after receipt of written notice from the Owner to commence and continue correction of such default or neglect with diligence and promptness, the Owner may, after seven days following receipt by the Contractor of an additional written notice and without prejudice to any other remedy the Owner may have, make good such deficiencies. In such case an appropriate Change Order shall be issued deducting from the payments then or thereafter due the Contractor the cost of correcting such deficiencies, including compensation for the Architect's and the Construction Manager's additional services made necessary by such default, neglect or failure. Such action by the Owner and the amount charged to the Contractor are both subject to the prior approval of the Architect, after consultation with the Construction Manager. If the payments then or thereafter due the Contractor are not sufficient to cover such amount, the Contractor shall pay the difference to the Owner.

ARTICLE 4
CONTRACTOR

4.1   DEFINITION

4.1.1   The Contractor is the person or entity identified as such in the Owner-Contractor Agreement. The term Contractor means the Contractor or the Contractor's authorized representative.

4.2   REVIEW OF CONTRACT DOCUMENTS

4.2.1   The Contractor shall carefully study and compare the Contract Documents and shall at once report to the Architect and the Construction Manager any error, inconsistency or omission that may be discovered. The Contractor shall not be liable to the Owner, the Architect or the Construction Manager for any damage resulting from any such errors, inconsistencies or omissions in the Contract Documents. The Contractor shall perform no portion of the Work at any time without Contract Documents or, where required, approved Shop Drawings, Product Data or Samples for such portion of the Work.

4.3   SUPERVISION AND CONSTRUCTION PROCEDURES

4.3.1   The Contractor shall supervise and direct the Work, using the Contractor's best skill and attention. The Contractor shall be solely responsible for all construction means, methods, techniques, sequences and procedures;

and shall coordinate all portions of the Work under the Contract, subject to the overall coordination of the Construction Manager.

4.3.2   The Contractor shall be responsible to the Owner for the acts and omissions of the Contractor's employees, Subcontractors and their agents and employees, and any other persons performing any of the Work under a contract with the Contractor.

4.3.3   The Contractor shall not be relieved from the Contractor's obligations to perform the Work in accordance with the Contract Documents either by the activities or duties of the Construction Manager or the Architect in their administration of the Contract, or by inspections, tests or approvals required or performed under Paragraph 7.7 by persons other than the Contractor.

4.4   LABOR AND MATERIALS

4.4.1   Unless otherwise provided in the Contract Documents, the Contractor shall provide and pay for all labor, materials, equipment, tools, construction equipment and machinery, water, heat, utilities, transportation, and other facilities and services necessary for the proper execution and completion of the Work, whether temporary or permanent and whether or not incorporated or to be incorporated in the Work.

4.4.2   The Contractor shall at all times enforce strict discipline and good order among the Contractor's employees and shall not employ on the Work any unfit person or anyone not skilled in the task assigned them.

4.5   WARRANTY

4.5.1   The Contractor warrants to the Owner, the Architect and the Construction Manager that all materials and equipment furnished under this Contract will be new unless otherwise specified, and that all Work will be of good quality, free from faults and defects and in conformance with the Contract Documents. All Work not conforming to these requirements, including substitutions not properly approved and authorized, may be considered defective. If required by the Architect or the Construction Manager, the Contractor shall furnish satisfactory evidence as to the kind and quality of materials and equipment. This warranty is not limited by the provisions of Paragraph 13.2.

4.6   TAXES

4.6.1   The Contractor shall pay all sales, consumer, use and other similar taxes for the Work or portions thereof provided by the Contractor which are legally enacted at the time bids are received, whether or not yet effective.

4.7   PERMITS, FEES AND NOTICES

4.7.1   Unless otherwise provided in the Contract Documents, the Owner shall secure and pay for the building permit and the Contractor shall secure and pay for all other permits and governmental fees, licenses and inspections necessary for the proper execution and completion of the Work which are customarily secured after execution of the Contract and which are legally required at the time bids are received.

4.7.2   The Contractor shall give all notices and comply with all laws, ordinances, rules, regulations and lawful orders of any public authority bearing on the performance of the Work.

AIA DOCUMENT A201/CM • GENERAL CONDITIONS OF THE CONTRACT FOR CONSTRUCTION CONSTRUCTION MANAGEMENT EDITION • JUNE 1980 EDITION • AIA® • © 1980 • THE AMERICAN INSTITUTE OF ARCHITECTS, 1735 NEW YORK AVE., N.W., WASHINGTON, D.C. 20006

4.7.3 It is not the responsibility of the Contractor to make certain that the Contract Documents are in accordance with applicable laws, statutes, building codes and regulations. If the Contractor observes that any of the Contract Documents are at variance therewith in any respect, the Contractor shall promptly notify the Architect and the Construction Manager in writing, and any necessary changes shall be accomplished by appropriate Modification.

4.7.4 If the Contractor performs any Work knowing it to be contrary to such laws, ordinances, rules and regulations, and without such notice to the Architect and the Construction Manager, the Contractor shall assume full responsibility therefor and shall bear all costs attributable thereto.

4.8   ALLOWANCES

4.8.1 The Contractor shall include in the Contract Sum all allowances stated in the Contract Documents. Items covered by these allowances shall be supplied for such amounts and by such persons as the Construction Manager may direct, but the Contractor will not be required to employ persons against whom the Contractor makes a reasonable objection.

4.8.2 Unless otherwise provided in the Contract Documents:

  .1 these allowances shall cover the cost to the Contractor, less any applicable trade discount, of the materials and equipment required by the allowance, delivered at the site, and all applicable taxes;

  .2 the Contractor's cost for unloading and handling on the site, labor, installation costs, overhead, profit and other expenses contemplated for the original allowance shall be included in the Contract Sum and not in the allowance;

  .3 whenever the cost is more or less than the allowance, the Contract Sum shall be adjusted accordingly by Change Order, the amount of which will recognize changes, if any, in handling costs on the site, labor, installation costs, overhead, profit and other expenses.

4.9   SUPERINTENDENT

4.9.1 The Contractor shall employ a competent superintendent and necessary assistants who shall be in attendance at the Project site during the progress of the Work. The superintendent shall represent the Contractor and all communications given to the superintendent shall be as binding as if given to the Contractor. Important communications shall be confirmed in writing. Other communications shall be so confirmed on written request in each case.

4.10   CONTRACTOR'S CONSTRUCTION SCHEDULE

4.10.1 The Contractor, immediately after being awarded the Contract, shall prepare and submit for the Construction Manager's approval a Contractor's Construction Schedule for the Work which shall provide for expeditious and practicable execution of the Work. This schedule shall be coordinated by the Construction Manager with the Project Construction Schedule. The Contractor's Construction Schedule shall be revised as required by the

conditions of the Work and the Project, subject to the Construction Manager's approval.

4.11   DOCUMENTS AND SAMPLES AT THE SITE

4.11.1 The Contractor shall maintain at the Project site, on a current basis, one record copy of all Drawings, Specifications, Addenda, Change Orders and other Modifications, in good order and marked currently to record all changes made during construction, and approved Shop Drawings, Product Data and Samples. These shall be available to the Architect and the Construction Manager. The Contractor shall advise the Construction Manager on a current basis of all changes in the Work made during construction.

4.12   SHOP DRAWINGS, PRODUCT DATA AND SAMPLES

4.12.1 Shop Drawings are drawings, diagrams, schedules and other data specially prepared for the. Work by the Contractor or any Subcontractor, manufacturer, supplier or distributor to illustrate some portion of the Work.

4.12.2 Product Data are illustrations, standard schedules, performance charts, instructions, brochures, diagrams and other information furnished by the Contractor to illustrate a material, product or system for some portion of the Work.

4.12.3 Samples are physical examples which illustrate materials, equipment or workmanship, and establish standards by which the Work will be judged.

4.12.4 The Contractor shall prepare, review, approve and submit through the Construction Manager, with reasonable promptness and in such sequence as to cause no delay in the Work or in the work of the Owner or any separate contractor, all Shop Drawings, Product Data and Samples required by the Contract Documents. The Contractor shall cooperate with the Construction Manager in the Construction Manager's coordination of the Contractor's Shop Drawings, Product Data and Samples with those of other separate contractors.

4.12.5 By preparing, approving and submitting Shop Drawings, Product Data and Samples, the Contractor represents that the Contractor has determined and verified all materials, field measurements and field construction criteria related thereto, or will do so with reasonable promptness, and has checked and coordinated the information contained within such submittals with the requirements of the Work, the Project and the Contract Documents.

4.12.6 The Contractor shall not be relieved of responsibility for any deviation from the requirements of the Contract Documents by the Architect's approval of Shop Drawings, Product Data or Samples under Subparagraph 2.3.18, unless the Contractor has specifically informed the Architect and the Construction Manager in writing of such deviation at the time of submission and the Architect has given written approval to the specific deviation. The Contractor shall not be relieved from responsibility for errors or omissions in the Shop Drawings, Product Data or Samples by the Architect's approval of them.

4.12.7 The Contractor shall direct specific attention, in writing or on resubmitted Shop Drawings, Product Data or Samples, to revisions other than those requested by the Architect on previous submittals.

4.12.8 No portion of the Work requiring submission of a Shop Drawing, Product Data or Sample shall be commenced until the submittal has been approved by the Architect as provided in Subparagraph 2.3.18. All such portions of the Work shall be in accordance with approved submittals.

4.13 USE OF SITE

4.13.1 The Contractor shall confine operations at the site to areas permitted by law, ordinances, permits and the Contract Documents, and shall not unreasonably encumber the site with any materials or equipment.

4.13.2 The Contractor shall coordinate all of the Contractor's operations with, and secure approval from, the Construction Manager before using any portion of the site.

4.14 CUTTING AND PATCHING OF WORK

4.14.1 The Contractor shall be responsible for all cutting, fitting or patching that may be required to complete the Work or to make its several parts fit together properly.

4.14.2 The Contractor shall not damage or endanger any portion of the Work or the work of the Owner or any separate contractors by cutting, patching or otherwise altering any work, or by excavation. The Contractor shall not cut or otherwise alter the work of the Owner or any separate contractor except with the written consent of the Owner and of such separate contractor. The Contractor shall not unreasonably withhold from the Owner or any separate contractor consent to cutting or otherwise altering the Work.

4.15 CLEANING UP

4.15.1 The Contractor shall at all times keep the premises free from accumulation of waste materials or rubbish caused by the Contractor's operations. At the completion of the Work, the Contractor shall remove all the Contractor's waste materials and rubbish from and about the Project as well as all the Contractor's tools, construction equipment, machinery and surplus materials.

4.15.2 If the Contractor fails to clean up at the completion of the Work, the Owner may do so as provided in Paragraph 3.4 and the cost thereof shall be charged to the Contractor.

4.16 COMMUNICATIONS

4.16.1 The Contractor shall forward all communications to the Owner and the Architect through the Construction Manager.

4.17 ROYALTIES AND PATENTS

4.17.1 The Contractor shall pay all royalties and license fees, shall defend all suits or claims for infringement of any patent rights and shall save the Owner and the Construction Manager harmless from loss on account thereof, except that the Owner, or the Construction Manager as the case may be, shall be responsible for all such loss when a particular design, process or the product of a particular manufacturer or manufacturers is selected by such person or such person's agent. If the Contractor, or the Construction Manager as the case may be, has reason to believe that the design, process or product selected is an infringement of a patent, that party shall be responsible for such loss unless such information is promptly given to the others and also to the Architect.

4.18 INDEMNIFICATION

4.18.1 To the fullest extent permitted by law, the Contractor shall indemnify and hold harmless the Owner, the Architect, the Construction Manager, and their agents and employees from and against all claims, damages, losses and expenses, including, but not limited to, attorneys' fees arising out of or resulting from the performance of the Work, provided that any such claim, damage, loss or expense (1) is attributable to bodily injury, sickness, disease or death, or to injury to or destruction of tangible property (other than the Work itself) including the loss of use resulting therefrom, and (2) is caused in whole or in part by any negligent act or omission of the Contractor, any Subcontractor, anyone directly or indirectly employed by any of them or anyone for whose acts any of them may be liable, regardless of whether or not it is caused in part by a party indemnified hereunder. Such obligation shall not be construed to negate, abridge or otherwise reduce any other right or obligation of indemnity which would otherwise exist as to any party or person described in this Paragraph 4.18.

4.18.2 In any and all claims against the Owner, the Architect, the Construction Manager or any of their agents or employees by any employee of the Contractor, any Subcontractor, anyone directly or indirectly employed by any of them or anyone for whose acts any of them may be liable, the indemnification obligation under this Paragraph 4.18 shall not be limited in any way by any limitation on the amount or type of damages, compensation or benefits payable by or for the Contractor or any Subcontractor under workers' or workmen's compensation acts, disability benefit acts or other employee benefit acts.

4.18.3 The obligations of the Contractor under this Paragraph 4.18 shall not extend to the liability of the Architect or the Construction Manager, their agents or employees, arising out of (1) the preparation or approval of maps, drawings, opinions, reports, surveys, Change Orders, designs or specifications, or (2) the giving of or the failure to give directions or instructions by the Architect or the Construction Manager, their agents or employees, provided such giving or failure to give is the primary cause of the injury or damage.

ARTICLE 5
SUBCONTRACTORS

5.1 DEFINITION

5.1.1 A Subcontractor is a person or entity who has a direct contract with the Contractor to perform any of the Work at the site. The term Subcontractor means a Subcontractor or a Subcontractor's authorized representative. The term Subcontractor does not include any separate contractor or any separate contractor's subcontractors.

5.1.2 A Sub-subcontractor is a person or entity who has a direct or indirect contract with a Subcontractor to perform any of the Work at the site. The term Sub-subcontractor means a Sub-subcontractor or an authorized representative thereof.

5.2 AWARDS OF SUBCONTRACTS AND OTHER CONTRACTS FOR PORTIONS OF THE WORK

5.2.1 Unless otherwise required by the Contract Docu-

AIA DOCUMENT A201/CM • GENERAL CONDITIONS OF THE CONTRACT FOR CONSTRUCTION CONSTRUCTION MANAGEMENT EDITION • JUNE 1980 EDITION • AIA® • © 1980 • THE AMERICAN INSTITUTE OF ARCHITECTS, 1735 NEW YORK AVE., N.W., WASHINGTON, D.C. 20006

ments or the Bidding Documents, the Contractor, as soon as practicable after the award of the Contract, shall furnish to the Construction Manager in writing for review by the Owner, the Architect and the Construction Manager, the names of the persons or entities (including those who are to furnish materials or equipment fabricated to a special design) proposed for each of the principal portions of the Work. The Construction Manager will promptly reply to the Contractor in writing stating whether or not the Owner, the Architect or the Construction Manager, after due investigation, has reasonable objection to any such proposed person or entity. Failure of the Construction Manager to reply promptly shall constitute notice of no reasonable objection.

5.2.2 The Contractor shall not contract with any such proposed person or entity to whom the Owner, the Architect or the Construction Manager has made reasonable objection under the provisions of Subparagraph 5.2.1. The Contractor shall not be required to contract with anyone to whom the Contractor has a reasonable objection.

5.2.3 If the Owner, the Architect or the Construction Manager has reasonable objection to any such proposed person or entity, the Contractor shall submit a substitute to whom the Owner, the Architect and the Construction Manager have no reasonable objection, and the Contract Sum shall be increased or decreased by the difference in cost occasioned by such substitution and an appropriate Change Order shall be issued; however, no increase in the Contract Sum shall be allowed for any such substitution unless the Contractor has acted promptly and responsively in submitting names as required by Subparagraph 5.2.1.

5.2.4 The Contractor shall make no substitution for any Subcontractor, person or entity previously selected if the Owner, the Architect or the Construction Manager makes reasonable objection to such substitution.

5.3 SUBCONTRACTUAL RELATIONS

5.3.1 By an appropriate agreement, written where legally required for validity, the Contractor shall require each Subcontractor, to the extent of the Work to be performed by the Subcontractor, to be bound to the Contractor by the terms of the Contract Documents, and to assume toward the Contractor all the obligations and responsibilities which the Contractor, by these Documents, assumes toward the Owner, the Architect and the Construction Manager. Said agreement shall preserve and protect the rights of the Owner, the Architect and the Construction Manager under the Contract Documents with respect to the Work to be performed by the Subcontractor so that the subcontracting thereof will not prejudice such rights, and shall allow to the Subcontractor, unless specifically provided otherwise in the Contractor-Subcontractor Agreement, the benefit of all rights, remedies and redress against the Contractor that the Contractor, by these Documents, has against the Owner. Where appropriate, the Contractor shall require each Subcontractor to enter into similar agreements with their Sub-subcontractors. The Contractor shall make available to each proposed Subcontractor, prior to the execution of the Subcontract, copies of the Contract Documents to which the Subcontractor will be bound by this Paragraph 5.3, and identify to the Subcontractor any terms and conditions of the pro-

posed Subcontract which may be at variance with the Contract Documents. Each Subcontractor shall similarly make copies of such Documents available to their Sub-subcontractors.

ARTICLE 6
WORK BY OWNER OR BY
SEPARATE CONTRACTORS

6.1 OWNER'S RIGHT TO PERFORM WORK AND TO AWARD SEPARATE CONTRACTS

6.1.1 The Owner reserves the right to perform work related to the Project with the Owner's own forces, and to award separate contracts in connection with other portions of the Project or other work on the site under these or similar Conditions of the Contract. If the Contractor claims that delay, damage or additional cost is involved because of such action by the Owner, the Contractor shall make such claim as provided elsewhere in the Contract Documents.

6.1.2 When separate contracts are awarded for different portions of the Project or other work on the site, the term Contractor in the Contract Documents in each case shall mean the Contractor who executes each separate Owner-Contractor Agreement.

6.1.3 The Owner will provide for the coordination of the work of the Owner's own forces and of each separate contractor with the Work of the Contractor, who shall cooperate therewith as provided in Paragraph 6.2.

6.2 MUTUAL RESPONSIBILITY

6.2.1 The Contractor shall afford the Owner, the Construction Manager and separate contractors reasonable opportunity for the introduction and storage of their materials and equipment and the execution of their work, and shall connect and coordinate the Work with theirs as required by the Contract Documents.

6.2.2 If any part of the Contractor's Work depends for proper execution or results upon the work of the Owner or any separate contractor, the Contractor shall, prior to proceeding with the Work, promptly report to the Construction Manager any apparent discrepancies or defects in such other work that render it unsuitable for such proper execution and results. Failure of the Contractor so to report shall constitute an acceptance of the Owner's or separate contractor's work as fit and proper to receive the Work, except as to defects which may subsequently become apparent in such work by others.

6.2.3 Any costs caused by defective or ill-timed work shall be borne by the party responsible therefor.

6.2.4 Should the Contractor wrongfully cause damage to the work or property of the Owner, or to other work or property on the site, the Contractor shall promptly remedy such damage as provided in Subparagraph 10.2.5.

6.2.5 Should the Contractor wrongfully delay or cause damage to the work or property of any separate contractor, the Contractor shall, upon due notice, promptly attempt to settle with such other contractor by agreement, or otherwise to resolve the dispute. If such separate contractor sues or initiates an arbitration proceeding against the Owner on account of any delay or damage alleged to have been caused by the Contractor, the Owner shall notify the Contractor who shall defend such proceedings at the Owner's expense, and if any judgment or award against the Owner arises therefrom, the Contractor shall pay or satisfy it and shall reimburse the Owner for all attorneys' fees and court or arbitration costs which the Owner has incurred.

6.3   OWNER'S RIGHT TO CLEAN UP

6.3.1   If a dispute arises between the Contractor and separate contractors as to their responsibility for cleaning up as required by Paragraph 4.15, the Owner may clean up and charge the cost thereof to the contractors responsible therefor as the Construction Manager shall determine to be just.

ARTICLE 7
MISCELLANEOUS PROVISIONS

7.1   GOVERNING LAW

7.1.1   The Contract shall be governed by the law of the place where the Project is located.

7.2   SUCCESSORS AND ASSIGNS

7.2.1   The Owner and the Contractor, respectively, bind themselves, their partners, successors, assigns and legal representatives to the other party hereto and to the partners, successors, assigns and legal representatives of such other party with respect to all covenants, agreements and obligations contained in the Contract Documents. Neither party to the Contract shall assign the Contract or sublet it as a whole without the written consent of the other.

7.3   WRITTEN NOTICE

7.3.1   Written notice shall be deemed to have been duly served if delivered in person to the individual or member of the firm or entity or to an officer of the corporation for whom it was intended, or if delivered at or sent by registered or certified mail to the last business address known to the party giving the notice.

7.4   CLAIMS FOR DAMAGES

7.4.1   Should either party to the Contract suffer injury or damage to person or property because of any act or omission of the other party or of any of the other party's employees, agents or others for whose acts such party is legally liable, claim shall be made in writing to such other party within a reasonable time after the first observance of such injury or damage.

**7.5   PERFORMANCE BOND AND LABOR AND
        MATERIAL PAYMENT BOND**

7.5.1   The Owner shall have the right to require the Contractor to furnish bonds covering the faithful performance of the Contract and the payment of all obligations arising thereunder if and as required in the Bidding Documents or the Contract Documents.

7.6   RIGHTS AND REMEDIES

7.6.1   The duties and obligations imposed by the Contract Documents and the rights and remedies available thereunder shall be in addition to, and not a limitation of, any duties, obligations, rights and remedies otherwise imposed or available by law.

7.6.2   No action or failure to act by the Owner, the Architect, the Construction Manager or the Contractor shall constitute a waiver of any right or duty afforded any of them under the Contract, nor shall any such action or failure to act constitute an approval of or acquiescence in any breach thereunder, except as may be specifically agreed in writing.

7.7   TESTS

7.7.1   If the Contract Documents, laws, ordinances, rules, regulations or orders of any public authority having jurisdiction require any portion of the Work to be inspected, tested or approved, the Contractor shall give the Architect and the Construction Manager timely notice of its readiness so the Architect and the Construction Manager may observe such inspection, testing or approval. The Contractor shall bear all costs of such inspections, tests or approvals conducted by public authorities. Unless otherwise provided, the Owner shall bear all costs of other inspections, tests or approvals.

7.7.2   If the Architect or the Construction Manager determines that any Work requires special inspection, testing or approval which Subparagraph 7.7.1 does not include, the Construction Manager will, upon written authorization from the Owner, instruct the Contractor to order such special inspection, testing or approval, and the Contractor shall give notice as provided in Subparagraph 7.7.1. If such special inspection or testing reveals a failure of the Work to comply with the requirements of the Contract Documents, the Contractor shall bear all costs thereof, including compensation for the Architect's and the Construction Manager's additional services made necessary by such failure; otherwise the Owner shall bear such costs, and an appropriate Change Order shall be issued.

7.7.3   Required certificates of inspection, testing or approval shall be secured by the Contractor and the Contractor shall promptly deliver them to the Construction Manager for transmittal to the Architect.

7.7.4   If the Architect or the Construction Manager wishes to observe the inspections, tests or approvals required by the Contract Documents, they will do so promptly and, where practicable, at the source of supply.

7.8   INTEREST

7.8.1   Payments due and unpaid under the Contract Documents shall bear interest from the date payment is due at such rate as the parties may agree upon in writing or, in the absence thereof, at the legal rate prevailing at the place of the Project.

7.9   ARBITRATION

7.9.1   All claims, disputes and other matters in question between the Contractor and the Owner arising out of or relating to the Contract Documents or the breach thereof, except as provided in Subparagraph 2.3.14 with respect to the Architect's decisions on matters relating to artistic effect, and except for claims which have been waived by the making or acceptance of final payment as provided by Subparagraphs 9.9.4 through 9.9.6, inclusive, shall be decided by arbitration in accordance with the Construction Industry Arbitration Rules of the American Arbitration Association then obtaining unless the parties mutually

AIA DOCUMENT A201/CM • GENERAL CONDITIONS OF THE CONTRACT FOR CONSTRUCTION CONSTRUCTION MANAGEMENT EDITION • JUNE 1980 EDITION • AIA® • © 1980 • THE AMERICAN INSTITUTE OF ARCHITECTS, 1735 NEW YORK AVE., N.W., WASHINGTON, D.C. 20006

182   The McGraw-Hill Construction Management Form Book

agree otherwise. No arbitration arising out of or relating to the Contract Documents shall include, by consolidation, joinder or in any other manner, the Architect, the Construction Manager, their employees or consultants except by written consent containing a specific reference to the Owner-Contractor Agreement and signed by the Architect, the Construction Manager, the Owner, the Contractor and any other person sought to be joined. No arbitration shall include by consolidation, joinder or in any other manner, parties other than the Owner, the Contractor and any other persons substantially involved in a common question of fact or law, whose presence is required if complete relief is to be accorded in the arbitration. No person other than the Owner or the Contractor shall be included as an original third party or additional third party to an arbitration whose interest or responsibility is insubstantial. Any consent to arbitration involving an additional person or persons shall not constitute consent to arbitration of any dispute not described therein or with any person not named or described therein. The foregoing agreement to arbitrate and any other agreement to arbitrate with an additional person or persons duly consented to by the parties to the Owner-Contractor Agreement shall be specifically enforceable under the prevailing arbitration law. The award rendered by the arbitrators shall be final, and judgment may be entered upon it in accordance with applicable law in any court having jurisdiction thereof.

7.9.2 Notice of the demand for arbitration shall be filed in writing with the other party to the Owner-Contractor Agreement and with the American Arbitration Association, and a copy shall be filed with the Architect and the Construction Manager. The demand for arbitration shall be made within the time limits specified in Subparagraph 2.3.15 where applicable, and in all other cases within a reasonable time after the claim, dispute or other matter in question has arisen; and in no event shall it be made after the date when institution of legal or equitable proceedings based on such claim, dispute or other matter in question would be barred by the applicable statute of limitations.

7.9.3 Unless otherwise agreed in writing, the Contractor shall carry on the Work and maintain its progress during any arbitration proceedings, and the Owner shall continue to make payments to the Contractor in accordance with the Contract Documents.

ARTICLE 8
TIME

8.1   DEFINITIONS

8.1.1 Unless otherwise provided, the Contract Time is the period of time allotted in the Contract Documents for Substantial Completion of the Work as defined in Subparagraph 8.1.3, including authorized adjustments thereto.

8.1.2 The date of commencement of the Work is the date established in a notice to proceed. If there is no notice to proceed, it shall be such other date as may be established in the Owner-Contractor Agreement or elsewhere in the Contract Documents.

8.1.3 The Date of Substantial Completion of the Work or designated portion thereof is the Date certified by the Architect when construction is sufficiently complete, in

accordance with the Contract Documents, so that the Owner or separate contractors can occupy or utilize the Work or a designated portion thereof for the use for which it is intended.

8.1.4 The Date of Substantial Completion of the Project or designated portion thereof is the Date certified by the Architect when construction is sufficiently complete so the Owner can occupy or utilize the Project or designated portion thereof for the use for which it was intended.

8.1.5 The term day as used in the Contract Documents shall mean calendar day unless specifically designated otherwise.

8.2   PROGRESS AND COMPLETION

8.2.1 All time limits stated in the Contract Documents are of the essence of the Contract.

8.2.2 The Contractor shall begin the Work on the date of commencement as defined in Subparagraph 8.1.2. The Contractor shall carry the Work forward expeditiously with adequate forces and shall achieve Substantial Completion of the Work within the Contract Time.

8.3   DELAYS AND EXTENSIONS OF TIME

8.3.1 If the Contractor is delayed at any time in the progress of the Work by any act or neglect of the Owner, the Architect, the Construction Manager, any of their employees, any separate contractor employed by the Owner, or by changes ordered in the Work, labor disputes, fire, unusual delay in transportation, adverse weather conditions not reasonably anticipatable, unavoidable casualties, any causes beyond the Contractor's control, delay authorized by the Owner pending arbitration, or by any other cause which the Construction Manager determines may justify the delay, then the Contract Time shall be extended by Change Order for such reasonable time as the Construction Manager may determine.

8.3.2 Any claim for extension of time shall be made in writing to the Construction Manager not more than twenty days after the commencement of the delay; otherwise it shall be waived. In the case of a continuing delay only one claim is necessary. The Contractor shall provide an estimate of the probable effect of such delay on the progress of the Work.

8.3.3 If no agreement is made stating the dates upon which interpretations as provided in Subparagraph 2.3.11 shall be furnished, then no claim for delay shall be allowed on account of failure to furnish such interpretations until fifteen days after written request is made for them, and not then unless such claim is reasonable.

8.3.4 This Paragraph 8.3 does not exclude the recovery of damages for delay by either party under other provisions of the Contract Documents.

ARTICLE 9
PAYMENTS AND COMPLETION

9.1   CONTRACT SUM

9.1.1 The Contract Sum is stated in the Owner-Contractor Agreement and, including authorized adjustments thereto, is the total amount payable by the Owner to the Contractor for the performance of the Work under the Contract Documents.

AIA DOCUMENT A201/CM • GENERAL CONDITIONS OF THE CONTRACT FOR CONSTRUCTION CONSTRUCTION MANAGEMENT EDITION • JUNE 1980 EDITION • AIA® • © 1980 • THE AMERICAN INSTITUTE OF ARCHITECTS, 1735 NEW YORK AVE., N.W., WASHINGTON, D.C. 20006

A201/CM — 1980   14

9.2 SCHEDULE OF VALUES

9.2.1 Before the first Application for Payment, the Contractor shall submit to the Construction Manager a schedule of values allocated to the various portions of the Work, prepared in such form and supported by such data to substantiate its accuracy as the Architect and the Construction Manager may require. This schedule, unless objected to by the Construction Manager or the Architect, shall be used only as a basis for the Contractor's Applications for Payment.

9.3 APPLICATIONS FOR PAYMENT

9.3.1 At least fifteen days before the date for each progress payment established in the Owner-Contractor Agreement, the Contractor shall submit to the Construction Manager an itemized Application for Payment, notarized if required, supported by such data substantiating the Contractor's right to payment as the Owner, the Architect or the Construction Manager may require, and reflecting retainage, if any, as provided elsewhere in the Contract Documents. The Construction Manager will assemble the Application with similar applications from other contractors on the Project into a combined Project Application for Payment and forward it with recommendations to the Architect within seven days.

9.3.2 Unless otherwise provided in the Contract Documents, payments will be made on account of materials or equipment not incorporated in the Work but delivered and suitably stored at the site and, if approved in advance by the Owner, payments may similarly be made for materials or equipment suitably stored at some other location agreed upon in writing. Payments for materials or equipment stored on or off the site shall be conditioned upon submission by the Contractor of bills of sale or such other procedures satisfactory to the Owner to establish the Owner's title to such materials or equipment or otherwise protect the Owner's interest, including applicable insurance and transportation to the site for those materials and equipment stored off the site.

9.3.3 The Contractor warrants that title to all Work, materials and equipment covered by an Application for Payment will pass to the Owner either by incorporation in the construction or upon receipt of payment by the Contractor, whichever occurs first, free and clear of all liens, claims, security interests or encumbrances, hereinafter referred to in this Article 9 as "liens"; and that no Work, materials or equipment covered by an Application for Payment will have been acquired by the Contractor, or by any other person performing Work at the site or furnishing materials and equipment for the Project, subject to an agreement under which an interest therein or an encumbrance thereon is retained by the seller or otherwise imposed by the Contractor or such other person.

9.4 CERTIFICATES FOR PAYMENT

9.4.1 The Architect will, within seven days after the receipt of the Project Application for Payment with the recommendations of the Construction Manager, review the Project Application for Payment and either issue a Project Certificate for Payment to the Owner with a copy to the Construction Manager for distribution to the Contractor for such amounts as the Architect determines are properly due, or notify the Construction Manager in writing of

the reasons for withholding a Certificate as provided in Subparagraph 9.6.1. Such notification will be forwarded to the Contractor by the Construction Manager.

9.4.2 The issuance of a Project Certificate for Payment will constitute a representation by the Architect to the Owner that, based on the Architect's observations at the site as provided in Subparagraph 2.3.4 and the data comprising the Project Application for Payment, the Work has progressed to the point indicated; that, to the best of the Architect's knowledge, information and belief, the quality of the Work is in accordance with the Contract Documents (subject to an evaluation of the Work for conformance with the Contract Documents upon Substantial Completion of the Work, to the results of any subsequent tests required by or performed under the Contract Documents, to minor deviations from the Contract Documents correctable prior to completion, and to any specific qualifications stated in the Certificate); and that the Contractor is entitled to payment in the amount certified. However, by issuing a Project Certificate for Payment, the Architect shall not thereby be deemed to represent that the Architect has made exhaustive or continuous on-site inspections to check the quality or quantity of the Work, has reviewed the construction means, methods, techniques, sequences or procedures, or has made any examination to ascertain how or for what purpose the Contractor has used the monies previously paid on account of the Contract Sum.

9.5 PROGRESS PAYMENTS

9.5.1 After the Architect has issued a Project Certificate for Payment, the Owner shall make payment in the manner and within the time provided in the Contract Documents.

9.5.2 The Contractor shall promptly pay each Subcontractor upon receipt of payment from the Owner, out of the amount paid to the Contractor on account of such Subcontractor's Work, the amount to which said Subcontractor is entitled, reflecting the percentage actually retained, if any, from payments to the Contractor on account of such Subcontractor's Work. The Contractor shall, by an appropriate agreement with each Subcontractor, require each Subcontractor to make payments to their Sub-subcontractors in similar manner.

9.5.3 The Architect may, on request and at the Architect's discretion, furnish to any Subcontractor, if practicable, information regarding the percentages of completion or the amounts applied for by the Contractor and the action taken thereon by the Architect on account of Work done by such Subcontractor.

9.5.4 Neither the Owner, the Architect nor the Construction Manager shall have any obligation to pay or to see to the payment of any monies to any Subcontractor except as may otherwise be required by law.

9.5.5 No certification of a progress payment, any progress payment, or any partial or entire use or occupancy of the Project by the Owner, shall constitute an acceptance of any Work not in accordance with the Contract Documents.

9.6 PAYMENTS WITHHELD

9.6.1 The Architect, following consultation with the Construction Manager, may decline to certify payment,

AIA DOCUMENT A201/CM • GENERAL CONDITIONS OF THE CONTRACT FOR CONSTRUCTION CONSTRUCTION MANAGEMENT EDITION • JUNE 1980 EDITION • AIA® • © 1980 • THE AMERICAN INSTITUTE OF ARCHITECTS, 1735 NEW YORK AVE., N.W., WASHINGTON, D.C. 20006

184   The McGraw-Hill Construction Management Form Book

and may withhold the Certificate in whole or in part to the extent necessary to reasonably protect the Owner, if, in the Architect's opinion, the Architect is unable to make representations to the Owner as provided in Subparagraph 9.4.2. If the Architect is unable to make representations to the Owner as provided in Subparagraph 9.4.2, and to certify payment in the amount of the Project Application, the Architect will notify the Construction Manager as provided in Subparagraph 9.4.1. If the Contractor and the Architect cannot agree on a revised amount, the Architect will promptly issue a Project Certificate for Payment for the amount for which the Architect is able to make such representations to the Owner. The Architect may also decline to certify payment or, because of subsequently discovered evidence or subsequent observations, the Architect may nullify the whole or any part of any Project Certificate for Payment previously issued to such extent as may be necessary, in the Architect's opinion, to protect the Owner from loss because of:

.1 defective Work not remedied;

.2 third party claims filed or reasonable evidence indicating probable filing of such claims;

.3 failure of the Contractor to make payments properly to Subcontractors, or for labor, materials or equipment;

.4 reasonable evidence that the Work cannot be completed for the unpaid balance of the Contract Sum;

.5 damage to the Owner or another contractor;

.6 reasonable evidence that the Work will not be completed within the Contract Time; or

.7 persistent failure to carry out the Work in accordance with the Contract Documents.

9.6.2 When the grounds in Subparagraph 9.6.1 above are removed, payment shall be made for amounts withheld because of them.

9.7   FAILURE OF PAYMENT

9.7.1 If the Construction Manager should fail to issue recommendations within seven days of receipt of the Contractor's Application for Payment, or if, through no fault of the Contractor, the Architect does not issue a Project Certificate for Payment within seven days after the Architect's receipt of the Project Application for Payment, or if the Owner does not pay the Contractor within seven days after the date established in the Contract Documents any amount certified by the Architect or awarded by arbitration, then the Contractor may, upon seven additional days' written notice to the Owner, the Architect and the Construction Manager, stop the Work until payment of the amount owing has been received. The Contract Sum shall be increased by the amount of the Contractor's reasonable costs of shut-down, delay and start-up, which shall be effected by appropriate Change Order in accordance with Paragraph 12.3.

9.8   SUBSTANTIAL COMPLETION

9.8.1 When the Contractor considers that the Work, or a designated portion thereof which is acceptable to the Owner, is substantially complete as defined in Subparagraph 8.1.3, the Contractor shall prepare for the Construction Manager a list of items to be completed or corrected. The failure to include any items on such list does

not alter the responsibility of the Contractor to complete all Work in accordance with the Contract Documents. When the Architect, on the basis of inspection and consultation with the Construction Manager, determines that the Work or designated portion thereof is substantially complete, the Architect will then prepare a Certificate of Substantial Completion of the Work which shall establish the Date of Substantial Completion of the Work, shall state the responsibilities of the Owner and the Contractor for security, maintenance, heat, utilities, damage to the Work and insurance, and shall fix the time within which the Contractor shall complete the items listed therein. The Certificate of Substantial Completion of the Work shall be submitted to the Owner and the Contractor for their written acceptance of the responsibilities assigned to them in such Certificate.

9.8.2 Upon Substantial Completion of the Work or designated portion thereof, and upon application by the Contractor and certification by the Architect, the Owner shall make payment, reflecting adjustment in retainage, if any, for such Work or portion thereof as provided in the Contract Documents.

9.8.3 When the Architect, on the basis of inspections, determines that the Project or designated portion thereof is substantially complete, the Architect will then prepare a Certificate of Substantial Completion of the Project which shall establish the Date of Substantial Completion of the Project and fix the time within which the Contractor shall complete any uncompleted items on the Certificate of Substantial Completion of the Work.

9.8.4 Warranties required by the Contract Documents shall commence on the Date of Substantial Completion of the Project or designated portion thereof unless otherwise provided in the Certificate of Substantial Completion of the Work or designated portion thereof.

9.9   FINAL COMPLETION AND FINAL PAYMENT

9.9.1 Following the Architect's issuance of the Certificate of Substantial Completion of the Work or designated portion thereof, and the Contractor's completion of the Work, the Contractor shall forward to the Construction Manager a written notice that the Work is ready for final inspection and acceptance, and shall also forward to the Construction Manager a final Application for Payment. Upon receipt, the Construction Manager will make the necessary evaluations and forward recommendations to the Architect who will promptly make such inspection. When the Architect finds the Work acceptable under the Contract Documents and the Contract fully performed, the Architect will issue a Project Certificate for Payment which will approve the final payment due the Contractor. This approval will constitute a representation that, to the best of the Architect's knowledge, information and belief, and on the basis of observations and inspections, the Work has been completed in accordance with the Terms and Conditions of the Contract Documents and that the entire balance found to be due the Contractor, and noted in said Certificate, is due and payable. The Architect's approval of said Project Certificate for Payment will constitute a further representation that the conditions precedent to the Contractor's being entitled to final payment as set forth in Subparagraph 9.9.2 have been fulfilled.

AIA DOCUMENT A201/CM • GENERAL CONDITIONS OF THE CONTRACT FOR CONSTRUCTION
CONSTRUCTION MANAGEMENT EDITION • JUNE 1980 EDITION • AIA® • © 1980 • THE
AMERICAN INSTITUTE OF ARCHITECTS, 1735 NEW YORK AVE., N.W., WASHINGTON, D.C. 20006

A201/CM — 1980   16

9.9.2 Neither the final payment nor the remaining retainage shall become due until the Contractor submits to the Architect, through the Construction Manager, (1) an affidavit that all payrolls, bills for materials and equipment, and other indebtedness connected with the Work for which the Owner or the Owner's property might in any way be responsible, have been paid or otherwise satisfied, (2) consent of surety, if any, to final payment, and (3) if required by the Owner, other data establishing payment or satisfaction of all such obligations, such as receipts, releases and waivers of liens arising out of the Contract, to the extent and in such form as may be designated by the Owner. If any Subcontractor refuses to furnish a release or waiver required by the Owner, the Contractor may furnish a bond satisfactory to the Owner to indemnify the Owner against any such lien. If any such lien remains unsatisfied after all payments are made, the Contractor shall refund to the Owner all monies that the latter may be compelled to pay in discharging such lien, including all costs and reasonable attorneys' fees.

9.9.3 If, after Substantial Completion of the Work, final completion thereof is materially delayed through no fault of the Contractor or by the issuance of Change Orders affecting final completion, and the Construction Manager so confirms, the Owner shall, upon application by the Contractor and certification by the Architect and without terminating the Contract, make payment of the balance due for that portion of the Work fully completed and accepted. If the remaining balance for Work not fully completed or corrected is less than the retainage stipulated in the Contract Documents, and if bonds have been furnished as provided in Paragraph 7.5, the written consent of the surety to the payment of the balance due for that portion of the Work fully completed and accepted shall be submitted by the Contractor to the Construction Manager prior to certification of such payment. Such payment shall be made under the Terms and Conditions governing final payments, except that it shall not constitute a waiver of claims.

9.9.4 The making of final payment shall, after the Date of Substantial Completion of the Project, constitute a waiver of all claims by the Owner except those arising from:

.1 unsettled liens;

.2 faulty or defective Work appearing after Substantial Completion of the Work;

.3 failure of the Work to comply with the requirements of the Contract Documents; or

.4 terms of any special warranties required by the Contract Documents.

9.9.5 The acceptance of final payment shall, after the Date of Substantial Completion of the Project, constitute a waiver of all claims by the Contractor except those previously made in writing and identified by the Contractor as unsettled at the time of the final Application for Payment.

9.9.6 All provisions of this Agreement, including without limitation those establishing obligations and procedures, shall remain in full force and effect notwithstanding the making or acceptance of final payment prior to the Date of Substantial Completion of the Project.

ARTICLE 10
PROTECTION OF PERSONS AND PROPERTY

10.1 SAFETY PRECAUTIONS AND PROGRAMS

10.1.1 The Contractor shall be responsible for initiating, maintaining and supervising all safety precautions and programs in connection with the Work.

10.2 SAFETY OF PERSONS AND PROPERTY

10.2.1 The Contractor shall take all reasonable precautions for the safety of, and shall provide all reasonable protection to prevent damage, injury or loss to:

.1 all employees on the Work and all other persons who may be affected thereby;

.2 all the Work and all materials and equipment to be incorporated therein, whether in storage on or off the site, under the care, custody or control of the Contractor or any of the Contractor's Subcontractors or Sub-subcontractors; and

.3 other property at the site or adjacent thereto, including trees, shrubs, lawns, walks, pavements, roadways, structures and utilities not designated for removal, relocation or replacement in the course of construction; and

.4 the Work of the Owner or other separate contractors.

10.2.2 The Contractor shall give all notices and comply with all applicable laws, ordinances, rules, regulations and lawful orders of any public authority bearing on the safety of persons or property or their protection from damage, injury or loss.

10.2.3 The Contractor shall erect and maintain, as required by existing conditions and the progress of the Work, all reasonable safeguards for safety and protection, including posting danger signs and other warnings against hazards, promulgating safety regulations and notifying owners and users of adjacent utilities.

10.2.4 When the use or storage of explosives or other hazardous materials or equipment is necessary for the execution of the Work, the Contractor shall exercise the utmost care and shall carry on such activities under the supervision of properly qualified personnel.

10.2.5 The Contractor shall promptly remedy all damage or loss (other than damage or loss insured under Paragraph 11.3) to any property referred to in Clauses 10.2.1.2 and 10.2.1.3 caused in whole or in part by the Contractor, any Subcontractor, any Sub-subcontractor, anyone directly or indirectly employed by any of them, or by anyone for whose acts any of them may be liable, and for which the Contractor is responsible under Clauses 10.2.1.2 and 10.2.1.3, except damage or loss attributable to the acts or omissions of the Owner, the Architect, the Construction Manager or anyone directly or indirectly employed by any of them, or by anyone for whose acts any of them may be liable, and not attributable to the fault or negligence of the Contractor. The foregoing obligations of the Contractor are in addition to the Contractor's obligations under Paragraph 4.18.

10.2.6 The Contractor shall designate a responsible member of the Contractor's organization at the site whose duty shall be the prevention of accidents. This person shall be the Contractor's superintendent unless

AIA DOCUMENT A201/CM • GENERAL CONDITIONS OF THE CONTRACT FOR CONSTRUCTION
CONSTRUCTION MANAGEMENT EDITION • JUNE 1980 EDITION • AIA® • © 1980 • THE
AMERICAN INSTITUTE OF ARCHITECTS, 1735 NEW YORK AVE., N.W., WASHINGTON, D.C. 20006

otherwise designated by the Contractor in writing to the Owner and the Construction Manager.

10.2.7 The Contractor shall not load or permit any part of the Work to be loaded so as to endanger its safety.

10.3   EMERGENCIES

10.3.1 In any emergency affecting the safety of persons or property the Contractor shall act, at the Contractor's discretion, to prevent threatened damage, injury or loss. Any additional compensation or extension of time claimed by the Contractor on account of emergency work shall be determined as provided in Article 12 for Changes in the Work.

ARTICLE 11
INSURANCE

11.1   CONTRACTOR'S LIABILITY INSURANCE

11.1.1 The Contractor shall purchase and maintain insurance for protection from the claims set forth below which may arise out of or result from the Contractor's operations under the Contract, whether such operations be by the Contractor or by any Subcontractor, or by anyone directly or indirectly employed by any of them, or by anyone for whose acts any of them may be liable:

.1 claims under workers' or workmen's compensation, disability benefit and other similar employee benefit acts;

.2 claims for damages because of bodily injury, occupational sickness or disease, or death of the Contractor's employees;

.3 claims for damages because of bodily injury, sickness or disease, or death of any person other than the Contractor's employees;

.4 claims for damages insured by usual personal injury liability coverage which are sustained (1) by any person as a result of an offense directly or indirectly related to the employment of such person by the Contractor, or (2) by any other person;

.5 claims for damages, other than to the Work itself, because of injury to or destruction of tangible property, including loss of use resulting therefrom; and

.6 claims for damages because of bodily injury or death of any person or property damage arising out of the ownership, maintenance or use of any motor vehicle.

11.1.2 The insurance required by Subparagraph 11.1.1 shall be written for not less than any limits of liability specified in the Contract Documents or required by law, whichever is greater.

11.1.3 The insurance required by Subparagraph 11.1.1 shall include contractual liability insurance applicable to the Contractor's obligations under Paragraph 4.18.

11.1.4 Certificates of Insurance acceptable to the Owner shall be submitted to the Construction Manager for transmittal to the Owner prior to commencement of the Work. These Certificates shall contain a provision that coverages afforded under the policies will not be canceled until at least thirty days' prior written notice has been given to the Owner.

11.2   OWNER'S LIABILITY INSURANCE

11.2.1 The Owner shall be responsible for purchasing and maintaining Owner's liability insurance and, at the Owner's option, may purchase and maintain insurance for protection against claims which may arise from operations under the Contract.

11.3   PROPERTY INSURANCE

11.3.1 Unless otherwise provided, the Owner shall purchase and maintain property insurance upon the entire Work at the site to the full insurable value thereof. This insurance shall include the interests of the Owner, the Construction Manager, the Contractor, Subcontractors and Sub-subcontractors in the Work, and shall insure against the perils of fire and extended coverage and shall include "all risk" insurance for physical loss or damage including, without duplication of coverage, theft, vandalism and malicious mischief. If the Owner does not intend to purchase such insurance for the full insurable value of the entire Work, the Owner shall inform the Contractor in writing prior to commencement of the Work. The Contractor may then effect insurance which will protect the interests of the Contractor, the Contractor's Subcontractors and the Sub-subcontractors in the Work, and by appropriate Change Order the cost thereof shall be charged to the Owner. If the Contractor is damaged by failure of the Owner to purchase or maintain such insurance and to so notify the Contractor, then the Owner shall bear all reasonable costs properly attributable thereto. If not covered under the all risk insurance or otherwise provided in the Contract Documents, the Contractor shall effect and maintain similar property insurance on portions of the Work stored off the site or in transit when such portions of the Work are to be included in an Application for Payment under Subparagraph 9.3.2.

11.3.2 The Owner shall purchase and maintain such boiler and machinery insurance as may be required by the Contract Documents or by law. This insurance shall include the interests of the Owner, the Construction Manager, the Contractor, Subcontractors and Sub-subcontractors in the Work.

11.3.3 Any loss insured under Subparagraph 11.3.1 is to be adjusted with the Owner and made payable to the Owner as trustee for the insureds, as their interests may appear, subject to the requirements of any applicable mortgagee clause and of Subparagraph 11.3.8. The Contractor shall pay each Subcontractor a just share of any insurance monies received by the Contractor, and by appropriate agreement, written where legally required for validity, shall require each Subcontractor to make payments to their Sub-subcontractors in similar manner.

11.3.4 The Owner shall file a copy of all policies with the Contractor before an exposure to loss may occur.

11.3.5 If the Contractor requests in writing that insurance for risks other than those described in Subparagraphs 11.3.1 and 11.3.2, or other special hazards, be included in the property insurance policy, the Owner shall, if possible, include such insurance, and the cost thereof shall be charged to the Contractor by appropriate Change Order.

11.3.6 The Owner and the Contractor waive all rights against (1) each other and the Subcontractors, Sub-subcontractors, agents and employees of each other, and (2) the Architect, the Construction Manager and separate contractors, if any, and their subcontractors, sub-subcontractors, agents and employees, for damages caused by fire or other perils to the extent covered by insurance obtained pursuant to this Paragraph 11.3 or any other property insurance applicable to the Work, except such rights as they may have to the proceeds of such insurance held by the Owner as trustee. The foregoing waiver afforded the Architect, the Construction Manager, their agents and employees shall not extend to the liability imposed by Subparagraph 4.18.3. The Owner or the Contractor, as appropriate, shall require of the Architect, the Construction Manager, separate contractors, Subcontractors and Sub-subcontractors by appropriate agreements, written where legally required for validity, similar waivers each in favor of all other parties enumerated in this Subparagraph 11.3.6.

11.3.7 If required in writing by any party in interest, the Owner as trustee shall, upon the occurrence of an insured loss, give bond for the proper performance of the Owner's duties. The Owner shall deposit in a separate account any money so received, and shall distribute it in accordance with such agreement as the parties in interest may reach, or in accordance with an award by arbitration in which case the procedure shall be as provided in Paragraph 7.9. If after such loss no other special agreement is made, replacement of damaged Work shall be covered by an appropriate Change Order.

11.3.8 The Owner, as trustee, shall have power to adjust and settle any loss with the insurers unless one of the parties in interest shall object, in writing, within five days after the occurrence of loss, to the Owner's exercise of this power, and if such objection be made, arbitrators shall be chosen as provided in Paragraph 7.9. The Owner as trustee shall, in that case, make settlement with the insurers in accordance with the directions of such arbitrators. If distribution of the insurance proceeds by arbitration is required, the arbitrators will direct such distribution.

11.3.9 If the Owner finds it necessary to occupy or use a portion or portions of the Work prior to Substantial Completion thereof, such occupancy shall not commence prior to a time mutually agreed to by the Owner and the Contractor and to which the insurance company or companies providing the property insurance have consented by endorsement to the policy or policies. This insurance shall not be canceled or lapsed on account of such partial occupancy. Consent of the Contractor and of the insurance company or companies to such occupancy or use shall not be unreasonably withheld.

11.4 **LOSS OF USE INSURANCE**

11.4.1 The Owner, at the Owner's option, may purchase and maintain insurance for protection against loss of use of the Owner's property due to fire or other hazards, however caused. The Owner waives all rights of action against the Contractor for loss of use of the Owner's property, including consequential losses due to fire or other hazards however caused, to the extent covered by insurance under this Paragraph 11.4.

ARTICLE 12
CHANGES IN THE WORK

12.1 **CHANGE ORDERS**

12.1.1 A Change Order is a written order to the Contractor signed to show the recommendation of the Construction Manager, the approval of the Architect and the authorization of the Owner, issued after execution of the Contract, authorizing a change in the Work or an adjustment in the Contract Sum or the Contract Time. The Contract Sum and the Contract Time may be changed only by Change Order. A Change Order signed by the Contractor indicates the Contractor's agreement therewith, including the adjustment in the Contract Sum or the Contract Time.

12.1.2 The Owner, without invalidating the Contract, may order changes in the Work within the general scope of the Contract consisting of additions, deletions or other revisions, the Contract Sum and the Contract Time being adjusted accordingly. All such changes in the Work shall be authorized by Change Order, and shall be performed under the applicable conditions of the Contract Documents.

12.1.3 The cost or credit to the Owner resulting from a change in the Work shall be determined in one or more of the following ways:

.1 by mutual acceptance of a lump sum properly itemized and supported by sufficient substantiating data to permit evaluation;

.2 by unit prices stated in the Contract Documents or subsequently agreed upon;

.3 by cost to be determined in a manner agreed upon by the parties and a mutually acceptable fixed or percentage fee; or

.4 by the method provided in Subparagraph 12.1.4.

12.1.4 If none of the methods set forth in Clauses 12.1.3.1, 12.1.3.2 or 12.1.3.3 is agreed upon, the Contractor, provided a written order signed by the Owner is received, shall promptly proceed with the Work involved. The cost of such Work shall then be determined by the Architect, after consultation with the Construction Manager, on the basis of the reasonable expenditures and savings of those performing the Work attributable to the change, including, in the case of an increase in the Contract Sum, a reasonable allowance for overhead and profit. In such case, and also under Clauses 12.1.3.3 and 12.1.3.4 above, the Contractor shall keep and present, in such form as the Owner, the Architect or the Construction Manager may prescribe, an itemized accounting together with appropriate supporting data for inclusion in a Change Order. Unless otherwise provided in the Contract Documents, cost shall be limited to the following: cost of materials, including sales tax and cost of delivery; cost of labor, including social security, old age and unemployment insurance, and fringe benefits required by agreement or custom; workers' or workmen's compensation insurance; bond premiums; rental value of equipment and machinery; and the additional costs of supervision and field office personnel directly attributable to the change. Pending final determination of cost to the Owner, payments on account shall be made on the Architect's approval of a Project Certificate for Payment.

AIA DOCUMENT A201/CM • GENERAL CONDITIONS OF THE CONTRACT FOR CONSTRUCTION CONSTRUCTION MANAGEMENT EDITION • JUNE 1980 EDITION • AIA® • © 1980 • THE AMERICAN INSTITUTE OF ARCHITECTS, 1735 NEW YORK AVE., N.W., WASHINGTON, D.C. 20006

The amount of credit to be allowed by the Contractor to the Owner for any deletion or change which results in a net decrease in the Contract Sum will be the amount of the actual net cost as confirmed by the Architect after consultation with the Construction Manager. When both additions and credits covering related Work or substitutions are involved in any one change, the allowance for overhead and profit shall be figured on the basis of the net increase, if any, with respect to that change.

12.1.5 If unit prices are stated in the Contract Documents or subsequently agreed upon, and if the quantities originally contemplated are so changed in a proposed Change Order that application of the agreed unit prices to the quantities of Work proposed will cause substantial inequity to the Owner or the Contractor, the applicable unit prices shall be equitably adjusted.

12.2   CONCEALED CONDITIONS

12.2.1 Should concealed conditions encountered in the performance of the Work below the surface of the ground or should concealed or unknown conditions in an existing structure be at variance with the conditions indicated by the Contract Documents, or should unknown physical conditions below the surface of the ground or should concealed or unknown conditions in an existing structure of an unusual nature, differing materially from those ordinarily encountered and generally recognized as inherent in work of the character provided for in this Contract, be encountered, the Contract Sum shall be equitably adjusted by Change Order upon claim by either party made within twenty days after the first observance of the conditions.

12.3   CLAIMS FOR ADDITIONAL COST

12.3.1 If the Contractor wishes to make a claim for an increase in the Contract Sum, the Contractor shall give the Architect and the Construction Manager written notice thereof within twenty days after the occurrence of the event giving rise to such claim. This notice shall be given by the Contractor before proceeding to execute the Work, except in an emergency endangering life or property in which case the Contractor shall proceed in accordance with Paragraph 10.3. No such claim shall be valid unless so made. If the Owner and the Contractor cannot agree on the amount of the adjustment in the Contract Sum, it shall be determined by the Architect after consultation with the Construction Manager. Any change in the Contract Sum resulting from such claim shall be authorized by Change Order.

12.3.2 If the Contractor claims that additional cost is involved because of, but not limited to, (1) any written interpretation pursuant to Subparagraph 2.3.11, (2) any order by the Owner to stop the Work pursuant to Paragraph 3.3 where the Contractor was not at fault, or any such order by the Construction Manager as the Owner's agent, (3) any written order for a minor change in the Work issued pursuant to Paragraph 12.4, or (4) failure of payment by the Owner pursuant to Paragraph 9.7, the Contractor shall make such claim as provided in Subparagraph 12.3.1.

12.4   MINOR CHANGES IN THE WORK

12.4.1 The Architect will have authority to order minor changes in the Work not involving an adjustment in the

Contract Sum or extension of the Contract Time and not inconsistent with the intent of the Contract Documents. Such changes shall be effected by written order issued through the Construction Manager, and shall be binding on the Owner and the Contractor. The Contractor shall carry out such written orders promptly.

ARTICLE 13
UNCOVERING AND CORRECTION OF WORK

13.1   UNCOVERING OF WORK

13.1.1 If any portion of the Work should be covered contrary to the request of the Architect or the Construction Manager, or to requirements specifically expressed in the Contract Documents, it must, if required in writing by either, be uncovered for their observation and shall be replaced at the Contractor's expense.

13.1.2 If any other portion of the Work has been covered which the Architect or the Construction Manager has not specifically requested to observe prior to its being covered, either may request to see such Work and it shall be uncovered by the Contractor. If such Work be found in accordance with the Contract Documents, the cost of uncovering and replacement shall, by appropriate Change Order, be charged to the Owner. If such Work be found not in accordance with the Contract Documents, the Contractor shall pay such costs unless it be found that this condition was caused by the Owner or a separate contractor as provided in Article 6, in which event the Owner shall be responsible for the payment of such costs.

13.2   CORRECTION OF WORK

13.2.1 The Contractor shall promptly correct all Work rejected by the Architect or the Construction Manager as defective or as failing to conform to the Contract Documents whether observed before or after Substantial Completion of the Project and whether or not fabricated, installed or completed. The Contractor shall bear all costs of correcting such rejected Work, including compensation for the Architect's and the Construction Manager's additional services made necessary thereby.

13.2.2 If, within one year after the Date of Substantial Completion of the Project or designated portion thereof, or within one year after acceptance by the Owner of designated equipment, or within such longer period of time as may be prescribed by the terms of any applicable special warranty required by the Contract Documents, any of the Work is found to be defective or not in accordance with the Contract Documents, the Contractor shall correct it promptly after receipt of a written notice from the Owner to do so unless the Owner has previously given the Contractor a written acceptance of such condition. This obligation shall survive both final payment for the Work or designated portion thereof and termination of the Contract. The Owner shall give such notice promptly after discovery of the condition.

13.2.3 The Contractor shall remove from the site all portions of the Work which are defective or nonconforming and which have not been corrected under Subparagraphs 4.5.1, 13.2.1 and 13.2.2, unless removal is waived by the Owner.

13.2.4 If the Contractor fails to correct defective or nonconforming Work as provided in Subparagraphs 4.5.1,

AIA DOCUMENT A201/CM • GENERAL CONDITIONS OF THE CONTRACT FOR CONSTRUCTION
CONSTRUCTION MANAGEMENT EDITION • JUNE 1980 EDITION • AIA® • © 1980 • THE
AMERICAN INSTITUTE OF ARCHITECTS, 1735 NEW YORK AVE., N.W., WASHINGTON, D.C. 20006

A201/CM — 1980   20

13.2.1 and 13.2.2, the Owner may correct it in accordance with Paragraph 3.4.

13.2.5 If the Contractor does not proceed with the correction of such defective or nonconforming Work within a reasonable time fixed by written notice from the Architect issued through the Construction Manager, the Owner may remove it and may store the materials or equipment at the expense of the Contractor. If the Contractor does not pay the cost of such removal and storage within ten days thereafter, the Owner may, upon ten additional days' written notice, sell such Work at auction or at private sale and shall account for the net proceeds thereof, after deducting all the costs that should have been borne by the Contractor, including compensation for the Architect's and the Construction Manager's additional services made necessary thereby. If such proceeds of sale do not cover all costs which the Contractor should have borne, the difference shall be charged to the Contractor and an appropriate Change Order shall be issued. If the payments then or thereafter due the Contractor are not sufficient to cover such amount, the Contractor shall pay the difference to the Owner.

13.2.6 The Contractor shall bear the cost of making good all work of the Owner or separate contractors destroyed or damaged by such correction or removal.

13.2.7 Nothing contained in this Paragraph 13.2 shall be construed to establish a period of limitation with respect to any other obligation which the Contractor might have under the Contract Documents, including Paragraph 4.5 hereof. The establishment of the time periods noted in Subparagraph 13.2.2, or such longer period of time as may be prescribed by law or by the terms of any warranty required by the Contract Documents, relates only to the specific obligation of the Contractor to correct the Work, and has no relationship to the time within which the Contractor's obligation to comply with the Contract Documents may be sought to be enforced, nor to the time within which proceedings may be commenced to establish the Contractor's liability with respect to the Contractor's obligations other than specifically to correct the Work.

13.3 ACCEPTANCE OF DEFECTIVE OR NONCONFORMING WORK

13.3.1 If the Owner prefers to accept defective or nonconforming Work, the Owner may do so instead of requiring its removal and correction, in which case a Change Order will be issued to reflect a reduction in the Contract Sum where appropriate and equitable. Such adjustment shall be effected whether or not final payment has been made.

ARTICLE 14
TERMINATION OF THE CONTRACT

14.1 TERMINATION BY THE CONTRACTOR

14.1.1 If the Work is stopped for a period of thirty days under an order of any court or other public authority

having jurisdiction, or as a result of an act of government such as a declaration of a national emergency making materials unavailable, through no act or fault of the Contractor or a Subcontractor or their agents or employees or any other persons performing any of the Work under a contract with the Contractor, or if the Work should be stopped for a period of thirty days by the Contractor because of the Construction Manager's failure to recommend or the Architect's failure to issue a Project Certificate for Payment as provided in Paragraph 9.7 or because the Owner has not made payment thereon as provided in Paragraph 9.7, then the Contractor may, upon seven additional days' written notice to the Owner, the Architect and the Construction Manager, terminate the Contract and recover from the Owner payment for all Work executed and for any proven loss sustained upon any materials, equipment, tools, construction equipment and machinery, including reasonable profit and damages.

14.2 TERMINATION BY THE OWNER

14.2.1 If the Contractor is adjudged a bankrupt, or makes a general assignment for the benefit of creditors, or if a receiver is appointed on account of the Contractor's insolvency, or if the Contractor persistently or repeatedly refuses or fails, except in cases for which extension of time is provided, to supply enough properly skilled workers or proper materials, or fails to make prompt payment to Subcontractors or for materials or labor, or persistently disregards laws, ordinances, rules, regulations or orders of any public authority having jurisdiction, or otherwise is guilty of a substantial violation of a provision of the Contract Documents, and fails within seven days after receipt of written notice to commence and continue correction of such default, neglect or violation with diligence and promptness, the Owner, upon certification by the Architect after consultation with the Construction Manager that sufficient cause exists to justify such action, may, after seven days following receipt by the Contractor of an additional written notice and without prejudice to any other remedy the Owner may have, terminate the employment of the Contractor and take possession of the site and of all materials, equipment, tools, construction equipment and machinery thereon owned by the Contractor and may finish the Work by whatever methods the Owner may deem expedient. In such case the Contractor shall not be entitled to receive any further payment until the Work is finished.

14.2.2 If the unpaid balance of the Contract Sum exceeds the costs of finishing the Work, including compensation for the Architect's and the Construction Manager's additional services made necessary thereby, such excess shall be paid to the Contractor. If such costs exceed the unpaid balance, the Contractor shall pay the difference to the Owner. The amount to be paid to the Contractor or to the Owner, as the case may be, shall be certified by the Architect, upon application, in the manner provided in Paragraph 9.4, and this obligation for payment shall survive the termination of the Contract.

AIA DOCUMENT A201/CM • GENERAL CONDITIONS OF THE CONTRACT FOR CONSTRUCTION CONSTRUCTION MANAGEMENT EDITION • JUNE 1980 EDITION • AIA® • © 1980 • THE AMERICAN INSTITUTE OF ARCHITECTS, 1735 NEW YORK AVE., N.W., WASHINGTON, D.C. 20006

AIA Document A311/CM, *Performance Bond and Labor and Material Payment Bond* (CM Edition).   The only difference in this bond form from the traditional bond form is that suit may be brought under A311/CM within one year after substantial completion of the project, even if this time period exceeds the traditional period of two years from final payment (as to the performance bond) or one year from ceasing work (as to the labor and material payment bond).

The purpose of this change is to permit enforcement of the bond through the one-year corrective period following completion of the entire project. Where a project is undertaken with separate contractors and phased construction, the traditional time limitations, which begin to run with reference to completion or cessation of the individual contractor's work, could expire before the entire project was complete.

THE AMERICAN INSTITUTE OF ARCHITECTS

AIA Document A311/CM

CONSTRUCTION MANAGEMENT EDITION

Performance Bond

KNOW ALL MEN BY THESE PRESENTS: that

(Here insert full name and address or legal title of Contractor)

as Principal, hereinafter called Contractor, and,

(Here insert full name and address or legal title of Surety)

as Surety, hereinafter called Surety, are held and firmly bound unto

(Here insert full name and address or legal title of Owner)

as Obligee, hereinafter called Owner, in the amount of

Dollars ($),

for the payment whereof Contractor and Surety bind themselves, their heirs, executors, administrators, successors and assigns, jointly and severally, firmly by these presents.

WHEREAS,

Contractor has by written agreement dated   19   , entered into a contract with Owner for
(Here insert full name, address and description of project)

in accordance with Drawings and Specifications prepared by

(Here insert full name and address or legal title of Architect)

which contract is by reference made a part hereof, and is hereinafter referred to as the Contract.

PERFORMANCE BOND

NOW, THEREFORE, THE CONDITION OF THIS OBLIGATION is such that, if Contractor shall promptly and faithfully perform said Contract, then this obligation shall be null and void; otherwise it shall remain in full force and effect.

The Surety hereby waives notice of any alteration or extension of time made by the Owner.

Whenever Contractor shall be and declared by Owner to be in default under the Contract, the Owner having performed Owner's obligations thereunder, the Surety may promptly remedy the default, or shall promptly

1) Complete the Contract in accordance with its terms and conditions, or

2) Obtain a bid or bids for completing the Contract in accordance with its terms and conditions, and upon determination by Surety of the lowest responsible bidder, or, if the Owner elects, upon determination by the Owner and the Surety jointly of the lowest responsible bidder, arrange for a contract between such bidder and Owner, and make available as Work progresses (even though there should be a default or a succession of defaults under the contract or contracts of completion arranged under this paragraph)

sufficient funds to pay the cost of completion less the balance of the contract price; but not exceeding, including other costs and damages for which the Surety may be liable hereunder, the amount set forth in the first paragraph hereof. The term "balance of the contract price," as used in this paragraph, shall mean the total amount payable by Owner to Contractor under the Contract and any amendments thereto, less the amount properly paid by Owner to Contractor.

Any suit under this bond must be instituted before the expiration of two (2) years from the date on which final payment under the Contract falls due or before the expiration of one (1) year from the Date of Substantial Completion of the Project, whichever is later.

No right of action shall accrue on this bond to or for the use of any person or corporation other than the Owner named herein or the heirs, executors, administrators or successors of the Owner.

SAMPLE

Signed and sealed this   day of   , 19

_____
(Witness)

{   _____
                     (Principal)   (Seal)

         _____
                       (Title)

_____
(Witness)

{   _____
                     (Principal)   (Seal)

         _____
                       (Title)

AIA DOCUMENT A311/CM • PERFORMANCE BOND AND LABOR AND MATERIAL PAYMENT BOND • CONSTRUCTION MANAGEMENT EDITION
JUNE 1980 EDITION • AIA® • THE AMERICAN INSTITUTE OF ARCHITECTS, 1735 NEW YORK AVENUE, N.W., WASHINGTON, D.C. 20006   2 of 4

THE AMERICAN INSTITUTE OF ARCHITECTS

AIA Document A311/CM

CONSTRUCTION MANAGEMENT EDITION

Labor and Material Payment Bond

THIS BOND IS ISSUED SIMULTANEOUSLY WITH PERFORMANCE BOND IN FAVOR OF THE
OWNER CONDITIONED ON THE FULL AND FAITHFUL PERFORMANCE OF THE CONTRACT

KNOW ALL MEN BY THESE PRESENTS: that

(Here insert full name and address or legal title of Contractor)

as Principal, hereinafter called Principal, and

(Here insert full name and address or legal title of Surety)

as Surety, hereinafter called Surety, are held and firmly bound unto

(Here insert full name and address or legal title of Owner)

as Obligee, hereinafter called Owner, for the use and benefit of claimants as hereinbelow defined, in the

amount of

(Here insert a sum equal to at least one-half of the contract price)   Dollars ($),

for the payment whereof Principal and Surety bind themselves, their heirs, executors, administrators,
successors and assigns, jointly and severally, firmly by these presents.

WHEREAS,

Principal has by written agreement dated   , 19   , entered into a contract with Owner for
(Here insert full name, address and description of project)

in accordance with Drawings and Specifications prepared by

(Here insert full name and address or legal title of Architect)

which contract is by reference made a part hereof, and is hereinafter referred to as the Contract.

LABOR AND MATERIAL PAYMENT BOND

NOW, THEREFORE, THE CONDITION OF THIS OBLIGATION is such that, if Principal shall promptly make payment to all claimants as hereinafter defined, for all labor and material used or reasonably required for use in the performance of the Contract, then this obligation shall be void; otherwise it shall remain in full force and effect, subject, however, to the following conditions:

1. A claimant is defined as one having a direct contract with the Principal or with a Subcontractor of the Principal for labor, material, or both, used or reasonably required for use in the performance of the Contract, labor and material being construed to include that part of water, gas, power, light, heat, oil, gasoline, telephone service or rental of equipment directly applicable to the Contract.

2. The above named Principal and Surety hereby jointly and severally agree with the Owner that every claimant as herein defined, who has not been paid in full before the expiration of a period of ninety (90) days after the date on which the last of such claimant's work or labor was done or performed, or materials were furnished by such claimant, may sue on this bond for the use of such claimant, prosecute the suit to final judgment for such sum or sums as may be justly due claimant, and have execution thereon. The Owner shall not be liable for the payment of any costs or expenses of any such suit.

3. No suit or action shall be commenced hereunder by any claimant:

a) Unless claimant, other than one having a direct contract with the Principal, shall have given written notice to any two of the following: the Principal, the Owner or the Surety above named, within ninety (90) days after such claimant did or performed the last of the work or labor, or furnished the last of the materials for which said claim is made, stating with substantial accuracy the amount claimed and the name of the party to whom the materials were furnished, or for whom the work or labor was done or performed. Such notice shall be served by mailing the same by registered mail or certified mail, postage prepaid, in an envelope addressed to the Principal, Owner or Surety, at any place where an office is regularly maintained for the transaction of business, or served in any manner in which legal process may be served in the state in which the aforesaid project is located, save that such service need not be made by a public officer.

b) After the expiration of one (1) year following the date on which Principal ceased Work on said Contract or after the expiration of one (1) year following the Date of Substantial Completion of the Project, whichever is later, it being understood, however, that if any limitation embodied in this bond is prohibited by any law controlling the construction hereof such limitation shall be deemed to be amended so as to be equal to the minimum period of limitation permitted by such law.

c) Other than in a state court of competent jurisdiction in and for the county or other political subdivision of the state in which the Project, or any part thereof, is situated, or in the United States District Court for the district in which the Project, or any part thereof, is situated, and not elsewhere.

4. The amount of this bond shall be reduced by and to the extent of any payment or payments made in good faith hereunder, inclusive of the payment by Surety of mechanics' liens which may be filed of record against said improvement, whether or not claim for the amount of such lien be presented under and against this bond.

Signed and sealed this   day of   , 19

| (Witness) | { | (Principal) | (Seal) |
| | | (Title) | |

| (Witness) | { | (Principal) | (Seal) |
| | | (Title) | |

AIA DOCUMENT A311/CM • PERFORMANCE BOND AND LABOR AND MATERIAL PAYMENT BOND • CONSTRUCTION MANAGEMENT EDITION
JUNE 1980 EDITION • AIA® • THE AMERICAN INSTITUTE OF ARCHITECTS, 1735 NEW YORK AVENUE, N.W., WASHINGTON, D.C. 20006   4 of 4

AIA Document B141/CM *Standard Form of Agreement between Owner and Architect* **(CM edition).**   This agreement form is substantially the same as the traditional owner-architect agreement, AIA Document B141, except for adjustments necessary to properly coordinate with the owner–construction manager agreement (B801) and the construction phase responsibilities of both the architect and the construction manager as set forth in the general conditions (A201/CM). B141/CM is the agreement which should govern the relationships between the owner and the architect when a construction manager has been or will be engaged by the owner with the responsibilities set forth in the AIA Document B801.

Often, an owner will have already engaged an architect, using AIA Document B141, before determining that a construction manager should be engaged or the construction management approach to construction utilized. In such a case, the owner and architect should execute an amendment to their original agreement substituting a document such as B141/CM for the original agreement. Failure to do so may create conflicts between the roles and authority of the construction manager and the architect.

INSTRUCTION SHEET *AIA DOCUMENT B141/CMa*

FOR AIA DOCUMENT B141/CM, STANDARD FORM OF AGREEMENT BETWEEN OWNER AND ARCHITECT, CONSTRUCTION MANAGEMENT EDITION—JUNE 1980 EDITION

A. GENERAL INFORMATION

1. Purpose

This Document is intended as the basis for the agreement between the Owner and the Architect for architectural services when a Construction Manager is used on the Project. It provides a choice of several methods of compensation, and the parties may select the appropriate one.

2. Related Documents

This Document is intended to be used in conjunction with the following AIA Documents:

a) A101/CM, Owner-Contractor Agreement, Construction Management Edition
b) A201/CM, General Conditions of the Contract for Construction, Construction Management Edition
c) B801, Owner-Construction Manager Agreement

3. Use of Non-AIA Forms

If a series of non-AIA documents or a mixture of AIA Documents plus non-AIA documents are to be used, particular care must be taken to achieve consistency of language and intent. Certain owners require the use of owner-contractor agreements and other contract forms prepared by them. Such forms should be carefully compared to the standard AIA forms for which they are being substituted before executing an agreement. If there are any significant omissions, additions or variances from the terms of the related standard AIA forms, both legal and insurance counsel should be consulted. Of particular concern is the need for consistency between the Owner-Architect Agreement and the anticipated General Conditions of the Contract for Construction in the delineation of the Architect's Construction Phase services and responsibilities.

4. Letter Forms of Agreement

Letter forms of agreement are generally discouraged by the AIA, as is the performance of a part or the whole of professional services based on oral agreements or understandings. The standard AIA Agreement forms have been developed through more than sixty years of experience and have been tested repeatedly in the courts. In addition, the standard forms have been carefully coordinated with other AIA Documents, including the various Architect-Consultant Agreement forms, the Owner-Contractor Agreements and the General Conditions of the Contract for Construction. The necessity for specific and complete correlation between these documents and any Owner-Architect Agreement used is of paramount importance.

5. Use of Current Documents

Prior to using any AIA Document, the user should consult the AIA or an AIA component to determine the current edition of each Document.

6. Reproduction

AIA Document B141/CM is a copyrighted document, and may not be reproduced or excerpted from in substantial part without the express written permission of the AIA. Purchasers of B141/CM are hereby entitled to reproduce a maximum of ten copies of the completed or executed document for use only in connection with the particular Project. AIA will not permit the reproduction of this Document in blank, or the use of substantial portions of, or language from, this Document, except upon written request and after receipt of written permission from AIA.

B. CHANGES FROM THE PREVIOUS EDITION

1. Format Changes

The compensation provisions have been moved to a position near the end of the Document rather than preceding the Terms and Conditions. Also, the language dealing with alternative methods of computing compensation has been deleted from the Agreement form and is now provided as part of this Instruction Sheet for incorporation by the user.

2. Changes in Content

Numerous changes in content have been made in response to experience with the previous edition, the recognition of concerns of Owners, and the recommendations of AIA members, committees, legal and insurance counsel. The following are some of the significant changes made to the content of the 1975 Edition of B141/CM:

a) **Architect's Services:** Under the Schematic Design, Design Development and Construction Documents Phases, the requirement for the Architect to submit a Statement of Probable Construction Cost has been eliminated. The Architect now has the responsibility to provide the Construction Manager with drawings and other documents for the purpose of preparing an estimate of Construction Cost.

b) **Construction Administration:** Subparagraphs 1.5.7 and 1.5.8 have been modified utilizing the term "Project Certificate for Payment" and require the Construction Manager's recommendation prior to the Architect's acting on Applications for Payment.

Subparagraph 1.5.12, dealing with the Architect's authority to require testing and reject nonconforming Work, requires action be taken only after consultation with the Construction Manager.

Subparagraph 1.5.14, pertaining to Change Orders, requires the Architect to review and sign or take other appropriate action on Change Orders *prepared by the Construction Manager.*

Subparagraph 1.7.6 has been added making the providing of services in connection with alternative designs for cost estimating or bidding purposes additional services.

c) **The Owner's Responsibilities:** Paragraph 2.4 has been added stating that the Owner shall retain a Construction Manager and outlining the Construction Manager's duties and responsibilities.

d) **Construction Cost:** Subparagraph 3.1.2 outlines what the construction cost will include and specifically mentions the Construction Manager's compensation and costs.

Subparagraph 3.2.4 outlines the procedure to be followed in the event the Project budget or fixed limit of construction cost is exceeded by the sum of the lowest bona fide bids or negotiated proposals. Note that it is the Construction Manager's responsibility to prepare a detailed cost estimate (provided for in B801); and the Architect is required to modify the drawings and specifications as necessary to comply with the fixed limit, without additional cost to the Owner, *only* if the Architect has concurred in the Construction Manager's estimate of Construction Cost.

e) **Reimbursable Expense:** This edition provides for reimbursement for reproductions, postage and handling of all drawings, specifications and other documents except those for the office use of the Architect and the Architect's consultants.

Expense of any additional insurance coverage or limits requested by the Owner in excess of that normally carried by the Architect or the Architect's consultants will be reimbursable.

f) **Payments to the Architect:** The percentage of the total fee to be paid for each phase of the Work is removed from this Article and is included in Article 15.

Compensation for administration of the Construction Contracts beyond the period initially established shall be as set forth in Paragraph 14.4.

g) **Ownership and Use of Documents:** Revised to conform with the 1977 Edition of AIA Document B141.

C. COMPLETING THE B141/CM FORM

1. Modifications

As with all AIA Documents, users are encouraged to consult an attorney with respect to completing the form. Generally, any necessary modifications can be accomplished by describing in Article 15 changes to the printed text. Legal counsel should also be sought concerning the effect of state and local law on the terms of the Agreement, particularly with respect to registration laws, duties imposed by building codes, interest charges and arbitration.

2. Cover Page

Date: The date represents the date as of which the Agreement becomes effective. It may be the date that an oral agreement was reached, the date the Agreement was originally submitted to the Owner, the date authorizing action was taken or the date of actual execution. Professional services should not be performed prior to the effective date of the Agreement.

Identification of Parties: Parties to this Agreement should be identified using the full legal name under which the Agreement is to be executed, including a designation of the legal status of both parties (sole proprietorship, partnership, joint venture, unincorporated association, limited partnership or corporation [general, close or professional], etc.).

Project Description: The proposed Project should be described in sufficient detail to identify (1) the official name or title of the facility, (2) the location of the site, if known, (3) the proposed building type or usage, and (4) the size, capacity or scope of the Project, if known.

AIA DOCUMENT B141/CMa • INSTRUCTION SHEET FOR OWNER-ARCHITECT AGREEMENT CONSTRUCTION MANAGEMENT EDITION • 1980 EDITION • AIA® • THE AMERICAN INSTITUTE OF ARCHITECTS, 1735 NEW YORK AVENUE, N.W., WASHINGTON, D.C. 20006

3. Article 14—Basis of Compensation

Paragraph 14.1

Insert the amount of the retainer, if any, and indicate whether it will be credited to the first, last, or proportionately to all of the payments on the Owner's account.

Paragraph 14.2, Subparagraph 14.2.1

Compensation—Multiple of Direct Personnel Expense

"Compensation for services rendered by Principals and employees shall be based on a Multiple of Direct Personnel Expense in the same manner as described in Subparagraph 14.4.1, and for the services of professional consultants as described in Subparagraph 14.4.2."

The Architect may wish to include a method of computing Direct Personnel Expense, such as by using a percentage of employee salaries.

Compensation—Professional Fee Plus Expenses

"Compensation shall be based on a Professional Fee of ——————————————— dollars ($) plus compensation for services rendered by Principals and employees in the same manner as described in Subparagraph 14.4.1, and for the services of professional consultants as described in Subparagraph 14.4.2."

Compensation—Stipulated Sum

"Compensation shall be a Stipulated Sum of ————————————— dollars ($)."

Compensation—Percentage of Construction Cost

"Compensation shall be based on one of the following Percentages of Construction Cost, as defined in Article 3:

For portions of the Project to be awarded under:

A single, stipulated sum construction contract:
——————————————————————————————percent (%)

Separate, stipulated sum construction contracts:
——————————————————————————————percent (%)

A single, cost-plus construction contract:
——————————————————————————————percent (%)

Separate, cost-plus construction contracts:
——————————————————————————————percent (%)."

Paragraph 14.2, Sugparagraph 14.2.2

If applicable, insert the percentages of compensation payable for each separate phase of services. Percentages contained in previous Owner-Architect Agreements would be expressed as follows:

Schematic Design Phase:	fifteen percent (15%)
Design Development Phase:	twenty percent (20%)
Construction Documents Phase:	forty percent (40%)
Bidding or Negotiation Phase:	five percent (5%)
Construction Phase:	twenty percent (20%)
Total:	one hundred percent (100%)

Because phases may overlap in time, these percentages have been expressed separately for each phase, rather than cumulatively. This facilitates billing when services are being provided during more than one phase at a time.

Paragraph 14.3

Attach AIA Document B352, Duties, Responsibilities and Limitations of Authority of the Architect's Project Representative, and the agreed compensation arrangement for such services. If this is to be determined at a later date, such as at the time of commencement of the Construction Phase, so indicate.

If a cost of living escalation is to be included, the amount or percentage increase should be stated if it can be determined in advance.

Paragraph 14.4, Subparagraph 14.4.1

If billing rates are used and employees are classified in accordance with the AIA publication *Compensation Guidelines for Architectural and Engineering Services*, insert:

"1. Principals' time at the fixed rate of ——————————— dollars ($) per hour. For the purposes of this Article, the Principals are:

2. Supervisory time at the fixed rate of ——————————— dollars ($) per hour. For the purposes of this Article, Supervisory personnel include those in the following positions:

3. Technical Level I time at the fixed rate of _____ dollars ($) per hour. For the purposes of this Article, Technical Level I personnel include those in the following positions:

4. Technical Level II time at the fixed rate of _____ dollars ($) per hour. For the purposes of this Article, Technical Level II personnel include those in the following positions:

5. Technical Level III time at the fixed rate of _____ dollars ($) per hour. For the purposes of this Article, Technical Level III personnel include those in the following positions:

These billing rates shall be adjusted annually (semi-annually), in accordance with the Architect's adjustments in compensation for Principals and employees." (Add agreed upon limitations.)

NOTE: The rates above will normally be the total compensation. For a more detailed explanation, refer to *Compensation Guidelines.*

If a Multiple of Direct Personnel Expense is used, insert:

"Principals' and employees' time at a multiple of _____ (   ) times their Direct Personnel Expense as defined in Article 4."

If a multiple of direct salaries is used, the term "direct salaries" should be substituted for Direct Personnel Expense above, and this should be noted in Article 15.

Paragraph 14.4, Subparagraph 14.4.2

Insert the multiplier applied to consultant billings used to cover the costs of administration, responsibility for consultants' work, coordination and profit.

Paragraph 14.5

Insert the multiplier, if any, applied to reimbursable expenses used to cover the costs of administration.

Paragraph 14.6

Establish a due date for payment and insert the percentage rate and basis (monthly, annual) and the time (such as a number of days) after the due date for payment on which interest charges will begin to run. This should be carefully checked against state usury laws which may set a limit on the rate of interest which may legally be charged. In addition, federal truth in lending and similar state and local consumer protection laws may require setting forth the annual percentage rate and other disclosures or waivers for certain types of transactions or with certain types of clients. Advice of legal counsel should be sought on such matters.

Paragraph 14.7, Subparagraph 14.7.2

Insert the amount of time after which the compensation shall be subject to renegotiation or adjustment. If the firm requires periodic adjustments in hourly rates and multiples, this should be stated, along with any limitations on the amount of upward adjustment which may be made.

4. Article 15—Other Conditions or Services

Here insert the following types of provisions:

Additional phases, such as Predesign, Site Analysis or Postconstruction, and the services provided in each
Identification of Additional Services, if any, provided under the Basic Compensation
Other Additional Services, and any special compensation arrangements for them
Description of consultants, if any, provided under the Basic Compensation
Procedure for award of Construction Contracts (i.e., bidding or negotiation)
Fixed Limit of Construction Cost
Fixed Time of Performance
Modifications to any services or conditions
Other additional conditions

Note that any changes in the duties of the Architect during the Construction Phase must be considered with extreme care and correlated with the terms of A201/CM, General Conditions of the Contract for Construction, Construction Management Edition.

D. EXECUTION OF THE AGREEMENT

Each person executing the Agreement should indicate the capacity in which they are acting (i.e., president, secretary, partner, etc.) and the authority under which they are executing the Agreement. Where appropriate, a copy of the resolution authorizing the individual to act on behalf of the firm or entity should be attached.

AIA DOCUMENT B141/CMa • INSTRUCTION SHEET FOR OWNER-ARCHITECT AGREEMENT
CONSTRUCTION MANAGEMENT EDITION • 1980 EDITION • AIA® • THE AMERICAN
INSTITUTE OF ARCHITECTS, 1735 NEW YORK AVENUE, N.W., WASHINGTON, D.C. 20006

THE AMERICAN INSTITUTE OF ARCHITECTS

AIA Document B141/CM

CONSTRUCTION MANAGEMENT EDITION

Standard Form of Agreement Between Owner and Architect

1980 EDITION

THIS DOCUMENT HAS IMPORTANT LEGAL CONSEQUENCES; CONSULTATION WITH AN ATTORNEY IS ENCOURAGED.

This document is intended to be used in conjunction with
AIA Documents B801, 1980; A101/CM, 1980; and A201/CM, 1980.

AGREEMENT

made as of the   day of   in the year of Nineteen
Hundred and

BETWEEN the Owner:

and the Architect:

For the following Project:
(Include detailed description of Project location and scope.)

the Construction Manager:

The Owner and the Architect agree as set forth below.

AIA DOCUMENT B141/CM • OWNER-ARCHITECT AGREEMENT • CONSTRUCTION MANAGEMENT EDITION • JUNE 1980 EDITION
AIA® • ©1980 • THE AMERICAN INSTITUTE OF ARCHITECTS, 1735 NEW YORK AVENUE, N.W., WASHINGTON, D.C. 20006   **B141/CM—1980 1**

| TERMS AND CONDITIONS OF AGREEMENT BETWEEN OWNER AND ARCHITECT |

ARTICLE 1
ARCHITECT'S SERVICES AND RESPONSIBILITIES

BASIC SERVICES

Unless modified by Article 15, the Architect's Basic Services shall be provided in conjunction with, and in reliance upon, the services of a Construction Manager as described in the Standard Form of Agreement Between Owner and Construction Manager, AIA Document B801, 1980 Edition. They shall consist of the five Phases described in Paragraphs 1.1 through 1.5, inclusive, and include normal structural, mechanical and electrical engineering services, and any other services included in Article 15 as part of Basic Services.

1.1 SCHEMATIC DESIGN PHASE

1.1.1 The Architect shall review the program furnished by the Owner to ascertain the requirements of the Project and shall review and confirm the understanding of these requirements and other design parameters with the Owner.

1.1.2 The Architect shall provide a preliminary evaluation of the program and the Project budget requirements, each in terms of the other, subject to the limitations set forth in Subparagraph 3.2.1.

1.1.3 The Architect shall review with the Owner and the Construction Manager site use and improvements; selection of materials, building systems and equipment; construction methods and methods of Project delivery.

1.1.4 Based on the mutually agreed upon program and the Project budget requirements, the Architect shall prepare, for approval by the Owner, Schematic Design Documents consisting of drawings, outline specifications and other documents illustrating the scale and relationship of Project components.

1.1.5 At intervals appropriate to the progress of the Schematic Design Phase, the Architect shall provide schematic design studies for the Construction Manager's review, which will be made so as to cause no delay to the Architect.

1.1.6 Upon completion of the Schematic Design Phase the Architect shall provide the drawings, outline specifications and other documents approved by the Owner for the Construction Manager's use in preparing an estimate of Construction Cost.

1.2 DESIGN DEVELOPMENT PHASE

1.2.1 Based on the approved Schematic Design Documents and any adjustments authorized by the Owner in the program or the Project budget, the Architect shall prepare, for approval by the Owner, the Design Development Documents consisting of drawings, outline specifications and other documents to fix and describe the size and character of the entire Project as to architectural, structural, mechanical and electrical systems, materials, and such other elements as may be appropriate.

1.2.2 At intervals appropriate to the progress of the Design Development Phase, the Architect shall provide de-

sign development documents for the Construction Manager's review, which will be made so as to cause no delay to the Architect.

1.2.3 Upon completion of the Design Development Phase, the Architect shall provide the Construction Manager with drawings, outline specifications and other documents approved by the Owner for use in preparing a further estimate of Construction Cost, and shall assist the Construction Manager in preparing such estimate of Construction Cost.

1.3 CONSTRUCTION DOCUMENTS PHASE

1.3.1 Based on the approved Design Development Documents, and any further adjustments in the scope or quality of the Project or in the Project budget authorized by the Owner, the Architect shall prepare, for approval by the Owner, Construction Documents consisting of Drawings and Specifications setting forth in detail the requirements for the construction of the Project.

1.3.2 The Architect shall keep the Construction Manager informed of any changes in requirements or in construction materials, systems or equipment as the Drawings and Specifications are developed so that the Construction Manager can adjust the estimate of Construction Cost appropriately.

1.3.3 The Architect shall assist the Owner and the Construction Manager in the preparation of the necessary bidding information, bidding forms, the Conditions of the Contracts, and the forms of Agreement between the Owner and the Contractors.

1.3.4 The Architect shall assist the Owner and the Construction Manager in connection with the Owner's responsibility for filing documents required for the approvals of governmental authorities having jurisdiction over the Project.

1.4 BIDDING OR NEGOTIATION PHASE

1.4.1 The Architect, following the Owner's approval of the Construction Documents and the latest estimate of Construction Cost, shall assist the Construction Manager in obtaining Bids or negotiated proposals by rendering interpretations and clarifications of the Drawings and Specifications in appropriate written form. The Architect shall assist the Construction Manager in conducting pre-award conferences with successful Bidders.

1.5 CONSTRUCTION PHASE-ADMINISTRATION OF THE CONSTRUCTION CONTRACT

1.5.1 The Construction Phase will commence with the award of the initial Contract for Construction and, together with the Architect's obligation to provide Basic Services under this Agreement, will end when final payment to all Contractors is due, or in the absence of a final Project Certificate for Payment or of such due date, sixty days after the Date of Substantial Completion of the Project whichever occurs first.

1.5.2 Unless otherwise provided in this Agreement and incorporated in the Contract Documents, the Architect, in cooperation with the Construction Manager, shall provide administration of the Contracts for Construction as set forth below and in the 1980 Edition of AIA Document A201/CM, General Conditions of the Contract for Construction, Construction Management Edition.

1.5.3 The Architect and the Construction Manager shall advise and consult with the Owner during the Construction Phase. All instructions to the Contractors shall be forwarded through the Construction Manager. The Architect and the Construction Manager shall have authority to act on behalf of the Owner only to the extent provided in the Contract Documents unless otherwise modified by written instrument in accordance with Subparagraph 1.5.18.

1.5.4 The Architect shall visit the site at intervals appropriate to the stage of construction, or as otherwise agreed by the Architect in writing, to become generally familiar with the progress and quality of Work and to determine in general if Work is proceeding in accordance with the Contract Documents. However, the Architect shall not be required to make exhaustive or continuous on-site inspections to check the quality or quantity of Work. On the basis of such on-site observations as an architect, the Architect shall keep the Owner informed of the progress and quality of Work, and shall endeavor to guard the Owner against defects and deficiencies in Work of the Contractors.

1.5.5 The Architect shall not be responsible for, nor have control or charge of, construction means, methods, techniques, sequences or procedures, or for safety precautions and programs in connection with the Project, and shall not be responsible for Contractors' failure to carry out Work in accordance with the Contract Documents. The Architect shall not be responsible for, nor have control over, the acts or omissions of the Contractors, Subcontractors, any of their agents or employees, or any other persons performing any Work, nor shall the Architect be responsible for the Construction Manager's obligations as an agent of the Owner.

1.5.6 The Architect shall at all times have access to Work wherever it is in preparation or progress.

1.5.7 Based on the Architect's observations at the site, the recommendations of the Construction Manager and an evaluation of the Project Application for Payment, the Architect shall determine the amounts owing to the Contractors and shall issue a Project Certificate for Payment in such amounts, as provided in the Contract Documents.

1.5.8 The issuance of a Project Certificate for Payment shall constitute a representation by the Architect to the Owner that, based on the Architect's observations at the site as provided in Subparagraph 1.5.4 and on the data comprising the Project Application for Payment, Work has progressed to the point indicated; that, to the best of the Architect's knowledge, information and belief, the quality of Work is in accordance with the Contract Documents (subject to an evaluation of Work for conformance with the Contract Documents upon Substantial Completion, to the results of any subsequent tests required by or performed under the Contract Documents, to minor deviations from the Contract Documents correctable prior to completion, and to any specific qualifications stated in the Project Certificate for Payment); and that the Contractors are entitled to payment in the amount certified. However, the issuance of a Project Certificate for Payment shall not be a representation that the Architect has made any examination to ascertain how or for what purpose the Contractors have used the monies paid on account of the Contract Sums.

1.5.9 The Architect shall be the interpreter of the requirements of the Contract Documents and the judge of the performance thereunder by both the Owner and the Contractors. The Architect shall render interpretations necessary for the proper execution or progress of Work, with reasonable promptness and in accordance with agreed upon time limits. The Architect shall render written decisions, within a reasonable time, on all claims, disputes and other matters in question between the Owner and the Contractors relating to the execution or progress of Work or the interpretation of the Contract Documents.

1.5.10 All interpretations and decisions of the Architect shall be consistent with the intent of, and reasonably inferable from, the Contract Documents, and shall be in writing or in graphic form. In the capacity of interpreter and judge, the Architect shall endeavor to secure faithful performance by both the Owner and the Contractors, shall not show partiality, and shall not be liable for the result of any interpretation or decision rendered in good faith in such capacity.

1.5.11 The Architect's decision in matters relating to artistic effect shall be final if consistent with the intent of the Contract Documents. The Architect's decisions on any other claims, disputes or other matters, including those in question between the Owner and the Contractor(s), shall be subject to arbitration as provided in this Agreement and in the Contract Documents.

1.5.12 The Architect shall have authority to reject Work which does not conform to the Contract Documents, and whenever, in the Architect's reasonable opinion, it is necessary or advisable for the implementation of the intent of the Contract Documents, the Architect shall have authority to require special inspection or testing of Work in accordance with the provisions of the Contract Documents, whether or not such Work be then fabricated, installed or completed; but the Architect shall take such action only after consultation with the Construction Manager.

1.5.13 The Architect shall receive Contractors' submittals such as Shop Drawings, Product Data and Samples from the Construction Manager and shall review and approve or take other appropriate action upon them, but only for conformance with the design concept of the Project and with the information given in the Contract Documents. Such action shall be taken with reasonable promptness so as to cause no delay. The Architect's approval of a specific item shall not indicate approval of an assembly of which the item is a component.

1.5.14 The Architect shall review and sign or take other appropriate action on Change Orders prepared by the Construction Manager for the Owner's authorization in accordance with the Contract Documents.

1.5.15 The Architect shall have authority to order minor changes in Work not involving an adjustment in a Contract Sum or an extension of a Contract Time and which are not inconsistent with the intent of the Contract Documents. Such changes shall be effected by written order issued through the Construction Manager.

AIA DOCUMENT B141/CM • OWNER-ARCHITECT AGREEMENT • CONSTRUCTION MANAGEMENT EDITION • JUNE 1980 EDITION
3 B141/CM—1980 AIA® • ©1980 • THE AMERICAN INSTITUTE OF ARCHITECTS, 1735 NEW YORK AVENUE, N.W., WASHINGTON, D.C. 20006

1.5.16 The Architect, assisted by the Construction Manager, shall conduct inspections to determine the Dates of Substantial Completion and final completion and shall issue appropriate Project Certificates for Payment.

1.5.17 The Architect shall assist the Construction Manager in receiving and forwarding to the Owner for the Owner's review written warranties and related documents assembled by the Contractors.

1.5.18 The extent of the duties, responsibilities and limitations of authority of the Architect as a representative of the Owner during construction shall not be modified or extended without the written consent of the Owner, the Contractors, the Architect and the Construction Manager, which consent shall not be unreasonably withheld.

1.6 PROJECT REPRESENTATION BEYOND BASIC SERVICES

1.6.1 If the Owner and the Architect agree that more extensive representation at the site than is described in Paragraph 1.5 shall be provided, the Architect shall provide one or more Project Representatives to assist the Architect in carrying out such responsibilities at the site.

1.6.2 Such Project Representatives shall be selected, employed and directed by the Architect, and the Architect shall be compensated therefor as mutually agreed between the Owner and the Architect, as set forth in an exhibit appended to this Agreement, which shall describe the duties, responsibilities and limitations of authority of such Project Representatives.

1.6.3 Through the observations of such Project Representatives, the Architect shall endeavor to provide further protection for the Owner against defects and deficiencies in Work, but the furnishing of such Project representation shall not modify the rights, responsibilities or obligations of the Architect as described in Paragraph 1.5.

1.7 ADDITIONAL SERVICES
The following services are not included in Basic Services unless so identified in Article 15. They shall be provided if authorized or confirmed in writing by the Owner, and they shall be paid for by the Owner as provided in this Agreement, in addition to the compensation for Basic Services.

1.7.1 Providing analyses of the Owner's needs, and programming the requirements of the Project.

1.7.2 Providing financial feasibility or other special studies.

1.7.3 Providing planning surveys, site evaluations, environmental studies or comparative studies of prospective sites, and preparing special surveys, studies and submissions required for approvals of governmental authorities or others having jurisdiction over the Project.

1.7.4 Providing services relative to future facilities, systems and equipment which are not intended to be constructed during the Construction Phase.

1.7.5 Providing services to investigate existing conditions or facilities, or to make measured drawings thereof, or to verify the accuracy of drawings or other information furnished by the Owner.

1.7.6 Providing services in connection with alternative designs for cost estimating or bidding purposes.

1.7.7 Providing coordination of work performed by separate contractors or by the Owner's own forces.

1.7.8 Providing services in connection with the work of separate consultants, other than the Construction Manager, retained by the Owner.

1.7.9 Providing interior design and other similar services required for or in connection with the selection, procurement or installation of furniture, furnishings and related equipment.

1.7.10 Providing services for planning tenant or rental spaces.

1.7.11 Making revisions in Drawings, Specifications or other documents when such revisions are inconsistent with written approvals or instructions previously given, are required by the enactment or revision of codes, laws or regulations subsequent to the preparation of such documents, or are due to other causes not solely within the control of the Architect.

1.7.12 Preparing Drawings, Specifications and supporting data and providing other services in connection with Change Orders. If Basic Compensation is to be adjusted according to adjustments in Construction Cost, to the extent that any Change Order not required by causes solely within the control of the Architect results in an adjustment in the Basic Compensation not commensurate with the services required of the Architect, compensation shall be equitably adjusted.

1.7.13 Making investigations, surveys, valuations, inventories, detailed appraisals of existing facilities, and services required in connection with construction performed by the Owner.

1.7.14 Providing consultation concerning replacement of any Work damaged by fire or other cause during construction, and furnishing services as may be required in connection with the replacement of such Work.

1.7.15 Providing services made necessary by the failure of performance, the termination or default of the Construction Manager; by default of a Contractor; by major defects or deficiencies in the Work of any Contractor; or by failure of performance of either the Owner or any Contractor under the Contracts for Construction.

1.7.16 Preparing a set of reproducible record drawings showing significant changes in Work made during construction based on marked-up prints, drawings and other data furnished to the Architect.

1.7.17 Providing extensive assistance in the utilization of any equipment or system such as initial start-up or testing, adjusting and balancing, preparation of operation and maintenance manuals, training personnel for operation and maintenance, and consultation during operation.

1.7.18 Providing services after issuance to the Owner of the final Project Certificate for Payment, or in the absence of a final Project Certificate for Payment, more than sixty days after the Date of Substantial Completion of the Project.

1.7.19 Preparing to serve or serving as a witness in connection with any public hearing, arbitration proceeding or legal proceeding.

1.7.20 Providing services of consultants for other than the normal architectural, structural, mechanical and electrical engineering services for the Project.

1.7.21 Providing any other services not otherwise included in this Agreement or not customarily furnished in accordance with generally accepted architectural practice.

AIA DOCUMENT B141/CM • OWNER-ARCHITECT AGREEMENT • CONSTRUCTION MANAGEMENT EDITION • JUNE 1980 EDITION
AIA® • ©1980 • THE AMERICAN INSTITUTE OF ARCHITECTS, 1735 NEW YORK AVENUE, N.W., WASHINGTON, D.C. 20006 **B141/CM—1980 4**

1.8 TIME

1.8.1 The Architect shall perform Basic and Additional Services as expeditiously as is consistent with professional skill and care and the orderly progress of the Project. Upon request of the Owner, the Architect shall submit for the Owner's approval a schedule for the performance of the Architect's services which shall be adjusted as required as the Project proceeds, and which shall include allowances for periods of time required for the Owner's review and approval of submissions and for approvals of authorities having jurisdiction over the Project. The Architect shall consult with the Construction Manager to coordinate the Architect's time schedule with the Project Schedule. This schedule, when approved by the Owner, shall not, except for reasonable cause, be exceeded by the Architect.

ARTICLE 2
THE OWNER'S RESPONSIBILITIES

2.1 The Owner shall provide full information regarding requirements for the Project, including a program which shall set forth the Owner's design objectives, constraints and criteria, including space requirements and relationships, flexibility and expandability, special equipment and systems and site requirements.

2.2 The Owner shall provide a budget for the Project based on consultation with the Architect and the Construction Manager, which shall include contingencies for bidding, changes during construction and other costs which are the responsibility of the Owner. The Owner shall, at the request of the Architect, provide a statement of funds available for the Project and their source.

2.3 The Owner shall designate a representative authorized to act in the Owner's behalf with respect to the Project. The Owner, or such authorized representative, shall examine the documents submitted by the Architect and shall render decisions pertaining thereto promptly to avoid unreasonable delay in the progress of the Architect's services.

2.4 The Owner shall retain a construction manager to manage the Project. The Construction Manager's services, duties and responsibilities will be as described in the Agreement Between Owner and Construction Manager, AIA Document B801, 1980 Edition. The Terms and Conditions of the Owner-Construction Manager Agreement will be furnished to the Architect and will not be modified without written consent of the Architect, which consent shall not be unreasonably withheld. Actions taken by the Construction Manager as agent of the Owner shall be the acts of the Owner, and the Architect shall not be responsible for them.

2.5 The Owner shall furnish a legal description and a certified land survey of the site, giving, as applicable, grades and lines of streets, alleys, pavements and adjoining property; rights-of-way, restrictions, easements, encroachments, zoning, deed restrictions, boundaries and contours of the site; locations, dimensions and complete data pertaining to existing buildings, other improvements and trees; and full information concerning available service and utility lines both public and private, above and below grade, including inverts and depths.

2.6 The Owner shall furnish the services of soil engineers or other consultants when such services are deemed necessary by the Architect. Such services shall include test borings, test pits, soil bearing values, percolation tests, air and water pollution tests, ground corrosion and resistivity tests including necessary operations for determining subsoil, air and water conditions, with reports and appropriate professional recommendations.

2.7 The Owner shall furnish structural, mechanical, chemical and other laboratory tests, inspections and reports as required by law or the Contract Documents.

2.8 The Owner shall furnish such legal, accounting and insurance counseling services as may be necessary for the Project, including such auditing services as the Owner may require to verify the Project Applications for Payment or to ascertain how or for what purposes the Contractors have used the monies paid by or on behalf of the Owner.

2.9 The services, information, surveys and reports required by Paragraphs 2.5 through 2.8, inclusive, shall be furnished at the Owner's expense, and the Architect shall be entitled to rely upon their accuracy and completeness.

2.10 If the Owner observes or otherwise becomes aware of any fault or defect in the Project, or nonconformance with the Contract Documents, prompt written notice thereof shall be given by the Owner to the Architect and the Construction Manager.

2.11 The Owner shall furnish the required information and services and shall render approvals and decisions as expeditiously as necessary for the orderly progress of the Architect's services and Work of the Contractors.

ARTICLE 3
CONSTRUCTION COST

3.1 DEFINITION

3.1.1 The Construction Cost shall be the total cost or estimated cost to the Owner of all elements of the Project designed or specified by the Architect.

3.1.2 The Construction Cost shall also include at current market rates, including a reasonable allowance for overhead and profit, the cost of labor and materials furnished by the Owner and any equipment which has been designed, specified, selected or specially provided for by the Architect. It shall also include the Construction Manager's compensation for services, Reimbursable Costs and the cost of work provided by the Construction Manager.

3.1.3 Construction Cost does not include the compensation of the Architect and the Architect's consultants, the cost of the land, rights-of-way, or other costs which are the responsibility of the Owner as provided in Article 2.

3.2 RESPONSIBILITY FOR CONSTRUCTION COST

3.2.1 The Architect, as a design professional familiar with the construction industry, shall assist the Construction Manager in evaluating the Owner's Project budget and shall review the estimates of Construction Cost prepared by the Construction Manager. It is recognized, however, that neither the Architect, the Construction Manager nor the Owner has control over the cost of labor, materials or equipment, over the Contractors' methods of determining Bid prices, or over competitive bidding, market or negotiating conditions. Accordingly, the Architect cannot and does not warrant or represent that

Bids or negotiated prices will not vary from the Project budget proposed, established or approved by the Owner, if any, or from the estimate of Construction Cost or other cost estimate or evaluation prepared by the Construction Manager.

3.2.2 No fixed limit of Construction Cost shall be established as a condition of this Agreement by the furnishing, proposal or establishment of a Project budget under Subparagraph 1.1.2 or Paragraph 2.2, or otherwise, unless such fixed limit has been agreed upon in writing and signed by the parties to this Agreement. If such a fixed limit has been established, the Construction Manager will include contingencies for design, bidding and price escalation, and will consult with the Architect to determine what materials, equipment, component systems and types of construction are to be included in the Contract Documents, to make reasonable adjustments in the scope of the Project, and to include in the Contract Documents alternate Bids to adjust the Construction Cost to the fixed limit. Any such fixed limit shall be increased in the amount of any increase in the Contract Sums occurring after the execution of the Contracts for Construction.

3.2.3 If Bids are not received within the time scheduled at the time the fixed limit of Construction Cost was established, due to causes beyond the Architect's control, any fixed limit of Construction Cost established as a condition of this Agreement shall be adjusted to reflect any change in the general level of prices in the construction industry between the originally scheduled date and the date on which Bids are received.

3.2.4 If a fixed limit of Construction Cost (adjusted as provided in Subparagraph 3.2.3) is exceeded by the sum of the lowest figures from bona fide Bids or negotiated proposals, plus the Construction Manager's estimate of other elements of Construction Cost for the Project, the Owner shall (1) give written approval of an increase in such limit, (2) authorize rebidding or renegotiation of the Project or portions of the Project within a reasonable time, (3) if the Project is abandoned, terminate in accordance with Paragraph 10. 2, or (4) cooperate in revising the Project scope and quality as required to reduce the Construction Cost. In the case of item (4), the Architect shall modify the Drawings and Specifications as necessary to comply with the fixed limit, without additional cost to the Owner if the Architect has concurred in the Construction Manager's estimate of Construction Cost, but subject to compensation as an Additional Service under Subparagraph 1.7.11 if the Architect has not so concurred. The providing of such service shall be the limit of the Architect's responsibility arising from the establishment of such fixed limit, and having done so, the Architect shall be entitled to compensation for all services performed in accordance with this Agreement, whether or not the Construction Phase is commenced.

ARTICLE 4
DIRECT PERSONNEL EXPENSE

4.1 Direct Personnel Expense is defined as the direct salaries of all the Architect's personnel engaged on the Project, and the portion of the cost of their mandatory and customary contributions and benefits related thereto, such as employment taxes and other statutory employee benefits, insurance, sick leave, holidays, vacations, pensions and similar contributions and benefits.

ARTICLE 5
REIMBURSABLE EXPENSES

5.1 Reimbursable Expenses are in addition to the compensation for Basic and Additional Services and include actual expenditures made by the Architect and the Architect's employees and consultants in the interest of the Project for the expenses listed in the following Subparagraphs:

5.1.1 Expense of transportation in connection with the Project; living expenses in connection with out-of-town travel; long distance communications; and fees paid for securing approvals of authorities having jurisdiction over the Project.

5.1.2 Expense of reproductions, postage and handling of Drawings, Specifications and other documents, excluding reproductions for the office use of the Architect and the Architect's consultants.

5.1.3 Expense of data processing and photographic production techniques when used in connection with Additional Services.

5.1.4 If authorized in advance by the Owner, expense of overtime work requiring higher than regular rates.

5.1.5 Expense of renderings, models and mock-ups requested by the Owner.

5.1.6 Expense of any additional insurance coverage or limits, including professional liability insurance, requested by the Owner in excess of that normally carried by the Architect and the Architect's consultants.

ARTICLE 6
PAYMENTS TO THE ARCHITECT

6.1 **PAYMENTS ON ACCOUNT OF BASIC SERVICES**

6.1.1 An initial payment as set forth in Paragraph 14.1 is the minimum payment under this Agreement.

6.1.2 Subsequent payments for Basic Services shall be made monthly and shall be in proportion to services performed within each Phase of services, on the basis set forth in Article 14.

6.1.3 If and to the extent that the period initially established for the Construction Phase of the Project is exceeded or extended through no fault of the Architect, compensation for Basic Services required for such extended period of Administration of the Construction Contracts shall be computed as set forth in Paragraph 14.4 for Additional Services.

6.1.4 When compensation is based on a percentage of Construction Cost, and any portions of the Project are deleted or otherwise not constructed, compensation for such portions of the Project shall be payable to the extent services are performed on such portions, in accordance

AIA DOCUMENT B141/CM • OWNER-ARCHITECT AGREEMENT • CONSTRUCTION MANAGEMENT EDITION • JUNE 1980 EDITION
AIA® • ©1980 • THE AMERICAN INSTITUTE OF ARCHITECTS, 1735 NEW YORK AVENUE, N.W., WASHINGTON, D.C. 20006 **B141/CM—1980 6**

with the schedule set forth in Subparagraph 14.2.2, based on (1) the lowest figures from bona fide Bids or negotiated proposals, or (2) if no such Bids or proposals are received, the most recent estimate of Construction Cost for such portions of the Project.

6.2 PAYMENTS ON ACCOUNT OF ADDITIONAL SERVICES

6.2.1 Payments on account of the Architect's Additional Services, as defined in Paragraph 1.7, and for Reimbursable Expenses, as defined in Article 5, shall be made monthly upon presentation of the Architect's statement of services rendered or expenses incurred.

6.3 PAYMENTS WITHHELD

6.3.1 No deductions shall be made from the Architect's compensation on account of penalty, liquidated damages or other sums withheld from payments to Contractors, or on account of changes in Construction Cost other than those for which the Architect is held legally liable.

6.4 PROJECT SUSPENSION OR ABANDONMENT

6.4.1 If the Project is suspended or abandoned in whole or in part for more than three months, the Architect shall be compensated for all services performed prior to receipt of written notice from the Owner of such suspension or abandonment, together with Reimbursable Expenses then due and all Termination Expenses as defined in Paragraph 10.4. If the Project is resumed after being suspended for more than three months, the Architect's compensation shall be equitably adjusted.

ARTICLE 7
ARCHITECT'S ACCOUNTING RECORDS

7.1 Records of Reimbursable Expenses and expenses pertaining to Additional Services and services performed on the basis of a Multiple of Direct Personnel Expense shall be kept on the basis of generally accepted accounting principles and shall be available to the Owner or the Owner's authorized representative at mutually convenient times.

ARTICLE 8
OWNERSHIP AND USE OF DOCUMENTS

8.1 Drawings and Specifications as instruments of service are and shall remain the property of the Architect whether the Project for which they are made is executed or not. The Owner shall be permitted to retain copies, including reproducible copies, of Drawings and Specifications for information and reference in connection with the Owner's use and occupancy of the Project. The Drawings and Specifications shall not be used by the Owner on other projects, for additions to this Project, or for completion of this Project by others provided the Architect is not in default under this Agreement, except by agreement in writing and with appropriate compensation to the Architect.

8.2 Submission or distribution to meet official regulatory requirements or for other purposes in connection with the Project is not to be construed as publication in derogation of the Architect's rights.

ARTICLE 9
ARBITRATION

9.1 All claims, disputes and other matters in question between the parties to this Agreement arising out of or relating to this Agreement or the breach thereof, shall be decided by arbitration in accordance with the Construction Industry Arbitration Rules of the American Arbitration Association then obtaining unless the parties mutually agree otherwise. No arbitration arising out of or relating to this Agreement shall include, by consolidation, joinder or in any other manner, any additional person not a party to this Agreement except by written consent containing a specific reference to this Agreement and signed by the Architect, the Owner and any other person sought to be joined. Any consent to arbitration involving an additional person or persons shall not constitute consent to arbitration of any dispute not described therein or with any person not named or described therein. This agreement to arbitrate and any agreement to arbitrate with an additional person or persons duly consented to by the parties to this Agreement shall be specifically enforceable under the prevailing arbitration law.

9.2 Notice of the demand for arbitration shall be filed in writing with the other party to this Agreement and with the American Arbitration Association. The demand shall be made within a reasonable time after the claim, dispute or other matter in question has arisen. In no event shall the demand for arbitration be made after the date when institution of legal or equitable proceedings based on such claim, dispute or other matter in question would be barred by the applicable statute of limitations.

9.3 The award rendered by the arbitrators shall be final, and judgment may be entered upon it in accordance with applicable law in any court having jurisdiction thereof.

ARTICLE 10
TERMINATION OF AGREEMENT

10.1 This Agreement may be terminated by either party upon seven days' written notice should the other party fail substantially to perform in accordance with its terms through no fault of the party initiating the termination.

10.2 This Agreement may be terminated by the Owner upon at least seven days' written notice to the Architect in the event that the Project is permanently abandoned.

10.3 In the event of termination not the fault of the Architect, the Architect shall be compensated for all services performed to the termination date, together with Reimbursable Expenses then due and all Termination Expenses as defined in Paragraph 10.4.

10.4 Termination Expenses include expenses directly attributable to termination for which the Architect is not otherwise compensated, plus an amount computed as a percentage of the total Basic and Additional Compensation earned to the time of termination, as follows:

.1 20 percent if termination occurs during the Schematic Design Phase; or

.2 10 percent if termination occurs during the Design Development Phase; or

.3 5 percent if termination occurs during any subsequent Phase.

ARTICLE 11
MISCELLANEOUS PROVISIONS

11.1 Unless otherwise specified, this Agreement shall be governed by the law of the principal place of business of the Architect.

11.2 Terms in this Agreement shall have the same meaning as those in the 1980 Edition of AIA Document A201/CM, General Conditions of the Contract for Construction, Construction Management Edition.

11.3 As between the parties to this Agreement: as to all acts or failures to act by either party to this Agreement, any applicable statute of limitations shall commence to run and any alleged cause of action shall be deemed to have accrued in any and all events not later than the relevant Date of Substantial Completion of the Project, and as to any acts or failures to act occurring after the relevant Date of Substantial Completion of the Project, not later than the date of issuance of the final Project Certificate for Payment.

11.4 The Owner and the Architect waive all rights against each other, and against the contractors, consultants, agents and employees of the other, for damages covered by any property insurance during construction, as set forth in the 1980 Edition of AIA Document A201/CM, General Conditions of the Contract for Construction, Construction Management Edition. The Owner and the Architect shall each require appropriate similar waivers from their contractors, consultants and agents.

ARTICLE 12
SUCCESSORS AND ASSIGNS

12.1 The Owner and the Architect, respectively, bind themselves, their partners, successors, assigns and legal representatives to the other party to this Agreement, and to the partners, successors, assigns and legal representatives of such other party with respect to all covenants of this Agreement. Neither the Owner nor the Architect shall assign, sublet or transfer any interest in this Agreement without the written consent of the other.

ARTICLE 13
EXTENT OF AGREEMENT

13.1 This Agreement represents the entire and integrated agreement between the Owner and the Architect and supersedes all prior negotiations, representations or agreements, either written or oral. This Agreement may be amended only by written instrument signed by both the Owner and the Architect.

13.2 Nothing contained herein shall be deemed to create any contractual relationship between the Architect and the Construction Manager or any of the Contractors, Subcontractors or material suppliers on the Project; nor shall anything contained in this Agreement be deemed to give any third party any claim or right of action against the Owner or the Architect which does not otherwise exist without regard to this Agreement.

AIA DOCUMENT B141/CM • OWNER-ARCHITECT AGREEMENT • CONSTRUCTION MANAGEMENT EDITION • JUNE 1980 EDITION
AIA® • ©1980 • THE AMERICAN INSTITUTE OF ARCHITECTS, 1735 NEW YORK AVENUE, N.W., WASHINGTON, D.C. 20006 **B141/CM—1980 8**

ARTICLE 14

BASIS OF COMPENSATION

The Owner shall compensate the Architect for the Scope of Services provided, in accordance with Article 6, Payments to the Architect, and the other Terms and Conditions of this Agreement, as follows:

14.1   AN INITIAL PAYMENT of   dollars ($) shall be made upon execution of this Agreement and credited to the Owner's account as follows:

14.2   BASIC COMPENSATION

14.2.1   FOR BASIC SERVICES, as described in Paragraphs 1.1 through 1.5, and any other services included in Article 15 as part of Basic Services, Basic Compensation shall be computed as follows:
(Here insert basis of compensation, including fixed amounts, multiples or percentages, and identify Phases or parts of the Project to which particular methods of compensation apply, if necessary.)

14.2.2   Where compensation is based on a Stipulated Sum or Percentage of Construction Cost, payments for Basic Services shall be made as provided in Subparagraph 6.1.2, so that Basic Compensation for each Phase shall equal the following percentages of the total Basic Compensation payable:

(Include any additional Phases as appropriate.)

Schematic Design Phase:	percent (%)
Design Development Phase:	percent (%)
Construction Documents Phase:	percent (%)
Bidding or Negotiation Phase:	percent (%)
Construction Phase:	percent (%)

14.3   FOR PROJECT REPRESENTATION BEYOND BASIC SERVICES, as described in Paragraph 1.6, compensation shall be computed separately in accordance with Subparagraph 1.6.2.
(Here insert basis of compensation which may be a stipulated sum for a given period of time or a Multiple of Direct Personnel Expense as defined in Article 4. If a Multiple of Direct Personnel Expense is used, the Multiple should be clearly stated.)

AIA DOCUMENT B141/CM • OWNER-ARCHITECT AGREEMENT • CONSTRUCTION MANAGEMENT EDITION • JUNE 1980 EDITION
AIA® • ©1980 • THE AMERICAN INSTITUTE OF ARCHITECTS, 1735 NEW YORK AVENUE, N.W., WASHINGTON, D.C. 20006   **B141/CM—1980 9**

14.4 **COMPENSATION FOR ADDITIONAL SERVICES**

14.4.1 FOR ADDITIONAL SERVICES OF THE ARCHITECT, as described in Paragraph 1.7, and any other services included in Article 15 as part of Additional Services, but excluding Additional Services of consultants, compensation shall be computed as follows:

(Here insert basis of compensation, including rates and/or Multiples of Direct Personnel Expense for Principals and employees, and identify Principals and classify employees, if required. Identify specific services to which particular methods of compensation apply, if necessary.)

14.4.2 FOR ADDITIONAL SERVICES OF CONSULTANTS, including additional structural, mechanical and electrical engineering services and those provided under Subparagraph 1.7.21 or identified in Article 15 as part of Additional Services, a multiple of ( ) times the amounts billed to the Architect for such services.

(Identify specific types of consultants in Article 15, if required.)

14.5 FOR REIMBURSABLE EXPENSES, as described in Article 5, and any other items included in Article 15 as Reimbursable Expenses, a multiple of ( ) times the amounts expended by the Architect, the Architect's employees and consultants in the interest of the Project.

14.6 Payments due the Architect and unpaid under this Agreement shall bear interest from the date payment is due at the rate entered below, or in the absence thereof, at the legal rate prevailing at the principal place of business of the Architect.

(Here insert any rate of interest agreed upon.)

(Usury laws and requirements under the Federal Truth in Lending Act, similar state and local consumer credit laws and other regulations at the Owner's and Architect's principal places of business, the location of the Project and elsewhere may affect the validity of this provision. Specific legal advice should be obtained with respect to deletion, modification or other requirements such as written disclosures or waivers.)

14.7 The Owner and the Architect agree in accordance with the Terms and Conditions of this Agreement that:

14.7.1 IF THE SCOPE of the Project or the Architect's services is changed materially, the amounts of compensation shall be equitably adjusted.

14.7.2 IF THE SERVICES covered by this Agreement have not been completed within ( ) months of the date hereof, through no fault of the Architect, the amounts of compensation, rates and multiples set forth herein shall be equitably adjusted.

AIA DOCUMENT B141/CM • OWNER-ARCHITECT AGREEMENT • CONSTRUCTION MANAGEMENT EDITION • JUNE 1980 EDITION
AIA® • ©1980 • THE AMERICAN INSTITUTE OF ARCHITECTS, 1735 NEW YORK AVENUE, N.W., WASHINGTON, D.C. 20006 **B141/CM—1980 10**

ARTICLE 15
OTHER CONDITIONS OR SERVICES

SAMPLE

This Agreement entered into as of the day and year first written above.

OWNER

_____

_____

_____

_____

ARCHITECT

_____

_____

_____

_____

AIA DOCUMENT B141/CM • OWNER-ARCHITECT AGREEMENT • CONSTRUCTION MANAGEMENT EDITION • JUNE 1980 EDITION
AIA® • ©1980 • THE AMERICAN INSTITUTE OF ARCHITECTS, 1735 NEW YORK AVENUE, N.W., WASHINGTON, D.C. 20006 **B141/CM—1980 11**

AIA Document G701/CM, *Change Order* (CM edition). This edition of the change order form is designed to reflect the change order approval process set forth in the construction management edition of the general conditions (A201/CM), in which the recommendation of the construction manager and the approval of the architect are required before a change order is submitted to the owner for authorization.

AIA Document G722, *Project Application and Project Certificate for Payment.* This form is designed to supplement AIA Document G702, the application and certificate for payment which will be used by individual contractors. The construction manager will prepare an overall project application for payment on Document G711, based on all the individual applications for payment collected by the construction manager. The architect then certifies that the entire amount is due and owing to the contractors on the project, as provided for in the general conditions (A201/CM).

AIA Document G723, *Project Application Summary.* This form supplements AIA Document G722 by breaking out the total disbursement to be made by the owner into the amounts due each individual contractor and noting changes in the contract sums, the value of stored materials, retainage, etc.

CHANGE ORDER
CONSTRUCTION MANAGEMENT EDITION
AIA DOCUMENT G701/CM

Distribution to:
OWNER ☐
ARCHITECT ☐
CONSTRUCTION MANAGER ☐
CONTRACTOR ☐
FIELD ☐
OTHER ☐

PROJECT:
(name, address)

TO (Contractor):

CHANGE ORDER NUMBER:

INITIATION DATE:

ARCHITECT'S PROJECT NO:

CONSTRUCTION MANAGER'S
PROJECT NO:

CONTRACT FOR:

CONTRACT DATE:

You are directed to make the following changes in this Contract:

SAMPLE

Not valid until signed by the Owner, the Architect and the Construction Manager.
Signature of the Contractor indicates agreement herewith, including any adjustment in the Contract Sum or the Contract Time.

The original (Contract Sum) (Guaranteed Maximum Cost) was . . . . $.
Net change by previously authorized Change Orders . . . . $.
The (Contract Sum) (Guaranteed Maximum Cost) prior to this Change Order was . . . . $.
The (Contract Sum) (Guaranteed Maximum Cost) will be (increased) (decreased) (unchanged)
     by this Change Order . . . . $.
The new (Contract Sum) (Guaranteed Maximum Cost) including this Change Order will be . . . $.
The Contract Time will be (increased) (decreased) (unchanged) by   (   ) Days.
The Date of Substantial Completion as of the date of this Change Order therefore is   .

Recommended: Approved:

_____   _____
CONSTRUCTION MANAGER   ARCHITECT

_____   _____
ADDRESS   ADDRESS

_____   _____
BY   DATE   BY   DATE

Agreed To: Authorized:

_____   _____
CONTRACTOR   OWNER

_____   _____
ADDRESS   ADDRESS

_____   _____
BY   DATE   BY   DATE

AIA DOCUMENT G701/CM • CHANGE ORDER • CONSTRUCTION MANAGEMENT EDITION • JUNE 1980 EDITION • AIA®
© 1980 • THE AMERICAN INSTITUTE OF ARCHITECTS, 1735 NEW YORK AVENUE, N.W., WASHINGTON, D.C. 20006   **G701/CM — 1980**

INSTRUCTION SHEET
AIA DOCUMENT G722a

A. GENERAL INFORMATION:

AIA Document G722 is a new document to be used in conjunction with AIA Document G723, Project Application Summary. These documents are designed to be used on Projects where a Construction Manager is employed and where multiple Contractors have separate direct Agreements with the Owner. Procedures for their use are covered in AIA Document A201/CM, General Conditions of the Contract, Construction Management Edition, 1980 Edition.

B. COMPLETING THE G722 FORM:

After the Construction Manager has completed AIA Document G723, Project Application Summary, summary information should be transferred to the G722 form.

The Construction Manager should sign the form, have it notarized and submit it, together with G723 and a separate G702, Application, from each Contractor, to the Architect.

The Architect should review it and, if acceptable, complete the lower Project Certificate for Payment on this form. DO NOT SIGN A CERTIFICATION ON EACH G702 SUBMITTED BY THE CONTRACTORS.

The completed form should be forwarded to the Owner. The Owner will make payment directly to each separate Contractor based on the amount due each as noted in Line J of G723.

C. COMPLETING THE G723 FORM:

Each separate Contractor on the Project should complete and sign AIA Document G702, Application and Certificate for Payment, and forward it to the Construction Manager. The Construction Manager will review each separate Contractor's Application for Payment and, if it is acceptable, complete one vertical column for each separate Contractor.

If the Construction Manager does not agree with the amounts requested by any Contractor, the Construction Manager should note the corrected amount in the appropriate location on G723.

One vertical column should be completed for each application period for all Contractors involved in the Project whether or not any amount is due the particular Contractor for the period in question. Each page should be summarized horizontally and all pages summarized once to provide Project totals.

Project totals should be transferred to AIA Document G722, Project Application and Project Certificate for Payment.

AIA Document G702 from each of the separate Contractors should be attached to the G723 and submitted with G722 to the Architect for review and appropriate action.

G722a — 1980

AIA DOCUMENT G722a • INSTRUCTION SHEET FOR PROJECT APPLICATION AND PROJECT CERTIFICATE FOR PAYMENT
1980 EDITION • AIA® • THE AMERICAN INSTITUTE OF ARCHITECTS, 1735 NEW YORK AVE., N.W., WASHINGTON, D.C. 20006

PROJECT APPLICATION AND PROJECT CERTIFICATE FOR PAYMENT
AIA Document G722 (Instructions on reverse side)

PAGE ONE OF ____ PAGES

TO (Owner):

PROJECT:

APPLICATION NO:

PERIOD FROM:
TO:

Distribution to:
☐ OWNER
☐ ARCHITECT
☐ CONSTRUCTION MANAGER
☐

ATTENTION:

CONSTRUCTION MANAGER:

ARCHITECT'S
PROJECT NO:

Application is made for Payment, as shown below, in connection with the Project.
AIA Document G723, **Project Application Summary**, is attached.

The present status of the account for all Contractors for this Project is as follows:

TOTAL CONTRACT SUMS (Item A Totals) . . . . $

Total Net changes by Change Orders (Item B Totals) . . . $

TOTAL CONTRACT SUM TO DATE (Item C Totals) . . . . . $

TOTAL COMPLETED & STORED TO DATE . . . . $
(Item D Totals)

RETAINAGE (Item H Totals) . . . . $

LESS PREVIOUS CERTIFICATES FOR PAYMENTS (Item I Totals) . . . $

CURRENT PAYMENT DUE (Item J Totals) . . . . $

TOTAL OF AMOUNTS CERTIFIED . . . . $
(Attach explanation if amount certified differs from the amount applied for.)
ARCHITECT:

By: _____ Date: _____

This Certificate is not negotiable. The AMOUNTS CERTIFIED are payable only to the Contractors named in AIA Document G723, attached. Issuance, payment and acceptance of payment are without prejudice to any rights of the Owner or the Contractor under this Contract.

PROJECT APPLICATION FOR PAYMENT

The undersigned Construction Manager certifies that to the best of the Construction Manager's knowledge, information and belief Work covered by this Project Application for Payment has been completed in accordance with the Contract Documents, that all amounts have been paid by the Contractors for Work for which previous Project Certificates for Payment were issued and payments received from the Owner, and that current payment shown herein is now due.

CONSTRUCTION MANAGER

By: _____ Date: _____

State of: _____ County of: _____
Subscribed and sworn to before me this ____ day of ____ 19__
Notary Public:
My Commission expires:

ARCHITECT'S PROJECT CERTIFICATE FOR PAYMENT

In accordance with the Contract Documents, based on on-site observations and the data comprising the above Application, the Architect certifies to the Owner that Work has progressed as indicated; that to the best of the Architect's knowledge, information and belief the quality of the Work is in accordance with the Contract Documents; and that the Contractors are entitled to payment of the AMOUNTS CERTIFIED.

G722—1980

INSTRUCTION SHEET
AIA DOCUMENT G723a

A. GENERAL INFORMATION:

AIA Document G723 is a new document to be used in conjunction with AIA Document G722, Project Application and Project Certificate for Payment. These documents are designed to be used on Projects where a Construction Manager is employed and where multiple Contractors have separate direct Agreements with the Owner. Procedures for their use are covered in AIA Document A201/CM, General Conditions of the Contract, Construction Management Edition, 1980 Edition.

B. COMPLETING THE G723 FORM:

Each separate Contractor on the Project should complete and sign AIA Document G702, Application and Certificate for Payment, and forward it to the Construction Manager. The Construction Manager will review each separate Contractor's Application for Payment and, if it is acceptable, complete one vertical column for each separate Contractor.

If the Construction Manager does not agree with the amounts requested by any Contractor, the Construction Manager should note the corrected amount in the appropriate location on G723.

One vertical column should be completed for each application period for all Contractors involved in the Project whether or not any amount is due the particular Contractor for the period in question. Each page should be summarized horizontally and all pages summarized once to provide Project totals.

Project totals should be transferred to AIA Document G722, Project Application and Project Certificate for Payment.

AIA Document G702 from each of the separate Contractors should be attached to this G723 and submitted with G722 to the Architect for review and appropriate action.

C. COMPLETING THE G722 FORM:

After the Construction Manager has completed AIA Document G723, Project Application Summary, summary information should be transferred to the G722 form.

The Construction Manager should sign the form, have it notarized and submit it, together with G723 and a separate G702, Application, from each Contractor to the Architect.

The Architect should review it and, if acceptable, complete the lower Project Certificate for Payment on this form. DO NOT SIGN A CERTIFICATION ON EACH G702 SUBMITTED BY THE CONTRACTORS.

The completed form should be forwarded to the Owner. The Owner will make payment directly to each separate Contractor based on the amount due each as noted in Line J of G723.

PROJECT APPLICATION SUMMARY
AIA Document G723 (Instructions on reverse side)

A separate AIA Document G702, APPLICATION AND CERTIFICATE FOR PAYMENT, for each Contractor's signed Certification is attached.

In tabulations below, amounts are stated to the nearest dollar.

PAGE ONE OF ___ PAGES

APPLICATION NO:
APPLICATION DATE:
PERIOD FROM:
                    TO:
ARCHITECT'S PROJECT NO:

CONTRACTOR'S NAME					TOTALS THIS PAGE OR ALL PAGES
PORTION OF WORK					
A ORIGINAL CONTRACT SUM					
B NET CHANGE ORDERS TO DATE					
C CONTRACT SUM TO DATE					
D WORK IN PLACE TO DATE					
E STORED MATERIALS (Not in D or I)					
F TOTAL COMPLETED & STORED TO DATE (D+E)					
G RETAINAGE PERCENTAGE					
H RETAINAGE AMOUNT					
I PREVIOUS PAYMENTS					
J CURRENT PAYMENT DUE (F-H-I)					
K BALANCE TO FINISH (C-F)					
L PERCENT COMPLETE (F÷C)					

SAMPLE

G723 — 1980

AIA DOCUMENT G723 • PROJECT APPLICATION SUMMARY • JUNE 1980 EDITION • AIA® • © 1980
THE AMERICAN INSTITUTE OF ARCHITECTS, 1735 NEW YORK AVE., N.W., WASHINGTON, D.C. 20006

AGC Document 8a, *Amendment to Owner–Construction Manager Contractor (Guaranteed Maximum Price Option).*   As mentioned above in the discussion of the contractor construction management approach, this document supplements AGC Document 8 by establishing a guaranteed maximum price for the project. This amendment is designed to be executed by the contractor–construction manager who is performing and subcontracting out substantially all the work. Presumably, this form is not executed until the design documents are relatively fixed and the construction manager can either accurately estimate or secure reliable prices on the project. The amendment form also features an optional shared-savings clause and a space to set forth the completion date for the project.

THE ASSOCIATED GENERAL CONTRACTORS

AMENDMENT TO OWNER-
CONSTRUCTION MANAGER CONTRACT

Pursuant to Article 6 of the original Agreement, AGC Form No. 8, dated _____

between _____ (Owner)

and _____ (the Construction Manager),

for _____ (the Project),

the Owner desires to fix a Guaranteed Maximum Price for the Project and the Construction Manager agrees that the design, plans and specifications are sufficiently complete for such purpose. Therefore, the Owner and Construction Manager agree as set forth below.

ARTICLE I

Guaranteed Maximum Price

The Construction Manager's Guaranteed Maximum Price for the Project, including the Cost of the Work as defined in Article 8 and the Construction Manager's Fee as defined in Article 7 is _____ Dollars ($ _____). This price is for the performance of the Work in accordance with the documents listed and attached to this Amendment and marked Amendment Exhibit A.

(OPTIONAL SAVINGS CLAUSE) It is further agreed that if, upon completion of the work, the actual cost of the work plus the Construction Manager's Fee is less than the Guaranteed Maximum Price as set forth herein and as adjusted by approved change orders that the Owner agrees to pay to the Construction Manager an amount equal to _____% of such savings, as additional compensation.

ARTICLE II

Time Schedule

The Construction Completion date established by this Amendment is:

SAMPLE

OWNER:

ATTEST:

By: _____

_____

Date: _____

CONSTRUCTION MANAGER:

_____

ATTEST:

By: _____

_____

Date: _____

AGC DOCUMENT NO. 8A • AMENDMENT TO OWNER CONSTRUCTION MANAGER CONTRACT • JUNE 1977
©ASSOCIATED GENERAL CONTRACTORS OF AMERICA 1977

AGC Document 8b, *General Conditions for Trade Contractors under Construction Management Agreements.* AGC Documents 8b, like AIA Document A201/ CM, is based on and is quite similar to the AIA A201 form of general conditions. As stated in the instructions, "These conditions primarily govern the obligations of the Trade Contractors. . . . They have been drafted to cover trade contracts with either the Owner or the Construction Manager." Thus, they are designed to govern the responsibility of the separate contractors actually performing the work to their principal, be it construction manager or owner. It is important to note that the form is written giving obligations to both the owner and the construction manager, but that only the one who is an actual party to the contract is intended to be bound. Because this form attempts to be two different things at the same time, it should be carefully reviewed and modified to properly serve either as a subcontract (when the construction manager is acting as a general contractor) or as a prime contract (when the construction manager is only coordinating).

Some provisions of the AIA general conditions do not appear in the AGC form, including paragraph 3.2.1 (owner's evidence of financial capacity). The AGC general conditions form separately identifies the responsibilities of the architect/engineer from those of the construction manager, while the AIA document covers the responsibilities of both the architect and the construction manager in the same article.

Trade contractors must provide access to the construction for the A/E when directed to do so by the construction manager, thereby giving the construction manager control over access by the A/E. While the AIA general conditions document (A201/CM, par. 2.3.10) makes the architect the interpreter of all the contract documents (including the requirements of the general conditions), the AGC document limits the A/E's interpretations to the drawings and specifications (AGC Document 8b, par. 3.2.2). Although the AIA general conditions document provides for the architect to determine substantial completion (A201/CM, par. 9.8.3), the AGC document provides that the inspection is a joint responsibility of the architect and the construction manager (AGC Document 8b, 3.2.7), although the date of substantial completion is certified by the A/E (AGC Document 8b, par. 9.1.3) as in the AIA document (A201/CM, par. 8.1.3). Consistent with AGC's philosophy, the construction manager enjoys control over the trade contractors "as the owner's authorized representative" (AGC Document 8b, par. 4.2.1). Both the construction manager and the A/E have authority to reject nonconforming work (AGC Document 8b, par. 3.2.5 and 4.2.4); in the AIA document, the construction manager's right to reject work is subject to review by the architect (A201/CM, par. 2.3.16). A major difference between the AIA and AGC forms is that the power to issue change orders in the AGC form is vested in the construction manager (AGC Document 8b, par. 4.2.5 and 13.1.1); in the AIA general conditions, change orders are approved by the architect and specifically authorized by the owner, upon the recommendation of the construction manager (A201/CM, par. 2.3.19 and 12.1.1). Also, the AGC form gives both the owner and the construction manager the right to stop the work (AGC Document 8b, par. 4.3.1) and gives the construction manager the right to carry out the work if a trade contractor fails to perform (AGC Document 8b, par. 4.3.2). Both these rights are vested exclusively in the owner in the AIA document (A201/CM par. 3.3.1 and 3.4.1).

The responsibilities of the trade contractor under AGC Document 8b (art. 6) and AIA Document A201/CM (art. 5) are essentially the same. The rights of the trade contractor in the AGC document are limited, however. For instance, it is not entitled to additional compensation if it must make a substitute for an unacceptable subcontractor (AGC Document 8b, par. 6.2.3), while it is so entitled under the AIA form (A201/CM, par. 5.2.3).

The AGC document omits the reservation by the owner of the right to perform work and to award separate contracts (found in par. 6.1 of the AIA form), as well as the owner's right to clean up (found in par. 6.3 of the AIA form). Presumably, this is based on the assumption that such matters are the construction manager's responsibility. The arbitration provisions of AGC Document 8b (par. 8.9) parallel those of AGC's owner–construction manager agreement, omitting the limitations on joinder and consolidation which appear in the AIA document (A201/CM, par. 7.9.1). The AGC general conditions document contains in paragraph 9.3.4 a no-damages-for-acceleration-or-delay clause not found in the AIA document.

In article 10, the AGC general conditions document omits a payment certification process found in paragraphs 9.4 and 9.9 of the AIA document for progress payments and final payment; the AGC form also omits any role for the A/E in reviewing or approving payments. While the architect controls the certification and withholding of payment under the AIA form, this control is vested in the construction manager in the AGC form.

The AGC form, like the AIA form, places the responsibility for safety programs on the trade contractors, but unlike the AIA form, gives the construction manager a right to take over such safety precautions (par. 11.1.2). A significant difference in insurance requires sixty days' notice of cancellation of a contractor's coverage from the insurer to the construction manager (par. 12.1.4). The AGC document also omits language found in the AIA general conditions document (A201/CM, par. 11.3.1) permitting a contractor to purchase property insurance to cover the project (at the owner's expense) if the owner fails to provide adequate coverage. Early occupancy by the owner under the AIA agreement form is subject to the mutual agreement of the owner and the individual contractor (A201/CM, par. 11.3.9); in the AGC form, the owner and construction manager may decide (AGC Document 8b, par. 12.2.7). The AGC form also omits reference found in paragraph 11.4 of the AIA form to loss-of-use insurance.

Where work which is found defective has been caused by another contractor's improper work, the AGC general conditions document shifts responsibility to the "innocent" contractor from the owner (as in A201/CM, par. 13.1.2) to the offending contractor (AGC Document 8b, par. 14.1.2).

Although this form of general conditions could reasonably serve, with appropriate modifications, when the trade contractor is working either for the contractor–construction manager or for the owner, it clearly does not serve well as a document which could govern the relationship between the owner and the contractor–construction manager with respect to the work performed by or under the construction manager. Because AGC Document 8, the owner–construction manager agreement form, does not cover all the subject areas of the general conditions, there will be a need for additional contract preparation.

THE ASSOCIATED GENERAL CONTRACTORS

GENERAL CONDITIONS FOR TRADE CONTRACTORS UNDER CONSTRUCTION MANAGEMENT AGREEMENTS

INSTRUCTIONS FOR CONSTRUCTION MANAGER

1. These conditions primarily govern the obligations of the Trade Contractors and in addition establish the general procedures for the administration of construction. They have been drafted to cover Trade Contracts with either the Owner or the Construction Manager.

2. In all cases your attorney should be consulted to advise you on their use and any modifications.

3. Nothing contained herein is intended to conflict with local, state or federal laws or regulations.

4. It is recommended all insurance matters be reviewed with your insurance consultant and carrier such as implications of errors and omission liability, completed operations, and waiver of subrogation.

5. Each article should be reviewed by the Construction Manager as to the applicability to a given project and contractual conditions.

6. Special conditions and terms for the project or the Trade Contractor Agreements should cover the following:

   — trade contractor retainages
   — payment schedules
   — insurance limits
   — owner's protective insurance if required of trade contractors
   — builder's risk deductible, if any.

7. If the Owner does not provide Builder's Risk Insurance, Paragraph 12.2 will need to be modified.

AGC DOCUMENT NO. 520 • GENERAL CONDITIONS FOR TRADE CONTRACTORS UNDER CONSTRUCTION MANAGEMENT AGREEMENTS • JULY 1980
©1980 Associated General Contractors of America

THE ASSOCIATED GENERAL CONTRACTORS

GENERAL CONDITIONS FOR TRADE CONTRACTORS UNDER CONSTRUCTION MANAGEMENT AGREEMENTS

TABLE OF CONTENTS

AGC DOCUMENT NO. 520 • GENERAL CONDITIONS FOR TRADE CONTRACTORS UNDER CONSTRUCTION MANAGEMENT AGREEMENTS • JULY 1980
© 1980 Associated General Contractors of America

ARTICLE 1

CONTRACT DOCUMENTS

1.1 DEFINITIONS

1.1.1 THE CONTRACT DOCUMENTS

The Contract Documents consist of the Agreement between the Owner or Construction Manager, as the case may be, and the Trade Contractor, the Conditions of the Contract (General, Supplementary and other Conditions), the Drawings (and criteria if the drawings are not complete), the Specifications, all Addenda issued prior to execution of the Contract, and all Modifications issued after the execution of the contract. A modification is (1) a written amendment to the Contract signed by both parties, (2) a Change Order, (3) a written interpretation issued by the Architect/Engineer pursuant to Subparagraph 3.2.2, or (4) a written order for a minor change in the Work issued on the Owner's behalf pursuant to Paragraph 13.4. The Contract Documents do not include Bidding or Proposal Documents such as the Advertisement or Invitation To Bid, Requests for Proposals, sample forms, Trade Contractors Bid or Proposal, or portions of Addenda relative to any of these, or any other documents other than those set forth in this subparagraph unless specifically set forth in the Agreement with the Trade Contractor. In the event of an inconsistency between the Agreement and the other Contract Documents, the provisions of the Agreement will control.

1.1.2 THE CONTRACT

The Contract Documents form the Contract with the Trade Contractor. This Contract represents the entire and integrated agreement and supersedes all prior negotiations, representations or agreements, either written or oral. The Contract may be amended or modified only by a Modification as defined in Subparagraph 1.1.1.

1.1.3 THE WORK

The Work comprises the completed construction performed by the Construction Manager with his own forces or required by a Trade Contractor's contract and includes all labor necessary to produce such construction required of the Construction Manager or a Trade Contractor, and all materials and equipment incorporated or to be incorporated in such construction.

1.1.4 THE PROJECT

The Project is the total construction to be performed under the Agreement between the Owner and Construction Manager of which the Work is a part.

1.2 EXECUTION, CORRELATION AND INTENT

1.2.1 By executing his Agreement, each Trade Contractor represents that he has visited the site, familiarized himself with the local conditions under which the Work is to be performed and correlated his observations with the requirements of the Contract Documents.

1.2.2 The intent of the Contract Documents is to include all items necessary for the proper execution and completion of the Work. The Contract Documents are complementary, and what is required by any one shall be as binding as if required by all. Work not covered in the Contract Documents will not be required unless it is consistent therewith and is reasonably inferable therefrom as being necessary to produce the intended results. Words and abbreviations in the Contract Documents which have well-known technical or trade meanings are used in accordance with such recognized meanings.

1.2.3 The organization of the Specifications into divisions, sections and articles, and the arrangements of Drawings shall not control the Construction Manager in dividing the Work among Trade Contractors or in establishing the extent of Work to be performed by any trade.

1.3 OWNERSHIP AND USE OF DOCUMENTS

1.3.1 Unless otherwise provided in the Contract Documents, the Trade Contractor will be furnished, free of charge, all copies of Drawings and Specifications reasonably necessary for the execution of the Work.

1.3.2 All Drawings, Specifications and copies thereof furnished by the Architect/Engineer are and shall remain his property. They are to be ued only with respect to this Project and are not to be used on any other project. With the exception of one contract set for each party, such documents are to be returned or suitably accounted for to the Architect/Engineer on request at the completion of the Work. Submission or distribution to meet official regulatory requirements or for other purposes in connection with the Project is not to be construed as publication in derogation of the Architect/Engineer's common law copyright or other reserved rights.

ARTICLE 2

OWNER

2.1 DEFINITION

2.1.1 The Owner is the person or entity identified as such in the Agreement between the Owner and Construction Manager and is referred to throughout the Contract Documents as if singular in number and masculine in gender. The term Owner means the Owner or his authorized representative.

2.2 INFORMATION AND SERVICES FURNISHED BY THE OWNER

2.2.1 The Owner will furnish all surveys describing the physical characteristics, legal limitations and utility locations for the site of the Project, and a legal description of the site.

2.2.2 Except as provided in Subparagraph 5.7.1 the Owner will secure and pay for necessary approvals, easements, assessments and charges required for the construction, use or occupancy of permanent structures or for permanent changes in existing facilities.

2.2.3 Information or services under the Owner's control will be furnished by the Owner with reasonable promptness to avoid delay in the orderly progress of the Work.

2.2.4 The Owner shall forward all instructions to the Trade Contractors through the Construction Manager even when the Owner has direct contracts with Trade Contractors.

ARTICLE 3

ARCHITECT/ENGINEER

3.1 DEFINITION

3.1.1 The Architect/Engineer is the person lawfully licensed to practice architecture or engineering or an entity lawfully practicing architecture or engineering and identified as such in the Agreement between the Owner and Construction Manager and is referred to throughout the Contract Documents as if singular in number and masculine in gender. The term Architect/Engineer means the Architect/Engineer or his authorized representative.

3.1.2 Nothing contained in the Contract Documents shall create any contractual relationship between the Architect/Engineer and any Trade Contractor.

3.2 ARCHITECT/ENGINEER'S DUTIES DURING CONSTRUCTION

3.2.1 The Architect/Engineer shall at all times have access to the Work wherever it is in preparation and progress. When directed by the Construction Manager, the Trade Contractor shall provide facilities for such access so the Architect/Engineer may perform his functions under the Contract Documents.

3.2.2 The Architect/Engineer will be the interpreter of the requirements of the Drawings and Specifications. The Architect/Engineer will, within a reasonable time, render such interpretations as are necessary for the proper execution of the progress of the Work.

3.2.3   All interpretations of the Architect/Engineer shall be consistent with the intent of and reasonably inferable from the Contract Documents and will be in writing or in the form of drawings. All requests for interpretations shall be directed through the Construction Manager. The Architect/Engineer shall not be liable to the Trade Contractor for the result of any interpretation or decision rendered in good faith in such capacity.

3.2.4   The Architect/Engineer's decisions in matters relating to artistic effect will be final if consistent with the intent of the Contract Documents.

3.2.5   The Architect/Engineer will have authority to reject Work which does not conform to the Contract Documents. Whenever, in his opinion, he considers it necessary or advisable for the implementation of the intent of the Contract Documents, he will have authority to require special inspection or testing of the Work in accordance with Subparagraph 8.7.2 whether or not such Work be then fabricated, installed or completed. However, neither the Architect/Engineer's authority to act under this Subparagraph 3.2.5, nor any decision made by him in good faith either to exercise or not to exercise such authority, shall give rise to any duty or responsibility of the Architect/Engineer to the Trade Contractor, any Trade Subcontractor, any of their agents or employees, or any other person performing any of the Work.

3.2.6   The Architect/Engineer will review and approve or take other appropriate action upon Trade Contractor's submittals such as Shop Drawings, Product Data and Samples, but only for conformance with the design concept of the Work and with the information given in the Contract Documents. Such action shall be taken with reasonable promptness so as to cause no delay. The Architect/Engineer's approval of a specific item shall not indicate approval of an assembly of which the item is a component.

3.2.7   The Architect/Engineer along with the Construction Manager will conduct inspections to determine the dates of Substantial Completion and final completion, will receive and review written warranties and related documents required by the Contract and assembled by the Trade Contractor.

3.2.8   The Architect/Engineer will communicate with the Trade Contractors through the Construction Manager.

ARTICLE 4
CONSTRUCTION MANAGER

4.1   DEFINITION

4.1.1   The Construction Manager is the person or entity who has entered into an agreement with the Owner to serve as Construction Manager and is referred to throughout the Contract Documents as if singular in number and masculine in gender. The term Construction Manager means the Construction Manager acting through his authorized representative.

4.1.2   Whether the Trade Contracts are between the Owner and Trade Contractors, or the Construction Manager and Trade Contractors, it is the intent of these General Conditions to allow the Construction Manager to direct and schedule the performance of all Work and the Trade Contractors are expected to follow all such directions and schedules.

4.2   ADMINISTRATION OF THE CONTRACT

4.2.1   The Construction Manager will provide, as the Owner's authorized representative, the general administration of the Project as herein described.

4.2.2   The Construction Manager will be the Owner's construction representative during construction until final payment and shall have the responsibility to supervise and coordinate the work of all Trade Contractors.

4.2.3   The Construction Manager shall prepare and update all Construction Schedules and shall direct the Work with respect to such schedules.

4.2.4   The Construction Manager shall have the authority to reject Work which does not conform to the Contract Documents and to require any Special Inspection and Testing in accordance with Subparagraph 8.7.2.

4.2.5   The Construction Manager will prepare and issue Change Orders to the Trade Contractors in accordance with Article 13.

4.2.6 The Construction Manager along with the Architect/Engineer will conduct inspections to determine the dates of Substantial Completion and final completion, and will receive and review written warranties and related documents required by the Contract and assembled by the Trade Contractor.

4.2.7 Nothing contained in the Contract Documents between a Trade Contractor and the Owner shall create any contractual relationship between the Construction Manager and any Trade Contractor.

4.3 OWNER'S AND CONSTRUCTION MANAGER'S RIGHT TO STOP WORK

4.3.1 If the Trade Contractor fails to correct defective Work as required by Paragraph 14.2 or persistently fails to carry out the Work in accordance with the Contract Documents, the Construction Manager or the Owner through the Construction Manager may order the Trade Contractor to stop the Work, or any portion thereof, until the cause for such order has been eliminated.

4.3.2 If the Trade Contractor defaults or neglects to carry out the Work in accordance with the Contract Documents and fails within seven days after receipt of written notice from the Construction Manager to commence and continue correction of such default or neglect with diligence and promptness, the Construction Manager may, by written notice, and without prejudice to any other remedy he or the Owner may have, make good such deficiencies. In such case an appropriate Change Order shall be issued deducting from the payments then or thereafter due the Trade Contractor the cost of correcting such deficiencies, including compensation for the Architect/Engineer's and Construction Manager's additional services made necessary by such default, neglect or failure.

ARTICLE 5

TRADE CONTRACTORS

5.1 DEFINITION

5.1.1 A Trade Contractor is the person or entity identified as such in the Agreement between the Owner or Construction Manager and a Trade Contractor and is referred to throughout the Contract Document as if singular in number and masculine in gender. The term Trade Contractor means the Trade Contractor or his authorized representative.

5.1.2 The Agreements with the Trade Contractors may either be with the Owner or with the Construction Manager. These conditions in several instances make reference to obligations and rights of the ''Owner or Construction Manager'' to cover both possibilities. Such references are only to cover either possibility and such use does not create a joint obligation on the Owner and Construction Manager to the Trade Contractor. The contract obligation with the Trade Contractor is solely with the person or entity with whom he has his Agreement.

5.1.3 If the Trade Contracts are with the Construction Manager, the Trade Contractor assumes toward the Construction Manager all the obligations and responsibilities which the Construction Manager assumes toward the Owner under the Agreement between the Owner and the Construction Manager. A copy of the pertinent parts of this Agreement will be made available on request.

5.2 REVIEW OF CONTRACT DOCUMENTS

5.2.1 The Trade Contractor shall carefully study and compare the Contract Documents and shall at once report to the Construction Manager any error, inconsistency or omission he may or reasonably should discover. The Trade Contractor shall not be liable to the Owner or the Architect/Engineer or the Construction Manager for any damage resulting from any such errors, inconsistencies or omissions.

5.3 SUPERVISION AND CONSTRUCTION PROCEDURES

5.3.1 The Trade Contractor shall supervise and direct the Work, using his best skill and attention. He shall be solely responsible for all construction means, methods, techniques, sequences and procedures and for coordinating all portions of the Work under the Contract subject to the overall coordination of the Construction Manager.

5.3.2 The Trade Contractor shall be responsible to the Owner and the Construction Manager for the acts and omissions of his employees and all his Trade Subcontractors and their agents and employees and other persons performing any of the Work under a contract with the Trade Contractor.

5.3.3 Neither observations nor inspections, tests or approvals by persons other than the Trade Contractor shall relieve the Trade Contractor from his obligations to perform the Work in accordance with the Contract Documents.

5.4 LABOR AND MATERIALS

5.4.1 Unless otherwise specifically provided in the Contract Documents, the Trade Contractor shall provide and pay for all labor, materials, equipment, tools, construction equipment and machinery, transportation, and other facilities and services necessary for the proper execution and completion of the Work.

5.4.2 The Trade Contractor shall at all times enforce strict discipline and good order among his employees and shall not employ on the Work any unfit person or anyone not skilled in the task assigned to him.

5.5 WARRANTY

5.5.1 The Trade Contractor warrants to the Owner and the Construction Manager that all materials and equipment furnished under this Contract will be new unless otherwise specified, and that all Work will be of good quality, free from faults and defects and in conformance with the Contract Documents. All Work not so conforming to these requirements, including substitutions not properly approved and authorized, may be considered defective. If required by the Construction Manager, the Trade Contractor shall furnish satisfactory evidence as to the kind and quality of materials and equipment. This warranty is not limited by the provisions of Paragraph 14.2.

5.6 TAXES

5.6.1 The Trade Contractor shall pay all sales, consumer, use and other similar taxes for the Work or portions thereof provided by the Trade Contractor which are legally enacted at the time bids or proposals are received, whether or not yet effective.

5.7 PERMITS, FEES AND NOTICES

5.7.1 Unless otherwise provided in the Contract Documents, the Trade Contractor shall secure and pay for all permits, governmental fees, licenses and inspections necessary for the proper execution and completion of his Work, which are customarily secured after execution of the contract and which are legally required at the time bids or proposals are received.

5.7.2 The Trade Contractor shall give all notices and comply with all laws, ordinances, rules, regulations and orders of any public authority bearing on the performance of the Work.

5.7.3 Unless otherwise provided in the Contract Documents, it is not the responsibility of the Trade Contractor to make certain that the Contract Documents are in accordance with applicable laws, statutes, building codes and regulations. If the Trade Contractor observes that any of the Contract Documents are at variance therewith in any respect, he shall promptly notify the Construction Manager in writing, and any necessary changes shall be by appropriate Modification.

5.7.4 If the Trade Contractor performs any Work knowing it to be contrary to such laws, ordinances, rules and regulations, and without such notice to the Construction Manager, he shall assume full responsibility therefor and shall bear all costs attributable thereto.

5.8 ALLOWANCES

5.8.1 The Trade Contractor shall include in the Contract Sum as defined in 10.1.1 all allowances stated in the Contract Documents. Items covered by these allowances shall be supplied for such amounts and by such persons as the Construction Manager may direct, but the Trade Contractor will not be required to employ persons against whom he makes a reasonable objection.

5.8.2 Unless otherwise provided in the Contract Documents:

.1   These allowances shall cover the cost to the Trade Contractor, less applicable trade discount, of the materials and equipment required by the allowance delivered at the site, and all applicable taxes;

.2   The Trade Contractor's costs for unloading and handling on the site, labor, installation costs, overhead, profit and other expenses contemplated for the original allowance shall be included in the Contract Sum and not in the allowance;

.3   Whenever the cost is more than or less than the allowance , the Contract Sum shall be adjusted accordingly by Change Order, the amount of which will recognize changes, if any, in handling costs on the site, labor, installation costs, overhead, profit and other expenses.

5.9 SUPERINTENDENT

5.9.1   The Trade Contractor shall employ a competent superintendent and necessary assistants who shall be in attendance at the Project site during the progress of the Work. The superintendent shall be satisfactory to the Construction Manager, and shall not be changed except with the consent of the Construction Manager, unless the superintendent proves to be unsatisfactory to the Trade Contractor or ceases to be in his employ. The superintendent shall represent the Trade Contractor and all communications given to the superintendent shall be as binding as if given to the Trade Contractor. Important communications shall be confirmed in writing. Other communications shall be so confirmed on written request in each case.

5.10 PROGRESS SCHEDULE

5.10.1   The Trade Contractor, immediately after being awarded the Contract, shall prepare and submit for the Construction Manager's information an estimated progress schedule for the Work. The progress schedule shall be related to the entire Project to the extent required by the Contract Documents and shall provide for expeditious and practicable execution of the Work. This schedule shall indicate the dates for the starting and completion of the various stages of construction, shall be revised as required by the conditions of the Work, and shall be subject to the Construction Manager's approval.

5.11 DRAWINGS AND SPECIFICATIONS AT THE SITE

5.11.1   The Trade Contractor shall maintain at the site for the Construction Manager and Architect/Engineer two copies of all Drawings, Specifications, Addenda, Change Orders and other Modifications, in good order and marked currently to record all changes made during construction. These Drawings, marked to record all changes during construction, and approved Shop Drawings, Product Data and Samples shall be delivered to the Construction Manager for the Owner upon completion of the Work.

5.12 SHOP DRAWINGS, PRODUCT DATA AND SAMPLES

5.12.1   Shop Drawings are drawings, diagrams, schedules and other data especially prepared for the Work by the Trade Contractor or any Trade Subcontractor, manufacturer, supplier or distributor to illustrate some portion of the Work.

5.12.2   Product Data are illustrations, standard schedules, performance charts, instructions, brochures, diagrams and other information furnished by the Trade Contractor to illustrate a material, product or system for some portion of the Work.

5.12.3   Samples are physical examples which illustrate materials, equipment or workmanship and establish standards by which the Work will be judged.

5.12.4   The Trade Contractor shall review, approve and submit through the Construction Manager with reasonable promptness and in such sequence as to cause no delay in the Work or in the work of any separate contractor, all Shop Drawings, Product Data and Samples required by the Contract Documents.

5.12.5   By approving and submitting Shop Drawings, Product Data and Samples, the Trade Contractor represents that he has determined and verified all materials, field measurements, and field construction criteria related thereto, or will do so, and that he has checked and coordinated the information contained within such submittals with the requirements of the Work and of the Contract Documents.

5.12.6   The Construction Manager, if he finds such submittals to be in order, will forward them to the Architect/Engineer. If the Construction Manager finds them not to be complete or in proper form, he may return them to the Trade Contractor for correction or completion.

5.12.7   The Trade Contractor shall not be relieved of responsibility for any deviation from the requirements of the Contract Documents by the Construction Manager's forwarding them to the Architect/Engineer, or by the Architect/Engineer's approval of Shop Drawings, Product Data or Samples under Subparagraph 3.2.6 unless the Trade Contractor has specifically informed the Architect/Engineer and Construction Manager in writing of such deviation at the time of submission and the Architect/Engineer has given written approval to the specific deviation. The Trade Contractor shall not be relieved from responsibility for errors or omissions in the Shop Drawings, Product Data or Samples by the Construction Manager's forwarding or the Architect/Engineer's approval thereof.

5.12.8   The Trade Contractor shall direct specific attention, in writing or on resubmitted Shop Drawings, Product Data or Samples, to revisions other than those requested by the Architect/Engineer or Construction Manager on previous submittals.

5.12.9   No portion of the Work requiring submission of a Shop Drawing, Product Data or Sample shall be commenced until the submittal has been approved by the Architect/Engineer. All such portions of the Work shall be in accordance with approved submittals.

5.13   USE OF SITE

5.13.1   The Trade Contractor shall confine operations at the site to areas designated by the Construction Manager, permitted by law, ordinances, permits and the Contract Documents and shall not unreasonably encumber the site with any materials or equipment.

5.14   CUTTING AND PATCHING OF WORK

5.14.1   The Trade Contractor shall be responsible for all cutting, fitting or patching that may be required to complete the Work or to make its several parts fit together properly. He shall provide protection of existing Work as required.

5.14.2   The Trade Contractor shall not damage or endanger any portion of the Work or the work of the Construction Manager or any separate contractors by cutting, patching or otherwise altering any work, or by excavation. The Trade Contractor shall not cut or otherwise alter the work of the Construction Manager or any separate contractor except with the written consent of the Construction Manager and of such separate contractor. The Trade Contractor shall not unreasonably withhold from the Construction Manager or any separate contractor his consent to cutting or otherwise altering the Work.

5.15   CLEANING UP

5.15.1   The Trade Contractor at all times shall keep the premises free from accumulation of waste materials or rubbish caused by his operations. At the completion of the Work he shall remove all his waste materials and rubbish from and about the Project as well as all his tools, construction equipment, machinery and surplus materials.

5.15.2   If the Trade Contractor fails to clean up, the Construction manager may do so and the cost thereof shall be charged to the Trade Contractor.

5.16   COMMUNICATIONS

5.16.1   The Trade Contractor shall forward all communications to the Owner and Architect/Engineer through the Construction Manager.

5.17   ROYALTIES AND PATENTS

5.17.1   The Trade Contractor shall pay all royalties and license fees. He shall defend all suits or claims for infringement of any patent rights and shall save the Owner and Construction Manager harmless from loss on account thereof, except that the Owner shall be responsible for all such loss when a particular design, process or the product of a particular manufacturer or manufacturers is specified, but if the Trade Contractor has reason to believe that the design, process or product specified is an infringement of a patent, he shall be responsible for such loss unless he promptly gives such information to the Construction Manager.

5.18 INDEMNIFICATION

5.18.1 To the fullest extent permitted by law, the Trade Contractor shall indemnify and hold harmless the Owner, the Construction Manager and the Architect/Engineer and their agents and employees from and against all claims, damages, losses and expenses, including but not limited to attorneys' fees, arising out of or resulting from the performance of the Work, provided that any such claim, damage, loss or expense (1) is attributable to bodily injury, sickness, disease or death, or to injury to or destruction of tangible property (other than the Work itself) including the loss of use resulting therefrom, and (2) is caused in whole or in part by any negligent act or omission of the Trade Contractor, any Trade Subcontractor, anyone directly or indirectly employed by any of them or anyone for whose acts any of them may be liable, regardless of whether or not it is caused in part by a party indemnified hereunder. Such obligation shall not be construed to negate, abridge or otherwise reduce any other right or obligation of indemnity which would otherwise exist as to any party or person described in this Paragraph 5.18.

5.18.2 In any and all claims against the Owner, the Construction Manager or the Architect/Engineer or any of their agents or employees by any employee of the Trade Contractor, any Trade Subcontractor, anyone directly or indirectly employed by any of them or anyone for whose acts any of them may be liable, the indemnification obligation under this Paragraph 5.18 shall not be limited in any way by any limitation on the amount or type of damages, compensation or benefits payable by or for the Trade Contractor or any Trade Subcontractor under workers' or workmen's compensation acts, disability benefit acts or other employee benefit acts.

5.18.3 The obligations of the Trade Contractor under this Paragraph 5.18 shall not extend to the liability of the Architect/Engineer, his agents or employees arising out of (1) the preparation or approval of maps, drawings, opinions, reports, surveys, designs or specifications, or (2) the giving of or the failure to give directions or instruction by the Architect/Engineer, his agents or employees provided such giving or failure to give is the primary cause of the injury or damage.

ARTICLE 6

TRADE SUBCONTRACTORS

6.1 DEFINITION

6.1.1 A Trade Subcontractor is a person or entity who has a direct contract with a Trade Contractor to perform any of the Work at the site. The term Trade Subcontractor is referred to throughout the Contract Documents as if singular in number and masculine in gender and means a Trade Subcontractor or his authorized representative.

6.1.2 A Trade Subsubcontractor is a person or entity who has a direct or indirect contract with a Trade Subcontractor to perform any of the Work at the site. The term Trade Subsubcontractor is referred to throughout the Contract Documents as if singular in number and masculine in gender and means a Trade Subsubcontractor or an authorized representative thereof.

6.2 AWARD OF TRADE SUBCONTRACTS AND OTHER CONTRACTS FOR PORTIONS OF THE WORK

6.2.1 Unless otherwise required by the Contract Documents or in the Bidding or Proposal Documents, the Trade Contractor shall furnish to the Construction Manager in writing, for acceptance by the Owner and the Construction Manager in writing, the names of the persons or entities (including those who are to furnish materials or equipment fabricated to a special design) proposed for each of the principal portions of the Work. The Construction Manager will promptly reply to the Trade Contractor in writing if either the Owner or the Construction Manager, after due investigation, has reasonable objection to any such proposed person or entity. Failure of the Owner or Construction Manager to reply promptly shall constitute notice of no reasonable objection.

6.2.2 The Trade Contractor shall not contract with any such proposed person or entity to whom the Owner or the Construction Manager has made reasonable objection under the provisions of Subparagraph 6.2.1. The Trade Contractor shall not be required to contract with anyone to whom he has a reasonable objection.

6.2.3 If the Owner or Construction Manager refuses to accept any person or entity on a list submitted by the Trade Contractor in response to the requirements of the Contract Documents, the Trade Contractor shall submit an acceptable substitute; however, no increase in the Contract Sum shall be allowed for any such substitution.

6.2.4 The Trade Contractor shall me; however, no increase in the Contract Sum shall be allowed for any such substitution.

6.2.4 The Trade Contractor shall make no substitution for any Trade Subcontractor, person or entity previously selected if the Owner or Construction Manager makes reasonable objection to such substitution.

6.3 TRADE SUBCONTRACTUAL RELATIONS

6.3.1 By an appropriate agreement, written where legally required for validity, the Trade Contractor shall require each Trade Subcontractor, to the extent of the work to be performed by the Trade Subcontrator, to be bound to the Trade Contractor by the terms of the Contract Documents, and to assume toward the Trade Contractor all the obligations and responsibilities which the Trade Contractor, by these Documents, assumes toward the Owner, the Construction Manager, or the Architect/Engineer. Said agreement shall preserve and protect the rights of the Owner, the Construction Manager and the Architect/Engineer under the Contract Documents with respect to the Work to be performed by the Trade Subcontractor so that the subcontracting thereof will not prejudice such rights, and shall allow to the Trade Subcontractor, unless specifically provided otherwise in the Trade Contractor-Trade Subcontractor agreement, the benefit of all rights, remedies and redress against the Trade Contractor that the Trade Contractor, by these Documents, has against the Owner or Construction Manager. Where appropriate, the Trade Contractor shall require each Trade Subcontractor to enter into similar agreements with his Trade Subsubcontractors. The Trade Contractor shall make available to each proposed Trade Subcontractor, prior to the execution of the Trade Subcontract, copies of the Contract Documents to which the Trade Subcontractor will be bound by this Paragraph 6.3, and shall identify to the Trade Subcontractor any terms and conditions of the proposed Trade Subcontract which may be at variance with the Contract Documents. Each Trade Subcontractor shall similarly make copies of such Documents available to his Trade Subsubcontractors.

ARTICLE 7

SEPARATE TRADE CONTRACTS

7.1 MUTUAL RESPONSIBILITY OF TRADE CONTRACTORS

7.1.1 The Trade Contractor shall afford the Construction Manager and other trade contractors reasonable opportunity for the introduction and storage of their materials and equipment and the execution of their work, and shall connect and coordinate his Work with others under the general direction of the Construction Manager.

7.1.2 If any part of the Trade Contractor's Work depends, for proper execution or results, upon the work of the Construction Manager or any separate trade contractor, the Trade Contractor shall, prior to proceeding with the Work, promptly report to the Construction Manager any apparent discrepancies or defects in such work that render it unsuitable for such proper execution and results. Failure of the Trade Contractor so to report shall constitute an acceptance of the other trade contractor's or Construction Manager's work as fit and proper to receive his Work, except as to defects which may subsequently become apparent in such work by others.

7.1.3 Any costs caused by defective or ill-timed work shall be borne by the party responsible thereof.

7.1.4 Should the Trade Contractor wrongfully cause damage to the work or property of the Owner or to other work on the site, the Trade Contractor shall promptly remedy such damage as provided in Subparagraph 11.2.5.

7.1.5 Should the Trade Contractor wrongfully cause damage to the work or property of any separate trade contractor or other contractor, the Trade Contractor shall, upon due notice, promptly attempt to settle with the separate trade contractor or other contractor by agreement, or otherwise resolve the dispute. If such separate trade contractor or other contractor sues the Owner or the Construction Manager or initiates an arbitration proceeding against the Owner or Construction Manager on account of any damage alleged to have been caused by the Trade Contractor, the Owner or Construction Manager shall notify the Trade Contractor who shall defend such proceedings at the Trade Contractor's expense, and if any judgment or award against the Owner or Construction Manager arises therefrom, the Trade Contractor shall pay or satisfy it and shall reimburse the Owner or Construction Manager for all attorney's fees and court or arbitration costs which the Owner or Construction Manager has incurred.

7.2 CONSTRUCTION MANAGER'S RIGHT TO CLEAN UP

7.2.1 If a dispute arises between the separate Trade Contractors as to their responsibility for cleaning up as required by Paragraph 5.15, the Construction Manager may clean up and charge the cost thereof to the Trade Contractors responsible therefor as the Construction Manager shall determine to be just.

ARTICLE 8

MISCELLANEOUS PROVISIONS

8.1 GOVERNING LAW

8.1.1 The Contract shall be governed by the law of the place where the Project is located.

8.2 SUCCESSORS AND ASSIGNS

8.2.1 The Owner or Construction Manager (as the case may be) and the Trade Contractor each binds himself, his partners, successors, assigns and legal representatives to the other party hereto and to the partners, successors, assigns and legal representatives of such other party in respect to all covenants, agreements and obligations contained in the Contract Documents. Neither party to the Contract shall assign the Contract or sublet it as a whole without the written consent of the other.

8.3 WRITTEN NOTICE

8.3.1 Written notice shall be deemed to have been duly served if delivered in person to the individual or member of the firm or entity or to an officer of the corporation for whom it was intended, or if delivered at or sent by registered or certified mail to the last business address known to him who gives the notice.

8.4 CLAIMS FOR DAMAGES

8.4.1 Should either party to the Trade Contract suffer injury or damage to person or property because of any act or omission of the other party or of any of his employees, agents or others for whose acts he is legally liable, claim shall be made in writing to such other party within a reasonable time after the first observance of such injury or damage.

8.5 PERFORMANCE BOND AND LABOR AND MATERIAL PAYMENT BOND

8.5.1 The Owner or Construction Manager shall have the right to require the Trade Contractor to furnish bonds in a form and with a corporate surety acceptable to the Construction Manager covering the faithful performance of the Contract and the payment of all obligations arising thereunder if and as required in the Bidding or Proposal Documents or in the Contract Documents.

8.6 RIGHTS AND REMEDIES

8.6.1 The duties and obligations imposed by the Contract Documents and the rights and remedies available thereunder shall be in addition to and not a limitation of any duties, obligations, rights and remedies otherwise imposed or available by law.

8.6.2 No action or failure to act by the Construction Manager, Architect/Engineer or Trade Contractor shall constitute a waiver of any right or duty afforded any of them under the Contract Documents, nor shall any such action or failure to act constitute an approval of or acquiescence in any breach thereunder, except as may be specifically agreed in writing.

8.7 TESTS

8.7.1 If the Contract Documents, laws, ordinances, rules, regulations or orders of any public authority having jurisdiction require any portion of the Work to be inspected, tested or approved, the Trade Contractor shall give the Construction Manager timely notice of its readiness so the Architect/Engineer and Construction Manager may observe such inspection, testing or approval. The Trade Contractor shall bear all costs of such inspections, tests or approvals unless otherwise provided.

8.7.2 If the Architect/Engineer or Construction Manager determines that any Work requires special inspection, testing or approval which Subparagraph 8.7.1 does not include, he will, through the Construction Manager, instruct the Trade Contractor to order such special inspection, testing or approval and the Trade Contractor shall give notice as in Subparagraph 8.7.1. If such special inspection or testing reveals a failure of the Work to comply with the requirements of the Contract Documents, the Trade Contractor shall bear all costs thereof, including compensation for the Architect/Engineer's and Construction Manager's additional services made necessary by such failure. If the Work complies, the Owner or Construction Manager (as the case may be) shall bear such costs and an appropriate Change Order shall be issued.

8.7.3 Required certificates of inspection, testing or approval shall be secured by the Trade Contractor and promptly delivered by him through the Construction Manager to the Architect/Engineer.

8.7.4 If the Architect/Engineer or Construction Manager is to observe the inspections, tests or approvals required by the Contract Documents, he will do so promptly and, where practicable, at the source of supply.

8.8 INTEREST

8.8.1 Payments due and unpaid under the Contract Documents shall bear interest from the date payment is due at such rate upon which the parties may agree in writing or, in the absence thereof, at the legal rate prevailing at the place of the Project.

8.9 ARBITRATION

8.9.1 All claims, disputes and other matters in question arising out of, or relating to this Contract or the breach thereof, except as set forth in Subparagraph 3.2.4 with respect to the Architect/Engineer's decisions on matters relating to artistic effect, and except for claims which have been waived by the making or acceptance of final payment provided by Subparagraphs 10.8.4. and 10.8.5, shall be decided by arbitration in accordance with the Construction Industry Arbitration Rules of the American Arbitration Association then obtaining unless the parties mutually agree otherwise. This agreement to arbitrate shall be specifically enforceable under the prevailing arbitration law. The award rendered by the arbitrators shall be final, and judgment may be entered upon it in accordance with applicable law in any court having jurisdiction thereof.

8.9.2 Notice of the demand for arbitration shall be filed in writing with the other party to the Contract and with the American Arbitration Association. The demand for arbitration shall be made within a reasonable time after the claim, dispute or other matter in question has arisen, and in no event shall be made after the date when institution of legal or equitable proceedings based on such claim, dispute or other matter in question would be barred by the applicable statute of limitations.

8.9.3 The Trade Contractor shall carry on the Work and maintain the progress schedule during any arbitration proceedings, unless otherwise agreed by him and the Construction Manager in writing.

8.9.4 All claims which are related to or dependent upon each other shall be heard by the same arbitrator or arbitrators even though the parties are not the same unless a specific contract prohibits such consolidation.

ARTICLE 9

TIME

9.1 DEFINITIONS

9.1.1 Unless otherwise provided, the Contract Time is the period of time allotted in the Contract Documents for the Substantial Completion of the Work as defined in Subparagraph 9.1.3 including authorized adjustments thereto.

9.1.2 The date of commencement of the Work is the date established in a notice to proceed. If there is no notice to proceed, it shall be the date of the Trade Contractor Agreement or such other date as may be established therein.

9.1.3 The Date of Substantial Completion of the Work or designated portion thereof is the Date certified by the Architect/Engineer when construction is sufficiently complete, in accordance with the Contract Documents, so the Owner can occupy or utilize the Work or designated portion thereof for the use for which it is intended.

9.1.4 The term day as used in the Contract Documents shall mean calendar day unless otherwise specifically designated.

9.2 PROGRESS AND COMPLETION

9.2.1 All time limits stated in the Contract Documents are of the essence of the Contract.

9.2.2 The Trade Contractor shall begin the Work on the date of commencement as defined in Subparagraph 9.1.2. He shall carry the Work forward expeditiously with adequate forces and shall achieve Substantial Completion within the Contract Time.

9.3 DELAYS AND EXTENSIONS OF TIME

9.3.1 If the Trade Contractor is delayed at any time in the progress of the Work by any act or neglect of the Owner, Construction Manager, or the Architect/Engineer, or by any employee of either, or by any separate contractor employed by the Owner, or by changes ordered in the Work, or by labor disputes, fire, unusual delay in transportation, adverse weather conditions not reasonably anticipatable, unavoidable casualties or any causes beyond the Trade Contractor's control, or by delay authorized by the Owner or Construction Manager pending arbitration, or by any other cause which the Construction Manager determines may justify the delay, then the Contract Time shall be extended by Change Order for such reasonable time as the Construction Manager may determine.

9.3.2 Any claim for extension of time shall be made in writing to the Construction Manager not more than twenty (20) days after the commencement of the delay; otherwise, it shall be waived. In the case of a continuing delay only one claim is necessary. The Trade Contractor shall provide an estimate of the probable effect of such delay on the progress of the Work.

9.3.3 If no agreement is made stating the dates upon which interpretations as set forth in Subparagraph 3.2.2 shall be furnished, then no claim for delay shall be allowed on account of failure to furnish such interpretations until fifteen days after written request is made for them, and not then unless such claim is reasonable.

9.3.4 It shall be recognized by the Trade Contractor that he may reasonably anticipate that as the job progresses, the Construction Manager will be making changes in and updating Construction Schedules pursuant to the authority given him in Subparagraph 4.2.3. Therefore, no claim for an increase in the Contract Sum for either acceleration or delay will be allowed for extensions of time pursuant to this Paragraph 9.3 or for other changes in the Construction Schedules which are of the type ordinarily experienced in projects of similar size and complexity.

9.3.5 This Paragraph 9.3 does not exclude the recovery of damages for delay by either party under other provisions of the Contract Documents.

ARTICLE 10

PAYMENTS AND COMPLETION

10.1 CONTRACT SUM

10.1.1 The Contract Sum is stated in the Agreement between the Owner or Construction Manager and the Trade Contractor including adjustments thereto and is the total amount payable to the Trade Contractor for the performance of the Work under the Contract Documents.

10.2 SCHEDULE OF VALUES

10.2.1 Before the first Application for Payment, the Trade Contractor shall submit to the Construction Manager a schedule of values allocated to the various portions of the Work prepared in such form and supported by such data to substantiate its accuracy as the Construction Manager may require. This schedule, unless objected to by the Construction Manager, shall be used only as a basis for the Trade Contractor's Application for Payment.

10.3 APPLICATIONS FOR PAYMENT

10.3.1 At least ten days before the date for each progress payment established in the Trade Contractor's Agreement, the Trade Contractor shall submit to the Construction Manager an itemized Application for Payment, notarized if required, supported by such data substantiating the Trade Contractor's right to payment as the Owner or the Construction Manager may require, and reflecting retainage, if any, as provided elsewhere in the Contract Documents.

10.3.2 Unless otherwise provided in the Contract Documents, payments will be made on account of materials or equipment not incorporated in the Work but delivered and suitably stored at the site and, if approved in advance by the Construction Manager, payments may similarly be made for materials or equipment stored at some other location agreed upon in writing. Payments made for materials or equipment stored on or off the site shall be conditioned upon submission by the Trade Contractor of bills of sale or such other procedures satisfactory to the Construction Manager to establish the Owner's title to such materials or equipment or otherwise protect the Owner's interest, including applicable insurance and transportation to the site for those materials and equipment stored off the site.

10.3.3 The Trade Contractor warrants that title to all Work, materials and equipment covered by an Application for Payment will pass to the Owner either by incorporation in the construction or upon the receipt of payment by the Trade Contractor, whichever occurs first, free and clear of all liens, claims, security interests or encumbrances, hereinafter referred to in this Article 10 as "liens," and that no Work, materials or equipment covered by an Application for Payment will have been acquired by the Trade Contractor, or by any other person performing his Work at the site or furnishing materials and equipment for his Work, subject to an agreement under which an interest therein or an encumbrance thereon is retained by the seller or otherwise imposed by the Trade Contractor or such other person. All Trade Subcontractors and Trade Subsubcontractors agree that title will so pass upon their receipt of payment from the Trade Contractor.

10.4 PROGRESS PAYMENTS

10.4.1 If the Trade Contractor has made Application for Payment as above, the Construction Manager will, with reasonable promptness but not more than seven days after the receipt of the Application, review and process such Application for payment in accordance with the Contract.

10.4.2 No approval of an application for a progress payment, nor any progress payment, nor any partial or entire use or occupancy of the Project by the Owner, shall constitute an acceptance of any Work not in accordance with the Contract Documents.

10.4.3 The Trade Contractor shall promptly pay each Trade Subcontractor upon receipt of payment out of the amount paid to the Trade Contractor on account of such Trade Subcontractor's Work, the amount to which said Trade Subcontractor is entitled, reflecting the percentage actually retained, if any, from payments to the Trade Contractor on account of such Trade Subcontractor's Work. The Trade Contractor shall, by an appropriate agreement with each Trade Subcontractor, also require each Trade Subcontractor to make payments to his Trade Subsubcontractors in a similar manner.

10.5 PAYMENTS WITHHELD

10.5.1 The Construction Manager may decline to approve an Application for Payment if in his opinion the Application is not adequately supported. If the Trade Contractor and Construction Manager cannot agree on a revised amount, the Construction Manager shall process the Application for the amount he deems appropriate. The Construction Manager may also decline to approve any Applications for Payment or, because of subsequently discovered evidence or subsequent inspections, he may nullify in whole or in part any approval previously made to such extent as may be necessary in his opinion because of:

.1   defective work not remedied;

.2   third party claims filed or reasonable evidence indicating probable filing of such claims;

.3   failure of the Trade Contractor to make payments properly to Trade Subcontractors or for labor, materials or equipment;

.4   reasonable evidence that the Work cannot be completed for the unpaid balance of the Contract Sum;

.5   damage to the Construction Manager, the Owner, or another contractor working at the Project;

.6   reasonable evidence that the Work will not be completed within the Contract Time; or

.7   persistent failure to carry out the Work in accordance with the Contract Documents.

10.5.2 When the above grounds in Subparagraph 10.5.1 are removed, payment shall be made for amounts withheld because of them.

10.6   FAILURE OF PAYMENT

10.6.1   If the Trade Contractor is not paid within seven days after any amount is approved for payment by the Construction Manager and has become due and payable, then the Trade Contractor may, upon seven additional days' written notice to the Owner and Construction Manager, stop the Work until payment of the amount owing has been received. The Contract Sum shall be increased by the amount of the Trade Contractor's reasonable costs of shutdown, delay and start up, which shall be effected by appropriate Change Order in accordance with Paragraph 13.3.

10.7   SUBSTANTIAL COMPLETION

10.7.1   When the Trade Contractor considers that the Work, or a designated portion thereof which is acceptable to the Owner, is substantially complete as defined in Subparagraph 9.1.3, the Trade Contractor shall prepare for submission to the Construction Manager a list of items to be completed or corrected. The failure to include any items on such list does not alter the responsibility of the Trade Contractor to complete all Work in accordance with the Contract Documents. When the Construction Manager and Architect/Engineer on the basis of inspection determine that the Work or designated portion thereof is substantially complete, the Architect/Engineer will then prepare a Certificate of Substantial Completion which shall establish the Date of Substantial Completion, shall state the responsibilities of the Owner, the Construction Manager and the Trade Contractor for security, maintenance, heat, utilities, damage to the Work, and insurance, and shall fix the time within which the Trade Contractor shall complete the items listed therein. Warranties required by the Contract Documents shall commence on the Date of Substantial Completion of the Work or designated portion thereof unless otherwise provided in the Certificate of Substantial Completion. The Certificate of Substantial Completion shall be submitted to the Owner, the Construction Manager and the Trade Contractor for their written acceptance of the responsibilities assigned to them in such Certificate.

10.8   FINAL COMPLETION AND FINAL PAYMENT

10.8.1   Upon receipt of written notice that the Work is ready for final inspection and acceptance and upon receipt of a final Application for Payment, the Architect/Engineer and the Construction Manager will promptly make such inspection and, when they find the Work acceptable under the Contract Documents and the Contract fully performed, the Construction Manager will promptly approve final payment.

10.8.2   Neither the final payment nor the remaining retained percentage shall become due until the Trade Contractor submits to the Construction Manager (1) an affidavit that all payrolls, bills for materials and equipment, and other indebtedness connected with the Work for which the Owner or his property might in any way be responsible, have been paid or otherwise satisfied, (2) consent of surety, if any, to final payment, and (3) if required by the Owner, other data establishing payment or satisfaction of all such obligations, such as receipts, releases and waivers of liens arising out of the Contract, to the extent and in such form as may be designated by the Owner. If any Trade Subcontractor refuses to furnish a release or waiver required by the Owner or Construction Manager, the Trade Contractor may furnish a bond satisfactory to the Owner and Construction Manager to indemnify them against any such lien. If any such lien remains unsatisfied after all payments are made, the Trade Contractor shall refund to the Owner or Construction Manager all moneys that the latter may be compelled to pay in discharging such lien, including all costs and reasonable attorneys' fees.

10.8.3   If, after Substantial Completion of the Work, final completion thereof is materially delayed through no fault of the Trade Contractor or by the issuance of Change Orders affecting final completion, and the Construction Manager so confirms, the Owner or Construction Manager shall, upon certification by the Construction Manager, and without terminating the Contract, make payment of the balance due for that portion of the Work fully completed and accepted. If the remaining balance for Work not fully completed or corrected is less than the retainage stipulated in the Contract Documents, and if bonds have been furnished as provided in Paragraph 8.5, the written consent of the surety to the payment of the balance due for that portion of the Work fully completed and accepted shall be submitted by the Trade Contractor to the Construction Manager prior to such payment. Such payment shall be made under the terms and conditions governing final payment, except that it shall not constitute a waiver of claims.

10.8.4   The making of final payment shall constitute a waiver of all claims by the Owner or Construction Manager except those arising from:

.1   unsettled liens;

.2   faulty or defective Work appearing after Substantial Completion;

.3   failure of the Work to comply with the requirements of the Contract Documents; or

.4   terms of any special warranties required by the Contract Documents.

10.8.5   The acceptance of final payment shall constitute a waiver of all claims by the Trade Contractor except those previously made in writing and identified by the Trade Contractor as unsettled at the time of the Final Application for Payment.

ARTICLE 11

PROTECTION OF PERSONS AND PROPERTY

11.1 SAFETY PRECAUTIONS AND PROGRAMS

11.1.1   The Trade Contractor shall be responsible for initiating, maintaining and supervising all safety precautions and programs in connection with the Work.

11.1.2   If the Trade Contractor fails to maintain the safety precautions required by law or directed by the Construction Manager, the Construction Manager may take such steps as necessary and charge the Trade Contractor therefor.

11.1.3   The failure of the Construction Manager to take any such action shall not relieve the Trade Contractor of his obligations in Subparagraph 11.1.1.

11.2 SAFETY OF PERSONS AND PROPERTY

11.2.1   The Trade Contractor shall take all reasonable precautions for the safety of, and shall provide all reasonable protection to prevent damage, injury or loss to:

.1   all employees on the Work and all other persons who may be affected thereby;

.2   all the Work and all materials and equipment to be incorporated therein, whether in storage on or off the site, under the care, custody or control of the Trade Contractor or any of his Trade Subcontractors or Trade Subsubcontractors; and

.3   other property at the site or adjacent thereto, including trees, shrubs, lawns, walks, pavements, roadways, structures and utilities not designated for removal, relocation or replacement in the course of construction.

11.2.2   The Trade Contractor shall give all notices and comply with all applicable laws, ordinances, rules, regulations and lawful orders of any public authority bearing on the safety of persons or property or their protection from damage, injury or loss.

11.2.3   The Trade Contractor shall erect and maintain, as required by existing conditions and progress of the Work, all reasonable safeguards for safety and protection, including posting danger signs and other warnings against hazards, promulgating safety regulations and notifying owners and users of adjacent utilities. If the Trade Contractor fails to so comply he shall, at the direction of the Construction Manager, remove all forces from the Project without cost or loss to the Owner or Construction Manager, until he is in compliance.

11.2.4   When the use or storage of explosives or other hazardous materials or equipment is necessary for the execution of the Work, the Trade Contractor shall exercise the utmost care and shall carry on such activities under the supervision of properly qualified personnel.

11.2.5   The Trade Contractor shall promptly remedy all damage or loss (other than damage or loss insured under Paragraph 12.2) to any property referred to in Clauses 11.2.1.2 and 11.2.1.3 caused in whole or in part by the Trade Contractor, his Trade Subcontractors, his Trade Subsubcontractors, or anyone directly or indirectly employed by any of them, or by anyone for whose acts any of them may be liable and for which the Trade Contractor is responsible under Clauses 11.2.1.2 and 11.2.1.3, except damage or loss attributable to the acts or omissions of the Owner or Architect/Engineer or anyone directly or indirectly employed by either of them or by anyone for whose acts either of them may be liable, and not attributable to the fault or negligence of the Trade Contractor. The foregoing obligations of the Trade Contractor are in addition to his obligations under Paragraph 5.18.

11.2.6   The Trade Contractor shall designate a responsible member of his organization at the site whose duty shall be the prevention of accidents. This person shall be the Trade Contractor's superintendent unless otherwise designated by the Trade Contractor in writing to the Construction Manager.

11.2.7   The Trade Contractor shall not load or permit any part of the Work to be loaded so as to endanger its safety.

11.3 EMERGENCIES

11.3.1   In any emergency affecting the safety of persons or property, the Trade Contractor shall act, at his discretion, to prevent threatened damage, injury or loss. Any additional compensation or extension of time claimed by the Trade Contractor on account of emergency work shall be determined as provided in Article 13 for Changes in the Work.

ARTICLE 12

INSURANCE

12.1 TRADE CONTRACTOR'S LIABILITY INSURANCE

12.1.1   The Trade Contractor shall purchase and maintain such insurance as will protect him from claims set forth below which may arise out of or result from the Trade Contractor's operations under the Contract, whether such operations be by himself or by any of his Trade Subcontractors or by anyone directly or indirectly employed by any of them, or by anyone for whose acts any of them may be liable:

 .1 claims under workers' or workmen's compensation, disability benefit and other similar employee benefit acts which are applicable to the Work to be performed including the "Broad Form All States" Endorsement;

 .2 claims for damages because of bodily injury, occupational sickness or disease, or death of his employees under any employers liability law including, if applicable, those required under maritime or admiralty law for wages, maintenance, and cure;

 .3 claims for damages because of bodily injury, sickness or disease, or death of any person other than his employees;

 .4 claims for damages insured by usual personal injury liability coverage which are sustained (1) by any person as a result of an offense directly or indirectly related to the employment of such person by the Trade Contractor, or (2) by any other person;

 .5 claims for damages other than to the Work itself because of injury to or destruction of tangible property, including loss of use resulting therefrom; and

 .6 claims for damages because of bodily injury or death of any person or property damage arising out of the ownership, maintenance or use of any motor vehicle.

12.1.2   The insurance required by Subparagraph 12.1.1 shall be written for not less than any limits of liability specified in the Contract Documents, or required by law, whichever is greater.

12.1.3   The insurance required by Subparagraph 12.1.1 shall include premises-operations (including explosion, collapse and underground coverage), elevators, independent contractors, products and/or completed operations, and contractual liability insurance (on a "blanket basis" designating all written contracts), all including broad form property damage coverage. Liability insurance may be arranged under Comprehensive General Liability policies for the full limits required or by a combination of underlying policies for lesser limits with the remaining limits provided by an Excess or Umbrella Liability Policy.

12.1.4   The foregoing policies shall contain a provision that coverages afforded under the policies will not be cancelled until at least sixty days' prior written notice has been given to the Construction Manager. Certificates of Insurance acceptable to the Construction Manager shall be filed with the Construction Manager prior to commencement of the Work. Upon request, the Trade Contractor shall allow the Construction Manager to examine the actual policies.

12.2 PROPERTY INSURANCE AND WAIVER OF SUBROGATION

12.2.1 Unless otherwise provided, the Owner will purchase and maintain property insurance upon the entire Work at the site to the full insurable value thereof. This insurance shall include the interests of the Owner, the Construction Manager, the Trade Contractors, and Trade Subcontractors in the Work and shall insure against the perils of fire and extended coverage, and shall include "all risk" insurance for physical loss or damage.

12.2.2 The Owner will effect and maintain such boiler and machinery insurance as may be necessary and/or required by law. This insurance shall include the interest of the Owner, the Construction Manger, the Trade Contractors, and Trade Subcontractors in the Work.

12.2.3 Any loss insured under Paragraph 12.2 is to be adjusted with the Owner and Construction Manager and made payable to the Owner and Construction Manager as trustees for the insureds, as their interests may appear, subject to the requirements of any applicable mortgagee clause.

12.2.4 The Owner, the Construction Manager, the Architect/Engineer, the Trade Contractors, and the Trade Subcontractors waive all rights against each other and any other contractor or subcontractor engaged in the Project for damages caused by fire or other perils to the extent covered by insurance provided under Paragraph 12.2, or any other property or consequential loss insurance applicable to the Project, equipment used in the Project, or adjacent structures, except such rights as they may have to the proceeds of such insurance. If any policy of insurance requires an endorsement to maintain coverage with such waivers, the owner of such policy will cause the policy to be so endorsed. The Owner will require, by appropriate agreement, written where legally required for validity, similar waivers in favor of the Trade Contractors and Trade Subcontractors by any separate contractor and his subcontractors.

12.2.5 The Owner and Construction Manager shall deposit in a separate account any money received as trustees, and shall distribute it in accordance with such agreement as the parties in interest may reach, or in accordance with an award by arbitration in which case the procedure shall be as provided in Paragraph 8.9. If after such loss no other special agreement is made, replacement of damaged Work shall be covered by an appropriate Change Order.

12.2.6 The Owner and Construction Manager as trustees shall have power to adjust and settle any loss with the insurers unless one of the parties in interest shall object in writing within five days after the occurrence of loss to the Owner's and Construction Manager's exercise of this power; and if such objection be made, arbitrators shall be chosen as provided in Paragraph 8.9. The Owner and Construction Manager as trustees shall, in that case, make settlement with the insurers in accordance with the directions of such arbitrators. If distribution of the insurance proceeds by arbitration is required, the arbitrators will direct such distribution.

12.2.7 If the Owner finds it necessary to occupy or use a portion or portions of the Work prior to Substantial Completion thereof, such occupancy shall not commence prior to a time mutually agreed to by the Owner and Construction Manager and to which the insurance company or companies providing the property insurance have consented by endorsement to the policy or policies. This insurance shall not be cancelled or lapsed on account of such partial occupancy.

ARTICLE 13

CHANGES IN THE WORK

13.1 CHANGE ORDERS

13.1.1 A Change Order is a written order to the Trade Contractor signed by the Owner or Construction Manager, as the case may be, issued after the execution of the Contract, authorizing a Change in the Work or an adjustment in the Contract Sum or the Contract Time. The Contract Sum and the Contract Time may be changed only by Change Order. A Change Order signed by the Trade Contractor indicates his agreement therewith, including the adjustment in the Contract Sum or the Contract Time.

13.1.2 The Owner or Construction Manager, without invalidating the Contract, may order Changes in the Work within the general scope of the Contract consisting of additions, deletions or other revisions, the Contract Sum and the Contract Time being adjusted accordingly. All such changes in the Work shall be authorized by Change Order, and shall be performed under the applicable conditions of the Contract Documents.

13.1.3   The cost or credit to the Owner or Construction Manager resulting from a Change in the Work shall be determined in one or more of the following ways:

.1   by mutual acceptance of a lump sum properly itemized and supported by sufficient substantiating data to permit evaluation; or

.2   by unit prices stated in the Contract Documents or subsequently agreed upon; or

.3   by cost to be determined in a manner agreed upon by the parties and a mutually acceptable fixed or percentage fee; or

.4   by the method provided in Subparagraph 13.1.4.

13.1.4   If none of the methods set forth in Clauses 13.1.3.1, 13.1.3.2 or 13.1.3.3 is agreed upon, the Trade Contractor, provided he receives a written order signed by the Owner or the Construction Manager, shall promptly proceed with the Work involved. The cost of such Work shall be determined by the Construction Manager on the basis of the reasonable expenditures and savings of those performing the Work attributable to the change, including, in the case of an increase in the Contract Sum, a reasonable allowance for overhead and profit. In such case, and also under Clauses 13.1.3.3 and 13.1.3.4 above, the Trade Contractor shall keep and present, in such form as the Construction Manager may prescribe, an itemized accounting together with appropriate supporting data for inclusion in a Change Order. Unless otherwise provided in the Contract Documents, cost shall be limited to the following: cost of materials, including sales tax and cost of delivery; cost of labor, including social security, old age and unemployment insurance, and fringe benefits required by agreement or custom; workers' or workmen's compensation insurance; bond premiums; rental value of equipment and machinery; and the additional costs of supervision and field office personnel directly attributable to the change. Pending final determination of cost, payments on account shall be made as determined by the Construction Manager. The amount of credit to be allowed by the Trade Contractor for any deletion or change which results in a net decrease in the Contract Sum will be the amount of the actual net cost as confirmed by the Construction Manager. When both additions and credits covering related Work or substitutions are involved in any one change, the allowance for overhead and profit shall be figured on the basis of the net increase, if any, with respect to that change.

13.1.5   If unit prices are stated in the Contract Documents or subsequently agreed upon, and if the quantities originally contemplated are so changed in a proposed Change Order that application of the agreed unit prices to the quantities of Work proposed will cause substantial inequity to the Owner, the Construction Manager, or the Trade Contractor, the applicable unit prices shall be equitably adjusted.

13.2   CONCEALED CONDITIONS

13.2.1   Should concealed conditions encountered in the performance of the Work below the surface of the ground or should concealed or unknown conditions in an existing structure be at variance with the conditions indicated by the Contract Documents, or should unknown physical conditions below the surface of the ground or should concealed or unknown conditions in an existing structure of an unusual nature, differing materially from those ordinarily encountered and generally recognized as inherent in work of the character provided for in this Contract, be encountered, the Contract Sum shall be equitably adjusted by Change Order upon claim by either party made within twenty days after the first observance of the conditions.

13.3   CLAIMS FOR ADDITIONAL COST

13.3.1   If the Trade Contractor wishes to make a claim for an increase in the Contract Sum, he shall give the Construction Manager written notice thereof within twenty days after the occurrence of the event giving rise to such claim. This notice shall be given by the Trade Contractor before proceeding to execute the Work, except in an emergency endangering life or property in which case the Trade Contractor shall proceed in accordance with Paragraph 11.3. No such claim shall be valid unless so made. Any change in the Contract Sum resulting from such claim shall be authorized by Change Order.

13.3.2   If the Trade Contractor claims that additional cost is involved because of, but not limited to, (1) any written interpretation issued pursuant to Subparagraph 3.2.2, (2) any order by the Owner or Construction Manager to stop the Work pursuant to Paragraph 4.3 where the Trade Contractor was not at fault, or (3) any written order for a minor change in the Work issued pursuant to Paragraph 13.4, the Trade Contractor shall make such claim as provided in Subparagraph 13.3.1.

13.4 MINOR CHANGES IN THE WORK

13.4.1 The Architect/Engineer will have authority to order through the Construction Manager minor changes in the Work not involving an adjustment in the Contract Sum or an extension of the Contract Time and not inconsistent with the intent of the Contract Documents. Such changes shall be effected by written order and such changes shall be binding on the Owner, the Construction Manager, and the Trade Contractor. The Trade Contractor shall carry out such written orders promptly.

ARTICLE 14

UNCOVERING AND CORRECTION OF WORK

14.1 UNCOVERING OF WORK

14.1.1 If any portion of the Work should be covered contrary to the request of the Construction Manager or Architect/Engineer, or to requirements specifically expressed in the Contract Documents, it must, if required in writing by the Construction Manager, be uncovered for their observation and replaced, at the Trade Contractor's expense.

14.1.2 If any other portion of the Work has been covered which neither the Construction Manager nor the Architect/Engineer has specifically requested to observe prior to being covered, the Architect/Engineer or Construction Manager may request to see such Work and it shall be uncovered by the Trade Contractor. If such Work be found in accordance with the Contract Documents, the cost of uncovering and replacement shall, by appropriate Change Order, be charged to the Owner or Construction Manager, as the case may be. If such Work be found not in accordance with the Contract Documents, the Trade Contractor shall pay such costs unless it be found that this condition was caused by a separate trade contractor employed as provided in Article 7, and in that event the separate trade contractor shall be responsible for the payment of such costs.

14.2 CORRECTION OF WORK

14.2.1 The Trade Contractor shall promptly correct all Work rejected by the Architect/Engineer or the Construction Manager as defective or as failing to conform to the Contract Documents whether observed before or after Substantial Completion and whether or not fabricated, installed or completed. The Trade Contractor shall bear all costs of correcting such rejected Work, including compensation for the Architect/Engineer's and/or Construction Manager's additional services made necessary thereby.

14.2.2 If, within one year after the Date of Substantial Completion of Work or designated portion thereof, or within one year after acceptance by the Owner of designated equipment or within such longer period of time as may be prescribed by law or by the terms of any applicable special warranty required by the Contract Documents, any of the Work is found to be defective or not in accordance with the Contract Documents, the Trade Contractor shall correct it promptly after receipt of a written notice from the Owner or Construction Manager to do so unless the Owner or Construction Manager has previously given the Trade Contractor a written acceptance of such condition. This obligation shall survive the termination of the Contract. The Owner or Construction Manager shall give such notice promptly after discovery of the condition.

14.2.3 The Trade Contractor shall remove from the site all portions of the Work which are defective or non-conforming and which have not been corrected under Subparagraphs 5.5.1, 14.2.1 and 14.2.2, unless removal has been waived by the Owner.

14.2.4 If the Trade Contractor fails to correct defective or non-conforming Work as provided in Subparagraphs 5.5.1, 14.2.1 and 14.2.2, the Owner or Construction Manager may correct it in accordance with Subparagraph 4.3.2.

14.2.5 If the Trade Contractor does not proceed with the correction of such defective or non-conforming Work within a reasonable time fixed by written notice from the Construction Manager, the Owner or Construction Manager may remove it and may store the materials or equipment at the expense of the Trade Contractor. If the Trade Contractor does not pay the cost of such removal and storage within ten days thereafter, the Owner or Construction Manager may upon ten additional days' written notice sell such Work at auction or at private sale and shall account for the net proceeds thereof, after deducting all the costs that should have been borne by the Trade Contractor, including compensation for the Construction Manager's additional services made necessary thereby. If such proceeds of sale do not cover all costs which the Trade Contractor should have borne, the difference shall be charged to the Trade Contractor and an appropriate Change Order shall be issued. If the payments then or thereafter due the Trade Contractor are not sufficient to cover such amount, the Trade Contractor shall pay the difference to the Owner or Construction Manager.

14.2.6   The Trade Contractor shall bear the cost of making good all work of the Construction Manager or other contractors destroyed or damaged by such removal or correction.

14.3   ACCEPTANCE OF DEFECTIVE OR NONCONFORMING WORK

14.3.1   If the Owner or Construction Manager prefers to accept defective or non-conforming Work, he may do so instead of requiring its removal and correction, in which case a Change Order will be issued to reflect reduction in the Contract Sum where appropriate and equitable. Such adjustment shall be effected whether or not final payment has been made.

ARTICLE 15

TERMINATION OF THE CONTRACT

15.1   TERMINATION BY THE TRADE CONTRACTOR

15.1.1   If the Work is stopped for a period of thirty days under an order of any court or other public authority having jurisdiction, or as a result of an act of government, such as a declaration of a national emergency making materials unavailable, through no act or fault of the Trade Contractor or a Trade Subcontractor or their agents or employees or any other persons performing any of the Work under a contract with the Trade Contractor, or if the Work should be stopped for a period of thirty days by the Trade Contractor because of a failure to receive payment in accordance with the Contract, then the Trade Contractor may, upon seven additional days' written notice to the Construction Manager, terminate the Contract and recover from the Owner or Construction Manager, as the case may be, payment for all Work executed and for any proven loss sustained upon any materials, equipment, tools, construction equipment and machinery, including reasonable profit and damages.

15.2   TERMINATION BY THE OWNER OR CONSTRUCTION MANAGER

15.2.1   If the Trade Contractor is adjudged a bankrupt, or if he makes a general assignment for the benefit of his creditors, or if a receiver is appointed on account of his insolvency, or if he persistently or repeatedly refuses or fails, except in cases for which extension of time is provided, to supply enough properly skilled workmen or proper materials, or if he fails to make prompt payment to Trade Subcontractors or for materials or labor, or persistently disregards laws, ordinances, rules, regulations or orders of any public authority having jurisdiction, or otherwise is guilty of a substantial violation of a provision of the Contract Documents, then the Owner or Construction Manager may, without prejudice to any right or remedy and after giving the Trade Contractor and his surety, if any, seven days' written notice, terminate the employment of the Trade Contractor and take possession of the site and of all materials, equipment, tools, construction equipment and machinery thereon owned by the Trade Contractor and may finish the Work by whatever method he may deem expedient. In such case the Trade Contractor shall not be entitled to receive any further payment until the Work is finished.

15.2.2   If the unpaid balance of the Contract Sum exceeds the costs of finishing the Work, including compensation for the Construction Manager's additional services made necessary thereby, such excess shall be paid to the Trade Contractor. If such costs exceed the unpaid balance, the Trade Contractor shall pay the difference to the Owner or Construction Manager.

AGC Document 8f, *Change Order.* This change order form is essentially a mechanism for the owner and construction manager to modify the project requirements, project cost, construction manager's fee, and completion time. It is designed to be used in conjunction with the change order procedure set up in AGC Document 8b, *General Conditions for Trade Contractors under Construction Management Agreements.*

THE ASSOCIATED GENERAL CONTRACTORS

CHANGE ORDER

PROJECT:
[Name, Address]

TO: [Owner]

CHANGE ORDER NUMBER:
DATE:

PROJECT NO.:

CONTRACT FOR:

In accordance with the terms of this Contract, the following changes are approved:

PROJECT COST ADJUSTMENTS
Original Project Costs $_____
Previous Change Order #1 thru #_____
This Change Order #_____
New Project Costs $_____

CONSTRUCTION MANAGER FEE ADJUSTMENTS
Original Fee   _____
Previous Fee Changes #1 thru #____   _____
This Fee Change #_____   _____
New Fee $_____

New Project Costs and New Fees $_____

COMPLETION DATE ADJUSTMENT
Original Completion Date   _____
Previous Time Change #1 thru #_____   _____days
This Time Change #_____   _____days
New Completion Date   _____

The said Contract as hereby amended shall remain in full force and effect.

IN WITNESS WHEREOF the said parties have caused this agreement to be executed as of the day and year signed below.

_____   _____
CONSTRUCTION MANAGER   OWNER

_____   _____
Address   Address

_____   _____

BY _____   BY _____

DATE _____   DATE _____

AGC Document No. 8f   ©*Copyright by Associated General Contractors of America*

CHAPTER FIVE

SPECIAL CONSTRUCTION MANAGEMENT CONTRACT FORMS

CAVEAT

Construction management is a concept. This concept differs from contract to contract. Consequently the standard contract forms analyzed in the prior chapter must be modified and tailored to fit the facts and the needs of a particular transaction. The following special construction management forms serve the reader only as a guide to the different delivery approaches. They also must be modified to fit the facts and the requirements of each specific situation.

THE PROFESSIONAL CONSTRUCTION MANAGEMENT APPROACH

As noted previously, when the construction manager acts as a professional consultant to the owner as the owner's representative, coordinating, administering, and managing the project for a fee, the arrangement is generally referred to as professional construction management. Because the construction manager is receiving a fair monetary agreement at the outset, it is not working for a mark-up and acts only in the interest of the owner to effectuate the advantages of construction management heretofore reviewed.

This is considered by many as a major advantage of the construction management approach, for once compensation is agreed upon, the interests of the construction manager and the owner are identical—getting the job completed within the time and budgetary restraints. Gone are the battles of substituted materials and change orders. Present is a true team effort in which design input and cost concern are given equal consideration throughout the entire project.

In professional construction management the owner contracts with the trade contractors directly. This then leads to the major disadvantage to the owner in that contracting with multiple primes can lead to multiple problems, primarily problems of responsibility which can pertain to guarantees, warranties, damages, etc. There is no one guarantor of the cost, timeliness, or quality of construction.

Owner Services Agreement for a Large Private Project (Owner-Oriented)

On large projects mistakes of omission and commission can have a most severe impact. Experienced construction input is essential through the project. Knowledge of labor cost and productivity, as well as an understanding of materials and systems capabilities and costs, is a must.

The following example contract geared to the construction of a large manufacturing facility typifies construction management in its purest form. Because of the tremendous number of decisions and contemplated changes pertaining to design and engineering, only a most preliminary budget estimate can be rendered. It must be continually revised. Here the construction manager is engaged to work with the architect engineer from the inception and actively assist the owner with professional expertise throughout the planning, design, and construction phases as the owner's consultant.

The following construction management services agreement establishes an effective working relationship between the construction manager and the owner when it is desired that the construction manager act as the owner's consultant, be responsible for the production of the project, but still be answerable to the owner. It sets forth clearly the understandings as to the extent of the construction manager's authority. It is owner-oriented in that the owner maintains overview control.

Here the construction manager has no authority to bind the owner (who will contract directly with the trade contractors) except in certain specific situations that are clearly set forth in the construction management agreement, as, for example, entering into material purchase orders. The construction manager pays the trade contractors out of a special disbursement account to which the owner advances funds once the construction manager is satisfied that the trade contractors have complied with their obligations under their contracts with the owner.

It is intended in this agreement that the construction manager not perform work as a trade contractor except in rare situations. In those rare situations the construction manager must undertake all obligations of a trade contractor to the owner and charge only direct expense (no overhead or profit) unless the work exceeds, in the aggregate, a large sum.

CONSTRUCTION MANAGER AGREEMENT

THIS AGREEMENT, made this_____day of_____, 1982 between ABC MANU-FACTURING CORPORATION ("Owner"), and XYZ CONSTRUCTION COMPANY, INC. ("Construction Manager"), for services in connection with the design and construction of its UVW DIVISION GEAR AND ASSEMBLY PLANT, Norristown, Pennsylvania.

WITNESSETH:

WHEREAS, Owner is engaged in the construction of the UVW DIVISION GEAR AND ASSEMBLY PLANT, in Norristown, Pennsylvania, (the "Project"); and

WHEREAS, DEF Associates ("Architect") have been hired by Owner as its Architect and Engineer to prepare the plans and specifications and render other services for the Project; and

WHEREAS, GHI Design Associates ("Designer") have been hired by Architect to prepare plans and specifications for the interior design of portions of the Project; and

WHEREAS, Owner desires to have the Project well designed and constructed in an expeditious, economic and efficient manner; and

WHEREAS, Owner deems it advisable to have an independent organization undertake the management of construction and the coordination and expedition of the work of Architect, Designer, and each of the Trade Contractors engaged for the Project.

NOW, THEREFORE, in consideration of the mutual covenants and conditions herein contained, the parties hereto agree as follows:

1. *Agreement to Serve:* Owner hereby retains Construction Manager to serve as Owner's construction manager to perform the duties set forth herein in connection with the construction of the Project as generally described in Architect's Report dated December 15, 1981 to perform the services hereinafter described, on the terms and conditions specified herein, and Construction Manager hereby agrees so to serve.

2. *Services of Construction Manager:* The services of Construction Manager shall be performed under the general direction of Owner and shall consist of consulting with, and advising and making recommendations to Owner, Architect, Designer and Trade Contractors, as the case may be, and of coordinating all aspects of the planning and construction of the Project, in order to expedite the completion in accordance with the plans and specifications and the Construction Schedule. Such services shall include, but are not limited to, the following:

DESIGN PHASE

(a) *Consultation During Project Development:* Review site, foundation, structural, architectural, mechanical, electrical and plumbing plans and specifications and advise on the selection of systems and materials as contained in the preliminary studies, schematic design studies and the design development documents; review the preliminary construction cost estimate submitted by Architect and revise such estimate at periodic intervals in accordance with progress on the plans and specifications; review Architect's final construction estimate, commenting thereon; and make recommendations with respect to such factors as construction feasibility, suggested economies, availability of materials and labor, time requirements for installation and construction, and cost.

(b) *Scheduling:* Provide and periodically update a design time schedule as part of the Construction Schedule that coordinates and integrates Architect's services with the Construction Schedules.

(c) *Project Budget:* Prepare a Project budget for Owner's approval as soon as major Project requirements have been identified and update periodically. Prepare an estimate of construction cost based on a quantity survey of drawings and specifications at the end of the schematic design phase for approval by Owner. Update and refine this estimate for Owner's approval as the development of the drawings and specifications proceeds, and advise Owner and Architect if it appears that the Project budget will not be met and make recommendations for corrective action.

(d) *Coordination of Contract Documents:* Review the plans and specifications with Architect and make recommendations to Owner and Architect regarding the division of the work for the purpose of bidding and awarding of separate contracts by Owner, taking into consideration such factors as the type or scope of work to be performed, time of performance, availability of labor, community relations and other pertinent criteria, relating to the various trades involved.

(e) *Job Site Facilities and Safety:* Review the specifications to assure that they contain (1) provisions for all temporary facilities necessary to enable the Trade Contractors to properly perform their work and (2) provisions for all job-site facilities necessary to enable Construction Manager, Architect, Designer and Owner to perform their duties in connection with the construction; and verify that the requirements and assignment of responsibilities for safety precautions and programs, and for equipment, materials and services for common use of Trade Contractors are included in the proposed contract documents.

(f) *Review of Specifications:* Advise on the method to be used for selecting Trade Contractors and awarding contracts; review the drawings and specifications to (1) ascertain and advise Architect of areas of possible conflicts and overlapping jurisdictions among the Trade Contractors and subcontractors on the job and recommend solutions for the elimination of such areas so that the work on the Project may be advanced and completed as expeditiously as possible; (2) verify that all work has been included; and (3) provide for phased construction.

(g) *Equipment Schedule:* Investigate and recommend a schedule for purchase or lease by Owner of all materials and equipment requiring long lead time procurement, and coordinate the schedule with the early preparation of contract documents by Architect. Expedite and coordinate delivery of these purchases or leases.

(h) *Labor:* Make a labor survey, including an analysis of the types and quantity of manpower required for the Project and a forecast of the availability thereof as and when needed.

(i) *Equal Opportunity:* Determine applicable requirements for equal employment opportunity programs for inclusion in the proposed contract documents.

(j) *Labor Relations:* Make recommendations and render assistance for the development and administration of an effective labor relations program for the Project and the avoidance of labor disputes during construction.

(k) *Bidding:* Prepare pre-qualification criteria for bidders and develop trade contractor interest in the Project. Establish bidding schedules and conduct pre-bid conferences to familiarize bidders with the bidding documents and management techniques and with any special systems, materials or methods. Prepare with the assistance of Architect and Owner the bidding documents and submit bidding documents to Owner for its approval prior to distributing bid packages to bidders.

(l) *Bid Receipt:* Receive bids, prepare bid analyses and make recommendations to Owner for award of contracts or rejection of bids.

(m) *Value Engineering:* Prior to award of Trade Contracts, evaluate any trade contractor value engineering proposals submitted by bidder and recommend those worthy of consideration to Architect and Owner.

(n) *Contract Awards:* Conduct pre-award conferences with successful bidders. Prepare Trade Contracts substantially in accordance with Exhibit D for approval by Owner and advise Owner on the acceptability of subcontractors and material suppliers proposed by Trade Contractors.

CONSTRUCTION PHASE

(o) *Project Control:* Coordinate the work of the Trade Contractors with the activities and responsibilities of Owner and Architect and make recommendations to permit the completion of the Project in accordance with Owner's objectives on cost, time and quality; maintain a competent full-time staff at the Project site with authority to achieve these objectives. Establish an on-site organization and lines of authority to effectively carry out all phases of the Project on a totally coordinated basis. Provide for site offices for Owner and Architect including necessary utilities, telephone, and message service, the cost of which is to be reimbursed by Owner to Construction Manager as a Reimbursable Expense. It is expressly understood that Construction Manager does not guarantee the work or the price or time of completion of the work performed by Trade Contractors pursuant to Trade Contracts.

(p) *Organization Chart:* Prepare and submit to Owner an organization chart, which shall include a time schedule, Construction Manager's proposed job-site staff, and the job classification and salary of each member of said staff, which organization chart and any subsequent changes shall be subject to the prior written approval of Owner.

(q) *Meetings:* Schedule and conduct weekly pre-construction and construction and progress meetings at which Trade Contractors, Architect and Construction Manager can discuss jointly and coordinate such matters as procedures, progress, problems and scheduling. Establish and implement procedures for coordination of activities. Prepare minutes and furnish copies to all interested parties and to Owner and meet with Owner and Architect biweekly to discuss the construction progress, problems, and to review applications for payment.

(r) *Shop Drawings, Etc.:* Establish and implement procedures to be followed for expediting the processing of shop drawings, catalogs and samples, and the scheduling of material requirements.

(s) *Construction Schedule:* Within 60 calendar days of the date hereof, Construction Manager shall submit to Owner a network plan and schedule of the job (the "Construction Schedule") in the Critical Path Method (CPM) format which shall include a schedule for each of the Trade Contractors (including materialmen) and submit the CPM schedule as required, but no less than monthly, to incorporate approved scheduling data submitted by the Trade Contractors and Architect for the coordination of the work of the Trade Contractors and Architect. Owner reserves the right to require Construction Manager to modify any portion of the schedule judged impractical, unfeasible, or unreasonable. Schedules returned to the Construction Manager for revision or correction shall be resubmitted to Owner for approval within 5 calendar days.

(t) *Progress Records:* Keep accurate and detailed written records of the progress of the Project during all stages of planning and construction; submit weekly and monthly written progress reports to Owner and Architect including, but not limited to, information concerning the work of each of the various Trade Contractors, the percentage of completion and the number and amount of all change orders; and identify potential variances between scheduled and probable completion dates. Construct a Schedule for work not started or incomplete and recommend to Owner the Trade Contractor adjustments in the schedule to meet the probable completion date. Provide summary reports of each monitoring and document all changes in the schedule, and provide the reports set forth in Exhibit B attached hereto at the times specified therein.

(u) *Trade Contract Defaults:* Recommend courses of action to Owner when requirements of a Trade Contract are not being fulfilled. Determine the need for and recommend to Owner the institution of partial or complete default proceedings against Trade Contractors, the assessment of liquidated damages, and/or the termination of Trade Contracts, and assist Owner in selecting alternate contractors to perform defaulted work and assist in evaluating back charges or other penalties to be assessed.

(v) *Cost Control:* Revise and refine the approved estimate of construction cost, incorporate approved changes as they occur, and develop cash flow reports and forecasts as needed. Derive cost estimates on changes of scope to verify Trade Contractor estimates for negotiation assistance, after prior approval of scope change by Owner.

(w) *Cost Monitoring:* Provide regular monitoring of the approved estimate of construction cost, showing actual costs for activities in process and estimates for uncompleted tasks. Identify variances between actual and budgeted or estimated costs, and advise Owner and Architect whenever projected costs exceed budgets or estimates.

(x) *Cost Records:* With respect to all work to be performed on the Project, including, without limitation, all work performed on a time and material basis, unit cost, or similar basis, keep current computerized records and computations and summaries; including component summaries and summaries by Trade Contractor, and maintain computerized cost accounting records in accordance with procedures approved by Owner; and provide two copies of a printout to Owner of such information at least once a month.

(y) *Change Orders:* Review all requests for change orders, and implement the procedure set forth in the ABC General Conditions of the Contract for Construction attached hereto as a part of Exhibit D, for processing of change orders by obtaining proposals from Trade Contractors and reviewing and evaluating them; assist in negotiating cost of same and make recommendations to Owner and be prepared to substantiate all evaluations to Owner, and prepare change orders for Owner's signature with the assistance of Architect.

(z) *Change Recommendations:* Recommend necessary or desirable changes to Owner and Architect, review requests for changes and applications for extension of time from Trade Contractors, Owner and Architect, submit recommendations to Owner and Architect, and negotiate change orders.

(aa) *Payment Procedures:* Develop and implement a procedure for the review and processing of applications by Trade Contractors for progress and final payments. Make recommendations to Architect for certification to Owner for payment.

(bb) *Permits and Fees:* In consultation with Owner and Architect, obtain all required permits, licenses and certificates, including certificates of occupancy, and in addition, at its own cost, those permits, licenses and certificates, if any, which may be required for Construction Manager to provide the services hereunder; verify that all applicable fees and assessments for permanent facilities have been paid; and ensure the obtaining of approvals from all the authorities having jurisdiction.

(cc) *Owner's Consultants:* If required, assist Owner and Architect in selecting and retaining professional services of a surveyor, special consultants and testing laboratories, and coordinate these services.

(dd) *Inspection:* Inspect the work of Trade Contractors to determine that the work is being performed in accordance with the requirements of the contract documents, drawings and specifications; guard Owner against defects and deficiencies in the work; require any Trade Contractor to stop work or any portion thereof, and require special inspection or testing of any work not in accordance with the provisions of the contract documents whether or not such work be then fabricated, installed or completed; reject work which does not conform to the requirements of the contract documents. Inspect the Project jointly with Owner and Architect at least every two weeks.

(ee) *Tests:* Ascertain that all tests of cement, concrete, structural or reinforcing steel or of any other material or equipment required to be tested under the terms of the contract documents, are performed.

(ff) *Grades, Dimensions, Etc.:* Check grades, dimensions and measurements both in the field and in drawings as an added precaution against performance errors.

(gg) *Contract Performance:* Coordinate the work of the Trade Contractors on the Project until final completion and acceptance of the Project by Owner, assure that the materials furnished and work performed are in accordance with the plans and specifications and other contract documents, and that the work on the Project is progressing on schedule. In the event the interpretation of the meaning and intent of the plans and specifications becomes necessary during construction, Construction Manager shall, on behalf of Owner, consult with Architect, ascertain Architect's interpretation, and transmit such information to the appropriate Trade Contractor. Construction Manager shall inform Owner of an undue delay in obtaining such interpretations.

(hh) *Safety Programs and Insurance:* Review the safety programs as developed by each of the Trade Contractors, submit a comprehensive project safety program to Owner, and require each Trade Contractor to adhere to such program, review insurance requirements imposed on Trade Contractors, obtain insurance certificates and policies from Trade Contractors, and forward such to Owner, and not permit entry upon the site by Trade Contractors not in compliance with insurance requirements.

(ii) *Reports and Records:* Record the progress of the Project. Submit written progress reports to Owner and Architect including information on the Trade Contractors and work, the percentage of completion and the number and amounts of change orders. Keep a daily log available to Owner and Architect in which shall be described in detail all work accomplished on the preceding working day, the number of men employed at the site by each Trade Contractor and the number of hours worked, material shortages, labor difficulties, weather conditions, list of visiting officials and jurisdictions, daily activities, decisions, observations in general and specific observations as required. Assure that "as built" drawings are kept current and produced in accordance with contract requirements and maintain a complete set of such drawings.

(jj) *Records:* Maintain at the Project site, on a current basis, to be available for review by Owner and Architect at all times: records of all contracts, orderly files for correspondence, reports of job conferences, reproductions of original contract documents including all addenda, change orders, and supplemental drawings; shop drawings; as-built drawings; samples; purchases; materials; equipment; applicable handbooks; federal, commercial and technical standards and specifications; maintenance and operating manuals and instructions; and any other related documents and revisions which arise out of the contract or the work. Obtain data from Trade Contractors and maintain a current set of record drawings, specifications and operating manuals. At the completion of the Project, deliver all such records to Owner. Prepare when requested by Owner all replies to correspondence from the Trade Contractors, including letters and complaints, for signature by Owner.

(kk) *Owner-Purchased Items:* Accept delivery and arrange storage, protection and security for all Owner-purchased materials, systems and equipment which are a part of the work until such items are turned over to the Trade Contractors.

(ll) *Substantial Completion:* Inspect the Project, jointly with Architect and Owner between thirty to forty-five days prior to the time Owner is to take over, use, occupy or operate any part or all of the Project and, in addition, fifteen to twenty days prior to the time any Trade Contractor is expected to complete its work under a Trade Contract; furnish a detailed report to Owner and Architect of discrepancies and deficiencies in the work performed by any Trade Contractor or any materials provided by any material supplier, determine substantial completion or beneficial occupancy as appropriate, prepare all necessary punch lists and expedite execution of same, supervise the correction and completion of work.

(mm) *Start-up:* With Owner's maintenance personnel, direct the checkout of utilities, operational systems and equipment for readiness and assist in their initial start-up and testing.

(nn) *Final Completion:* Inspect the Project jointly with Architect and Owner upon the completion of construction and thereafter Construction Manager shall certify to Owner in writing that the Project is complete. Owner need not issue the Certificate of Final Acceptance until it has received certification of completion from Construction Manager and Architect. Collect guarantee, maintenance and operations manuals, keying schedules and other data required of the Trade Contractors, and maintain photographic records, material and equipment delivery records, visual aids, charts and graphs.

(oo) *Claims and Litigation:* In the event any claim is made or any action brought in any way relating to the design or construction of the Project, Construction Manager shall diligently render to Owner any and all assistance which Owner may require. Such services shall be rendered by Construction Manager without additional fee or other compensation except for the costs and expense of personnel who were assigned to the Project as job-site staff, or comparable personnel if those who were assigned to the Project are no longer employed by Construction Manager. Payment for the services of such personnel shall be in accordance with the provisions of Article 22(a) of this Agreement. In the event any legal action or arbitration is brought against Construction Manager, and Construction Manager is found to have been acting within the scope of its authority and obligations hereunder, then Owner shall indemnify Construction Manager for reasonable legal expenses and fees in connection with the defense of such action or arbitration. Owner at its option may, upon the instituting of any such action or arbitration, undertake to defend Construction Manager in lieu of payment of Construction Manager's legal expenses.

(pp) *Call Backs:* As part of the fee, supervise all call-backs for one year after the date of Final Completion.

(qq) *Guarantee Inspections:* Inspect the Project jointly with Architect and Designer between 30 to 45 days prior to the end of any one-year guarantee period provided in the Trade Contracts and furnish a detailed report to Owner of discrepancies and deficiencies applicable to any such guarantees.

3. *Duration of Contract:* The services of Construction Manager shall commence as of the date hereof and Construction Manager shall continue its services hereunder until the Final Acceptance. Construction Manager's obligation under Articles 3(oo), 3(pp), 3(qq), 27, 29, 30, 34, 42 and 43 shall survive Final Acceptance and final payment by Owner, and termination of this Agreement with or without cause.

4. *Time of the Essence:* In performing the services hereunder, Construction Manager shall place emphasis on considerations which will aid in expediting the construction of the Project. Construction Manager agrees to use its best efforts to effect completion of the Project on or before the Project Completion Date by the Trade Contractors and, to this end, but without limitation, it shall give constant attention to the adequacy of the Trade Contractors' planning, personnel, equipment and the availability of materials and supplies. Construction Manager acknowledges that time will be of the essence in completion of the project and that it will be its responsibility to prevent delays to the extent possible. Construction Manager will keep informed as to situations which may result in delays and take immediate steps to the extent possible to avoid the occurrence of such delays. If such a situation is not resolved immediately, Construction Manager shall bring it to the immediate attention of Owner.

5. *Communications in Writing:* All recommendations and communications by Construction Manager to Owner and Architect that will affect the cost of the Project shall be made or confirmed by it in writing. Owner may also require other recommendations and communications by Construction Manager to be made or confirmed by it in writing. All recommendations relating to proposed changes in the work, work schedules, and instructions to Trade Contractors, shall be made directly to Owner and Architect, unless otherwise directed by Owner. After approval by Owner and subject to the general supervision of Owner and Architect, Construction Manager shall issue instructions directly to the Trade Contractors.

6. *Services by Construction Manager's Own Staff:* The services to be performed hereunder shall be performed by Construction Manager's own staff, unless otherwise authorized by Owner. The employment of, contract with, or use of the services of any other person or firm by Construction Manager, as independent consultant or otherwise, shall be subject to prior written approval of Owner. No provision of this Agreement shall, however, be construed as constituting an agreement between Owner and any such person or firm.

7. *Construction Manager/Independent Contractor:* The relationship of Construction Manager to Owner shall be that of independent contractor, and Construction Manager shall have no authority to bind Owner in any way with third parties, except as specifically set forth herein where Owner may permit Construction Manager to execute agreements as an agent of Owner.

8. *No Third Party Beneficiaries:* Nothing contained herein shall be deemed to create a contractual relationship between or among Construction Manager, Architect, or any of the Trade Contractors, subcontractors or material suppliers for the Project, nor shall anything contained herein be deemed to give any such party or any third party any claim or right of action against Owner or Construction Manager beyond such as may otherwise exist without regard to this Agreement.

9. *Trade Contracts:*

(a)   Trade Contractors will enter into independent agreements with Owner for the provision of all work and materials for the Project substantially in the form attached hereto as Exhibit D and subject to the ABC Manufacturing Corporation General Conditions of the Contract for Construction attached thereto as a part of

Exhibit D. Unless otherwise directed by Owner in writing with reference to a specific contract, each Trade Contract will be between Trade Contractor and Owner. In either event, at Owner's request, Construction Manager will make payments to the Trade Contractors from the Special Account as provided in Article 10 below. Amendments to the Trade Contracts, that is change orders, must be approved in writing by Owner prior to execution of the change orders; however, in unusual emergency situations when after the exercise of due diligence by Construction Manager an authorized representative of Owner cannot be contacted, Construction Manager may execute on behalf of Owner as Owner's agent a change order in the amount of $3,000 or less.

(b)   In addition to the Trade Contracts for the provisions of major components of the labor or materials for the Project, Owner may from time to time deem it advisable to enter into short term agreements and purchase orders for the provision of labor, materials and services to Owner. Such purchase orders and short term agreements shall be deemed to be Trade Contracts and payment may be made from the Special Account directly to the provider of such labor or materials if:

(i)   The purchase order or agreement is in the form set forth in Exhibit E and contains the specific rider disclosing that Construction Manager is acting as the agent for the Owner thereunder.

(ii)   The Owner has previously agreed in writing to the specific contract or purchase order.

It is intended to the extent practicable that all such contracts or purchase orders shall be executed by Owner. It is intended to the extent possible and practical that all work for the Project shall be provided pursuant to Trade Contracts, and that Construction Manager will not undertake the role of a Trade Contractor as set forth in the following subparagraph (c).

(c)   Construction Manager hereby agrees, if requested by Owner, to act as Trade Contractor from time to time to perform work and provide materials to Owner in connection with the construction of the Project, which work and materials Construction Manager agrees shall be subject to the ABC Manufacturing Corporation General Conditions of the Contract for Construction, and Construction Manager agrees to undertake all the responsibilities and obligations of a Trade Contractor thereunder. It is expressly understood and agreed that Construction Manager will only charge Owner for its direct expenses in connection with such work or materials and that Construction Manager will not add any amounts for profit or overhead, except that should the cost of such work (excluding materials) exceed $100,000.00 in the aggregate, the Construction Manager shall be entitled to an additional fee of 10% of the cost of the work (excluding materials). Payment to Construction Manager for services and materials provided under this section may not be paid from the Special Account, but Construction Manager will submit a separate invoice to Owner together with all supporting information including the prior written authorization by Owner to undertake the work and services.

10.   *Special Account:* As a fiduciary agent for Owner, Construction Manager shall pay all Trade Contractors (except where Construction Manager is a Trade Contractor pursuant to a Trade Contract) out of funds to be advanced by Owner. Promptly after execution of this Agreement, Construction Manager and Owner shall establish a disbursement account (the "Special Account") from which funds may be drawn by an authorized agent of Construction Manager and by an authorized agent of Owner or either of them as the parties might agree. Any authorized agent of Construction Manager with permission to draw from the Special Account shall be bonded under a surety bond in form and substance satisfactory to Owner and in amount equal to the maximum expected balance to be deposited in such account. Funds from the Special Account may not be used to make payments of any

kind whatsoever to Construction Manager, including any fees to Construction Manager, reimbursement of expenses of Construction Manager or any other payments to Construction Manager. The bank at which such account is established shall be instructed not to pay upon any checks made to the name of Construction Manager and Construction Manager hereby waives the right to demand payment upon any check drawn on the Special Account. At least 15 days before any payments are required to be made to Trade Contractors, Construction Manager shall notify Owner, in writing, of the expected amounts of such payments, and within 15 days after receipt of such notice Owner shall deposit such amount in the Special Account. Provided that Owner makes the payments above, Construction Manager or its agent shall issue checks to the Trade Contractors, but only if Construction Manager first shall have determined that the respective Trade Contractors have complied in full with all of their obligations under their Trade Contracts and that such payments are then due to the respective Trade Contractors. Construction Manager shall act as a trustee for Owner in the disbursement of funds and shall be responsible to determine that such funds are paid to the proper recipients in accordance with the terms of their respective contracts. Construction Manager shall not be responsible for the bankruptcy or failure of a depository provided that such depository is a member of the Federal Reserve System, insured by the Federal Deposit Insurance Corporation and approved by Owner. Construction Manager shall not be responsible for any checks drawn on such account by Owner. It is not contemplated that Owner will draw checks on such account.

11.   *Construction Manager's Personnel:*

(a)   Construction Manager shall designate one person who, on its behalf, shall be responsible for the coordination of all the services to be rendered by it hereunder and to serve as its principal representative in its relationship with the Owner and the Architect. The designation of such person and his continuance in such capacity shall be subject to the prior written approval of Owner.

(b)   All personnel assigned by Construction Manager to its performance of this Agreement shall cooperate fully with personnel assigned to the Project by Owner and Architect; and in the event Owner determines that any personnel of Construction Manager have failed so to cooperate, or are not performing their duties satisfactorily, Construction Manager shall, at the request of Owner, replace such personnel. Designation of personnel assigned to the Project by Construction Manager and their continuance in such designation shall be subject to approval of Owner. Once assigned, personnel may not be transferred without prior approval of Owner.

12.   *Rights of Owner to Postpone and Terminate:* Owner shall have the right upon ten (10) days' written notice to Construction Manager, to postpone, delay, suspend or terminate all or any portion of the services to be performed by Construction Manager under the Agreement, or any additions hereto or modifications hereof, at any time and for any reason deemed to be in Owner's interest. In such event, Construction Manager shall be paid such part of the fee as shall have become due and payable hereunder so as to fairly compensate Construction Manager for the work done by him prior thereto, plus all unpaid reimbursable costs and expenses. Such postponement, delay, suspension or termination shall not give rise to any cause of action for damages or extra remuneration against Owner.

13.   *Termination for Cause:*

(a)   In the event that Construction Manager, for any reason or through any cause, (1) fails to perform materially any of the terms, covenants or provisions of this Agreement on his part to be performed; or (2) if Construction Manager shall violate any of the terms, covenants or provisions of this Agreement, then Owner may, upon the giving of a written notice (signed by the President or a Vice President of Owner) to Construction Manager, immediately terminate this Agreement for cause. Upon such termination, Construction Manager shall be entitled to payment of such amount, to be determined by Owner, as shall fairly compensate him for the work satisfactorily performed to the termination date; provided, however, that (i) no allowance shall be included for termination expenses; and (ii) Owner shall deduct from such amount and from any amount due and payable to Construction Manager to the termination date, but withheld or not paid, the total amount of additional expenses incurred by Owner in order satisfactorily to complete the services required to be performed by Construction Manager under this Agreement including the expense of engaging another construction manager for this purpose. If such additional expense shall exceed the amounts otherwise due and payable, Construction Manager shall pay Owner the full amount of such excess incurred by Owner.

(b)   In the event that the Owner violates any of the material terms, covenants or provisions of this Agreement, including failure to make payments within 20 days of the time any amount becomes due, then the Construction Manager may, upon written notice to Owner, terminate this Agreement, unless such violation is cured within ten (10) days after Owner's receipt of such notice. If this Agreement is so terminated, the Construction Manager shall recover from the Owner in lieu of the CM Fee a sum calculated at 2.0 times the Site Personnel Compensation incurred to the date of termination, less any fees paid to such date. Such sum together with any payments of parts of the CM Fee pursuant to Paragraph 24 shall not exceed the total CM Fee provided for in Article 21 and any other proven losses sustained.

14.   *Ownership of Working Papers:* All office diaries, daily records of labor, materials and equipment used, notes, designs, reports, drawings, tracings, and estimate schedules and other documents prepared by or for Construction Manager pursuant to this Agreement shall become the property of Owner. Upon the termination of Construction Manager's services for any reason whatsoever, Construction Manager shall deliver to Owner all the heretofore enumerated items which thereafter Owner may utilize in whole or in part or in modified form and in such manner or for such purposes or as many times as it may deem advisable without employment of or additional compensation to Construction Manager.

15.   *Responsibilities of Construction Manager:* Construction Manager specifically agrees that (a) its consultants, agents or employees shall possess the experience, knowledge and character necessary to qualify them individually for the particular duties they perform, and (b) it will comply with the provisions of the State of Pennsylvania's labor laws and all other Federal, state and local laws, ordinances and regulations that are applicable to the performance of the Agreement.

16.   *Standard of Execution of Work:* All work under this Agreement shall be performed in accordance with the highest standards of practice.

17.   *Inspection:* Owner shall have the right at all times to inspect the work of Construction Manager, Architect and Trade Contractors.

18.   *Construction Manager's Employees:* It is understood that all personnel and other consultants of Construction Manager are employees of Construction Manager and not of Owner, and Construction Manager alone is responsible for their work deportment.

19.   *Obligations of Owner:* Owner shall:

(a)   provide full information regarding its requirements for the project.

(b)   retain an Architect to design and to prepare construction documents for the Project. Architect's services, duties and responsibilities are described in the Agreement between Owner and Architect, pertinent parts of which will be furnished to Construction Manager and will not be modified without written notification to it.

(c)   furnish Construction Manager with a sufficient quantity of construction documents.

(d)   if Owner becomes aware of any fault or defect in the Project or nonconformance with the contract documents, give prompt written notice thereof to Construction Manager.

(e)   furnish the services, information, surveys and reports required by Paragraphs (b) and (c) inclusive at Owner's expense.

(f)   respond expeditiously to reasonable written requests from Construction Manager as to approvals by and Instructions from Owner.

20.   *Payment:* Owner shall pay to Construction Manager and Construction Manager shall accept from Owner, as compensation for all services to be rendered by Construction Manager pursuant to this Agreement, a fee for profit and overhead which shall be One Million Three Hundred Thousand dollars ($1,300,000.00) (the "CM Fee"), and reimbursement of Reimbursable Expenses as set forth below.

21.   *Construction Manager Fee:* The CM Fee shall be as set forth in Paragraph 20 even if there are substantial changes in the work ordered by Owner, and shall constitute full compensation to Construction Manager for the following:

(a)   compensation to officers or principals, home and regional office salaries (including, but not limiting to, the construction, accounting, purchasing, estimating, and cost control departments), services of a senior officer who shall have overall responsibility for the performance of this Agreement, home and regional office professional salaries, salaries of general supervisory employees who do not devote full time to the Project, and all payments mandated by law (including, e.g., taxes, pensions, and insurance) and all fringe benefits relating to these employees;

(b)   legal and accounting fees and bookkeeping expenses except for bookkeeping and accounting performed at the site;

(c)   cost of home office and regional general facilities including, but not limited to, rental cost or depreciation factor, light, heat and water, insurance related to home office, telephone, telegraph, sales, plans and estimating expenses, stationery, printing and postage, office and miscellaneous expenses, etc.;

(d)   miscellaneous travel and expediting costs (except as provided for in Article 22 (b) and (c);

(e)   taxes other than sales and use taxes;

(f)   interest expense;

(g)   advertising, dues and subscriptions;

(h)   contributions;

(i)   home office and regional office overhead;

(j)   profit; and

(k)   all site overhead and administrative expenses except as set forth in Article 22 (b) below.

22.   *Reimbursable Expenses:* Owner shall reimburse Construction Manager for the expenses set forth on Exhibit A attached hereto as set forth below (the "Reimbursable Expenses"). The Reimbursable Expenses shall include:

(a)   the salaries of the on-site employees, together with a 25% payroll burden which shall be deemed to include all amounts and allowances for vacation or holiday pay, social security, unemployment insurance, workmen's compensation or other fringe benefits. For on-site employees as described in Exhibit A to this Agreement, the compensation shall be substantially in accordance with the salary scale specified in said Exhibit (the "Site Personnel Compensation"), as such scale may be adjusted to reflect individual salaries and cost-of-living increases generally applicable to all employees of Construction Manager. Should Owner request Construction Manager to place additional personnel on the job-site staff over and above those indicated in Exhibit A, Owner shall reimburse Construction Manager for the cost of such personnel at the rates set forth on Exhibit A.

(b)   The following expenses by Construction Manager incurred on the site, substantially in accordance with the estimated amounts set forth on Exhibit A, with any variations to be approved in advance in writing by Owner:

(i)   Site Offices
(ii)   Site Utilities
(iii)   Site Equipment
(iv)   Site Supplies
(v)   Site Telephone
(vi)   Handling, shipping, mailing and reproduction of Project related needs.
(vii)   Fees paid for obtaining Governmental approvals.
(viii)   Any fines imposed on Construction Manager acting to perform its duties hereunder, when the acts could not have been avoided if Construction Manager was to have performed its duties are required hereunder.

(c)   Other costs and expenses for which reimbursement may be authorized by Owner in writing pursuant to this Article which may include, among other items, the following:

(i)   travel expenses of employees as and when approved by the owner.
(ii)   special consultants authorized by Owner.
(iii)   expenses to provide site offices, utilities, equipment and supplies for Architect and Owner at the site

It is expressly understood that this paragraph applies only to reimbursement of funds expended by Construction Manager to perform its duties and obligations hereunder, and does not apply to payment by Owner to Construction Manager for, *inter alia*, purchase orders and short form contracts deemed to be Trade Contracts or to compensate Construction Manager for any construction work performed by Construction Manager.

23.   *Payment of Fee and Expenses for Extra Work:* In the event the Project is not completed by the anticipated completion date, said date to be twenty-four (24) months from the date of this Agreement, by reason of any delay beyond the control of Construction Manager, by reason of a delay due solely to a design change during the construction phase, or if in the opinion of Owner, the circumstances and services performed by Construction Manager have so materially changed as to require in good faith a renegotiation of the CM Fee then Owner shall: (a) reimburse Construction Manager for Reimbursable Expenses incurred subsequent to the contemplated completion date; and (b) enter into good faith negotiations with Construction Manager for the renegotiation of the CM Fee.

24.   *Method of Payment of CM Fee:* The $1,300,000 CM Fee shall be paid in monthly installments in an amount equal to 1.4 times the Site Personnel Compensation for the preceding month, with any remainder payable by Owner upon completion of the Project. All payments for service are contingent upon the satisfactory performance of the Construction Manager's obligations hereunder.

25.   *Payment of Reimbursable Expenses:* Reimbursable Expenses shall be payable by Owner monthly on the basis of expenses incurred during the preceding month. Construction Manager may submit application for reimbursement of expenses together with supporting data as soon as incurred but not more often than once each month.

26.   *Delay Due to Strike:* In the event the completion of the Project is delayed by reason of strike or other circumstances not due to the fault of Construction Manager, Owner shall reimburse Construction Manager for the cost of its full job-site staff as provided by this Agreement up to the first 30 days of such delay. Construction Manager shall reduce the size of its job-site staff for the remainder of the delay

period after the first 30 days, as directed by Owner and, during such period, Owner shall reimburse Construction Manager for the costs of such reduced, rather than full, staff plus 25% to cover all fringe benefits. Upon the termination of the delay, Construction Manager shall restore its job-site staff to its former size, subject to the approval of Owner, whereupon the provisions of Articles 20, 21, 22, 23, 24, and 25 shall apply as though there has been no delay.

27. *Accounting Records:* Construction Manager shall maintain complete, detailed and accurate cost and accounting records as to all costs. During the term of this Agreement and at any time within three years thereafter, Construction Manager shall make such records available to Owner or its authorized representatives for review and audit at Construction Manager's place of business during normal business hours. In the event Owner authorizes Construction Manager to retain the services of consultants or subcontractors for which Construction Manager will be entitled to reimbursement hereunder, Construction Manager agrees to include in all its contracts with such consultants and subcontractors a requirement that they maintain complete, detailed and accurate cost and accounting records as to all their costs relating to the services and materials furnished by them under such contracts and that during the term of this Agreement and at any time within three years thereafter if required by Owner they will make such records available to Owner or its authorized representatives for review and audit at such places as may be designated by Owner. This paragraph is not intended to imply that the Trade Contractors are subcontractors of Construction Manager.

In the event all or any part of such records are not maintained by Construction Manager, its consultants or any of its subcontractors (not including Trade Contractors), or made available to Owner as provided herein, any item not supported by reason of the insufficiency or unavailability of such records shall, at the election of Owner be disallowed and, if payment therefor has already been made, Construction Manager, upon demand, shall refund to Owner the amounts so disallowed. Payments to Construction Manager or approval by Owner of any application for payment submitted by Construction Manager shall in no way affect Construction Manager's obligations hereunder or the rights of Owner to obtain a refund of any payment to Construction Manager which was in excess of that to which it was lawfully entitled.

28. *Insurance:* Construction Manager shall procure and maintain, during the term of this Agreement, policies of insurance with insurance companies authorized to do business in Pennsylvania and approved by Owner with minimum limits as set forth on Exhibit C hereof, covering all acts and omissions of Construction Manager and of Owner, as a named insured, and their agents, employees and consultants in their performance of this Agreement and the construction of the Project. The policies shall be endorsed to include Owner as an additional or named insured. In the event that Owner carries insurance which covers the same risks as are covered by Construction Manager's insurance policies, then Owner's insurance shall be considered to be primary to the Construction Manager's policy and will apply only if the limits under Owner's policy are exceeded. Before commencing its performance of this Agreement, Construction Manager shall furnish to Owner certificates in the form satisfactory to Owner showing that it has procured such insurance, which certificates shall provide that the policy shall not be changed or cancelled without thirty days' written notice to Owner and which certificates shall make reference to this Agreement and Articles 23 and 29 hereof.

29. *Indemnity:* To the fullest extent permitted by law, the Construction Manager shall indemnify and hold harmless Owner, Architect and their agents and employees from and against all claims, damages, losses and expenses arising out of or resulting from the performance of its services, provided that any such claim, damage, loss or expense is attributable to bodily injury, sickness, disease or death,

or to injury to or destruction of tangible property (other than the work itself) including the loss of use resulting therefrom, and is caused in whole or in part by any negligent act or omission of the Construction Manager or anyone for whose acts he may be liable, regardless of whether or not it is caused in part by a party indemnified hereunder.

30. *Copyrights and Patents:* The Construction Manager shall not, without the prior written approval of Owner, specify for the Project, or necessarily imply the required use of any article, product, material, fixture or form of construction, the use of which is covered by a patent, or which is otherwise exclusively controlled by a particular firm or group of firms. Construction Manager shall be liable and hereby agrees to defend, indemnify and hold harmless Owner against all claims against Owner for infringement of any copyright or patent rights or claim for unfair use of systems, graphs, charts, designs, drawings or specifications furnished or recommended by Construction Manager in the performance of this Agreement.

31. *Construction Manager to Perform Properly:* Construction Manager shall be liable to Owner for all losses, expenses and damage caused by the failure of Construction Manager properly to perform its obligations under this Agreement. Construction Manager shall not be entitled to any compensation for services or reimbursement for costs or expenses with respect to any such obligations not properly performed by it hereunder.

32. *Acceptance of Final Payment:* The acceptance of final payment under this Agreement by Construction Manager or any person or firm claiming under it shall operate as and shall be a release of Owner from all claims by and liability from or by Construction Manager and any such person or firm and his and its successors, legal representatives and assigns, for anything done or furnished pursuant to the provisions of this Agreement.

33. *Owner's Project Representative to Authorize Action:* Wherever under this Agreement action is to be taken or approval given by Owner, such action or approval may be taken or given only by the Owner's Project Representative or such person as may be designated in writing by the Owner's Project Representative to act on behalf of Owner for such purpose. The Construction Manager shall not act or rely upon any purported direction or approval by any other person on behalf of Owner, unless authorized in writing by a President or Vice President of Owner.

34. *Time to File Claim:* No action shall be maintained by Construction Manager, its successors or assigns against Owner on any claim based upon or arising out of this Agreement or out of anything done in connection with this Agreement unless such action shall be commenced within six (6) months after Architect has determined final payment is due.

35. *Assignment:* Since it is intended to secure the services of XYZ Construction Company, Inc. as Construction Manager, this Agreement shall not be assigned, sublet or transferred without the prior consent of Owner.

36. *Unlawful Provisions:* If this Agreement contains any unlawful provision not an essential part of the Agreement and which appears not to have been a controlling or material inducement to the making thereof, the same shall be deemed of no effect and shall upon the application of either party be stricken from the Agreement without affecting the binding force of the Agreement as it shall remain after omitting such provision.

37. *Errors:* If this Agreement contains any errors, inconsistencies, ambiguities or discrepancies, including typographical errors, Construction Manager shall request a clarification of same by writing to Owner whose decision shall be binding upon the parties.

38. *Workmen's Compensation:* Construction Manager shall procure and maintain during the life of this Agreement Workmen's Compensation Insurance and Disability Benefits Insurance in accordance with the laws of the State of Pennsylva-

nia and shall furnish to Owner two (2) certificates of such insurance upon execution of this Agreement.

39. *Notices:* Notices and invoices for payment to Owner shall be addressed as set forth below:

(a) All notices, demands and other communications hereunder shall be in writing and shall be deemed to have been duly given if personally delivered or mailed first class, postage prepaid:

    (i) If to Construction Manager: _____

        Attention: _____

   (ii) If to Owner: _____

        Attention: _____

    with a copy to the Owner's Project Representative: _____

(b) Construction Manager's invoices to Owner and Owner's payments to Construction Manager shall be personally delivered or mailed first class, postage prepaid to the address set forth above.

(c) Either party may change the addresses set forth for it herein upon written notice thereof to the other.

40. *Conflict of Interest:* Construction Manager covenants that neither it nor any of its officers has any interest, nor shall it or they acquire any interest, directly or indirectly, which would conflict in any manner or degree with the performance by Construction Manager of its services hereunder. Construction Manager further covenants that in the performance of this Agreement, no person having such interest shall be employed by it.

41. *Non-Discrimination:* Construction Manager shall comply with the requirements set forth in U.S. Department of Labor regulations dealing with (i) equal employment opportunity obligations of government contractors and subcontractors, 41 C.F.R. S60-1.4(a) (1)-(7), (ii) employment by government contractors and subcontractors of Vietnam-era and disabled veterans, 41 C.F.R. S60-250.4 (a)-(m), and (iii) employment of the physically handicapped by government contractors and subcontractors, 41 C.F.R. S60-741.4(a)-(f). All of the above referenced regulations are hereby incorporated herein and expressly made a part hereof. Inclusion in this Agreement of this paragraph 41 does not, and shall not be deemed to, constitute an acknowledgement or admission by Owner that it is a government contractor or first-tier subcontractor or that it is obligated to abide by the aforementioned regulations for the purposes contemplated thereby.

42. *Confidentiality:* During the confidential relationship established hereby, Owner may communicate to Construction Manager certain information to enable Construction Manager to render the services hereunder or may develop confidential information for Owner. Construction Manager agrees (i) to treat, and to obligate Construction Manager's employees, if any, to treat as secret and confidential, all information identified by Owner as confidential and any such information which constitutes a trade secret as a matter of law; (ii) not to disclose any such information or make available any reports, recommendations and/or conclusions which Construction Manager may make for Owner to any person, firm or corporation or use it in any manner whatsoever without first obtaining Owner's written approval; and (iii) not to disclose to Owner any information obtained by Construction Manager on a confidential basis from any third party unless (a) Construction Manager shall have first received written permission from such third party to disclose such information or (b) such information is presently in the public domain.

43. *Arbitration:*

(a) All claims, disputes and other matters in question between the parties to this Agreement, arising out of, or relating to this Agreement or the breach thereof, shall be decided by arbitration in accordance with the Construction Industry Arbitration Rules of the American Association then obtaining unless the parties mutually agree otherwise. No arbitration, arising out of, or relating to this Agreement, shall include, by consolidation, joinder or in any other manner, any additional party not a party to this Agreement except by written consent containing a specific reference to this Agreement and signed by all the parties hereto. Any consent to arbitration involving an additional party or parties shall not constitute consent to arbitration of any dispute not described therein or with any party not named or described therein. This Agreement to arbitrate and any agreement to arbitrate with an additional party or parties duly consented to by the parties hereto shall be specifically enforceable under the prevailing arbitration law. Notwithstanding the foregoing, if Owner has been made or becomes a party to an arbitration with the Architect or any of the Trade Contractors, Construction Manager hereby expressly consents upon the demand of Owner (but not Architect or a Trade Contractor) to become a party to such arbitration and to be bound thereby in accordance with this Paragraph 43.

(b) Notice of the demand for arbitration shall be filed in writing with the other party to this Agreement and with the American Arbitration Association. The demand shall be made within a reasonable time after the claim, dispute or other matter in question has arisen. In no event shall the demand for arbitration be made after the date when institution of legal or equitable proceedings based on such claim, dispute or other matter in question would be barred by the applicable statute of limitations.

(c) The award rendered by the arbitrators shall be final, and judgment may be entered upon it in accordance with applicable law in any court having jurisdiction thereof.

44. *Captions:* Captions are not part of the Agreement and are for reference purposes only and in no way define, limit or describe the scope or intent of the article or section of the Agreement nor in any other way affect this Agreement.

45. *ABC Manufacturing Corporation General Conditions of Contract:* The ABC Manufacturing Corporation General Conditions of Contract attached hereto as a part of Exhibit D are not incorporated herein by reference and are not deemed to be a part of this Agreement. The duties of Architect and Construction Manager set forth therein are merely descriptive for the information of Trade Contractors and to the extent there is any conflict between the obligations set forth herein and those set forth in the ABC Manufacturing Corporation General Conditions of Contract, those set forth herein shall govern. Owner may without permission of Construction Manager, amend or modify the ABC Manufacturing Corporation General Conditions of Contract.

IN WITNESS WHEREOF, the parties hereto have executed this Agreement in quadruplicate the day and year first above written, two copies to remain with Owner and two copies to be delivered to Construction Manager.

CONSTRUCTION MANAGER	OWNER
XYZ Construction Corporation, Inc.	ABC Manufacturing Corporation
By_____	By_____
Date:_____	Date:_____
Attest:	Attest:
_____	_____
Date:_____	Date:_____

Implementation Agreement between Owner, Construction Manager, and Trade Contractor

Although the construction manager agrees to guard the owner against defects and deficiencies in the work, the construction manager does not guarantee the cost, timeliness, or quality of construction. For this reason, particular attention must be paid to the agreement between the owner and the trade contractors to be certain that appropriate guarantees and warranties are rendered by the trade contractors and, in addition, that adequate control provisions are maintained.

AIA Document A101/CM (with appropriate general and supplemental conditions) can be used to accomplish and implement this purpose. Another form which is an example of an agreement implementing construction between the construction manager as the owner's representative and the trade contractor follows. It is also owner-oriented in terms of controlling the work. (This form is adapted from one found in Robert F. Cushman and William J. Palmer, *The Dow Jones Businessman's Guide to Construction,* Dow Jones Books, Princeton, N.J., 1981.)

IMPLEMENTATION AGREEMENT

This Agreement, made this_____day of_____, 1982, by and between DEF Trade Contractor ("Contractor") and XYZ Construction Company ("Construction Manager") for the supply and installation of _____ by Contractor in connection with the construction of the UVW Division Gear and Assembly Plant, Norristown, Pennsylvania for ABC Manufacturing Company ("Owner").

WITNESSETH

WHEREAS, Owner plans to construct a Gear and Assembly Plant and related facilities in Norristown, Pennsylvania on property located at _____, (hereinafter called the "Project"); and

WHEREAS, Owner and Managing Contractor desire to employ Contractor to perform certain work in the construction and/or equipping of the Project; and

WHEREAS, Contractor desires to perform such work;

NOW THEREFORE, in consideration of the mutual covenants herein contained, the parties hereto agree as follows:

SECTION 1. AGREEMENT FOR SERVICES

Subject to the terms and conditions of this Agreement:

1.01   Contractor agrees to perform the services set forth herein relating to the construction and/or equipping of Owner's Project.

1.02   In consideration of Contractor performing the services referred to in Section 1.01, Owner agrees to pay the Contractor the amount set forth in Section 3 hereof.

SECTION 2. SCOPE OF WORK

2.01   Contractor represents that it has visited the site of the Project, has examined the drawings, specifications, general conditions and schedules (hereinafter

called "Drawings and Specifications") prepared by _____,
A.I.A. Architect & Associates (hereinafter called the "Architect"), which Drawings
and Specifications, together with the bid documents are hereby incorporated by
reference and made a part of this Agreement, and has familiarized itself with local
conditions under which the work hereunder is to be performed and has correlated
its observations with the requirements of the bid documents. A list of specific
drawings applicable to the work to be performed by Contractor is contained in
EXHIBIT "A" attached hereto and made a part of this Agreement. Such listing
shall not limit the generality of the foregoing provisions.

2.02   Contractor agrees to perform the following work and/or deliver the
following materials in connection with the construction of the Project:

SECTION 3. CONTRACT PRICE

Owner agrees that it will pay to Contractor and Contractor agrees to accept as
full compensation for the work to be done and the material to be furnished as
specified in Section 2.

SECTION 4. PAYMENTS

4.01   Contractor shall be paid for its services rendered hereunder by Owner
as follows:

4.01.1   On or before the tenth day of each month, Owner will pay to Contractor ninety percent (90%) of the contract price of the material and/or labor delivered and erected at the job site up to the twentieth day of the previous month, as
estimated by Contractor and approved by Managing Contractor, less the aggregate of previous payments made by Owner to Contractor;

4.01.2   A final payment of ten percent (10%), shall be payable sixty (60) days
after completion of construction of the Project (for purposes of this Agreement,
"completion of construction," "completion date" or "date of completion" shall all
mean the date upon which the Project is first opened to the general public),
provided that all of Contractor's work hereunder (including Managing Contractor's "punch list") has been completed to the full satisfaction of Architect and
Managing Contractor.

4.02   As a prerequisite to any payments being made to it, Contractor shall
execute and deliver to Owner:

4.02.1   full and complete waivers and releases of liens from all persons furnishing labor and materials toward the performance hereof; and

4.02.2   such formal guarantees pertaining to the work as may be required by
Owner, and such other affidavits, receipts, waivers and other documents as
Owner shall reasonably require.

4.03   Prior to making any payment to Contractor hereunder, Owner may
demand proof of payment by Contractor to all people working on this job and
entitled to compensation from Contractor, and proof of payment to all materialmen who have delivered materials thereto. If such proof is not provided after
demand, Owner shall have the right to withhold payment of any monies otherwise
due to Contractor and shall not be deemed to be in default hereunder. Nothing in
this Agreement shall be construed to obligate Owner to any such person or materialmen. Owner shall have no obligation to demand such proof and may at any time
waive its right thereto.

4.04   The payment or payments made to Contractor during the progress of
the work or the occupancy of any part of the premises shall not be construed as an
approval or acceptance of Contractor's work by Owner or Managing Contractor.

4.05   The final payment by Owner to Contractor shall constitute a waiver of all claims by Owner against Contractor except those arising out of (a) unsettled liens, (b) faulty or defective work appearing after completion and acceptance of the Project by Owner, (c) failure of the work to comply with the requirements of the Drawings and Specifications, or (d) the terms of any special guarantees required by the Drawings and Specifications, bid documents or this Agreement. The acceptance of final payment by Contractor shall constitute its waiver of all claims against Owner, except those claims previously made in writing and still unsettled at the time of the final payment.

SECTION 5. CONTRACTOR'S SERVICES AND STATUS

5.01   Contractor covenants with Owner to use its best skills and judgment in forwarding the interests of Owner. Contractor agrees to provide efficient business administration and superintendence and to keep an adequate supply of competent workmen and materials on the job at all times and to secure its execution in the best, soundest, most expeditious and economical manner consistent with the interests of Owner. Managing Contractor shall have the right to reasonably request the replacement of supervisory personnel of Contractor and Contractor shall promptly comply with such request.

5.02   Contractor hereby warrants to Owner the proper quality and character of the materials and workmanship and the adequacy, suitability and workability of the materials and equipment to be furnished by Contractor hereunder. Contractor shall repair and make good, at no cost to Owner, any damages, deficiencies or faults that may occur as the result of any imperfect or defective work done or material furnished by Contractor for a period of one (1) year from the date of completion of the Project.

5.03   Contractor agrees to obtain (if applicable), for the benefit of Owner, a written guarantee from the manufacturer(s) of the materials referred to in Exhibit B hereto, warranting the durability of such materials for a period of one (1) year, from date of completion.

5.04   All work shall be done in a workmanlike manner and all materials shall be new and of superior quality. All work performed and all labor used by Contractor shall be acceptable to Owner. All work performed and material delivered shall also be subject to the approval of Managing Contractor (who shall be the sole judge thereof) and shall comply with all laws, regulations, requirements and rules of government authorities having jurisdiction over the Project. Contractor shall, at its own expense, promptly replace all work or material not so approved.

5.05   In the event a jurisdictional labor dispute should arise affecting any aspect of Contractor's work, Contractor shall abide by the decision of the Joint Conference Board, but such decision shall not relieve Contractor from fulfilling all of its obligations hereunder.

5.06   Contractor shall make all necessary field measurements at its expense.

5.07   The decision of the Managing Contractor or the Owner in interpreting the true intent and meaning of the Drawings and Specifications or explanations shall be final and binding upon Contractor.

5.08   All work covered by this Agreement done at the site of construction or in preparing or delivering materials to the site shall be at the sole risk of Contractor until final completion of the work and Owner's acceptance of it.

5.09   Contractor shall have the sole responsibility to protect its work from damage or dirt caused by it or other contractors during construction. Contractor shall maintain its work area in a broom-clean condition at all times during construction and Contractor shall, during and upon completion of its work, clean up and remove from the premises all debris which results from the work performed pursuant to this Agreement. It shall be Contractor's sole responsibility to clean its work area and repair any damage at its expense, or if the dirt or damage was caused by another contractor or other person, to collect the costs from such contractor or person. Owner shall have no obligation or liability to Contractor in such case and shall not be a party to any dispute which may arise therefrom. If Managing Contractor instructs Contractor to clean up and remove debris from its work area and Contractor does not promptly comply, Owner or Managing Contractor may have such services performed and deduct the cost thereof from any monies due Contractor hereunder.

5.10   Contractor shall, at its own expense, repair, restore or replace any personal or real property belonging to Owner which Contractor or his subcontractors, their employees, agents or suppliers, may damage or destroy while on or near Owner's premises.

5.11   Contractor shall not interfere in any manner with the normal operations of the Project in accomplishing its work hereunder.

5.12   Contractor shall be responsible for initiating, maintaining and supervising all health and safety precautions and programs in connection with its work and shall comply with all local, state and federal laws and regulations relating hereto, including but not limited to Williams-Steiger Occupational Safety and Health Act of 1970.

5.13   If any work is covered over by Contractor contrary to the prior request of Owner, Owner may require that it be uncovered for its inspection and replaced, all at Contractor's expense. If any work has been covered which Owner has not specifically requested to observe prior to its being covered, Owner may request to see such work and it shall be uncovered by Contractor. If the work is found to be in accordance with this Agreement, in the opinion of Managing Contractor, the cost of uncovering and replacing the work shall be charged to Owner. If the work is found by Managing Contractor not to be in accordance with this Agreement, Contractor shall pay such costs.

5.14   If the laws, ordinances, rules, regulations or orders of any public authority having jurisdiction over the Project require any work to be inspected, tested or approved, Contractor shall give Owner and Managing Contractor timely notice that it is ready for inspection and of the date of the inspection so Owner and Managing Contractor may observe such inspection or testing. Contractor shall bear all costs of such inspections, tests and approvals, unless otherwise provided herein.

5.15.   Owner may agree to accept defective, deficient or non-conforming work instead of requiring its removal and correction. In such case a Change Order will be issued to reflect an appropriate reduction in the compensation to be paid Contractor hereunder. Acceptance of such defective, deficient or non-conforming work shall not relieve Contractor of any of its obligations as set forth in the Drawings and Specifications.

5.16   Contractor is an independent contractor and is not authorized to represent, obligate or contract for or on behalf of Owner or Managing Contractor.

5.17   Contractor shall prepare all necessary shop drawings and samples as required by the General Conditions or the Architect, Owner or Managing Contractor. The approval by Architect or Owner or Managing Contractor of any of Contractor's shop drawings shall not relieve Contractor from responsibility for any deviations from the Drawings and Specifications or errors in such drawings unless the Contractor has, at the time of submission, called such deviations to the attention of Owner or Managing Contractor in writing, and has received the prior written approval for such deviations from Owner or Managing Contractor. Contractor hereby indemnifies Owner against any claims which may be asserted by anyone against Owner due to unapproved deviations or errors in Contractor's shop drawings.

5.18 Contractor shall not interfere with the work of other contractors at the job site or the storage of their material or equipment and shall connect and coordinate its work in accordance with the directions and schedules of the Managing Contractor.

5.19 All blueprints of the plans and copies of the specifications as may be required by Contractor for the construction of work specified in this Agreement will be paid for by Contractor.

SECTION 6. TIME OF COMPLETION

6.01 Time is of the essence of this Agreement and work to be performed and material to be furnished under this Agreement shall be scheduled and completed not later than _____ (_____) days after the execution hereof (hereinafter referred to as the "Time of Completion"). Contractor shall be responsible for maintaining such schedule and the failure to do so shall be a material breach of this Agreement. Any additional or unanticipated cost or expense required to maintain the schedule shall be Contractor's obligation.

6.02 Contractor shall coordinate the delivery of the material required and the work to be performed under this Agreement with the schedule determined by the Managing Contractor in accordance with the schedule set forth in Exhibit C hereto.

6.03 If Contractor is delayed at any time in the progress of its work by:
(a) any act or negligence of Owner,
(b) any other contractor employed by Owner,
(c) changes ordered by Owner in the work,
(d) labor disputes,
(e) delays authorized by Owner, or
(f) delays caused by casualties or causes beyond the Contractor's control,
then in such event, the Time of Completion shall, at Contractor's request, be extended by a Change Order to such reasonable time as Owner may determine. Contractor shall make such request for a time extension in writing to Owner within five (5) days after the commencement of the delay or such delay will be waived by Contractor.

6.04 Owner or Managing Contractor may require certain parts of Contractor's work hereunder to be prosecuted in preference to other parts. Owner shall have the right, at any time, to delay or suspend the whole or any part of the work for a reasonable time without additional compensation to Contractor and no delay, suspension or obstruction by Owner, or its agents, shall serve to terminate this Agreement except as otherwise provided in Section 13 hereafter. If Contractor shall delay the progress of its work so as to cause any loss or damage to Owner, Contractor shall indemnify Owner for any such loss, damage, claim or judgment resulting from such delay.

SECTION 7. PERFORMANCE BOND

Contractor shall furnish a Performance Bond and Payment Bond (with a provision for one (1) year guarantee against defects in materials and workmanship from the date of issuance of a final occupancy permit for the completed building), which Bonds shall be conditioned upon the full and faithful performance of this Agreement by Contractor, and upon payment by Contractor for all labor and materials used by it in the performance hereof, in form and with sureties acceptable to Owner. Said Bonds shall be furnished within ten (10) days of the date of this Agreement and the premiums for such Bonds shall be paid by Contractor.

SECTION 8. INSURANCE

8.01   Before starting its work or delivering any materials to the site of construction, Contractor shall furnish certificates of insurance evidencing that Contractor has placed in force valid insurance, with an insurer acceptable to Owner, as follows:

8.01.1   Workmen's Compensation Insurance covering its full liability under the Workmen's Compensation Laws of the State of _____ as required by the _____ Industrial Commission;

8.01.2   Comprehensive General Liability Insurance in the amount of _____ for injury or death to each person, _____ for each occurrence and for each claim of property damage; and _____.

Such insurance shall fully protect Owner, Architect and Managing Contractor and shall be maintained in full force until all of Contractor's obligations hereunder are completed.

8.02   The certificates of insurance required by Section 8.01 shall contain the following provisions:

(a)   name Owner, Architect and Managing Contractor as additional insureds;

(b)   in the event of any change in the limits of liability, or the cancellation of the insurance in its entirety, the insurer will give Owner, Architect and Managing Contractor written notice at least thirty (30) days prior to the effective date of such change or cancellation and the insurance coverage shall remain in force during said thirty (30) days; and

(c)   waive any right of subrogation of the insured against Owner, Architect and Managing Contractor.

8.03   If Contractor fails to carry any insurance provided for herein, Owner may, but shall not be obligated to, procure the same and charge the cost thereof to Contractor.

8.04   At Owner's request, Contractor will name any bank and/or financial institution designated by Owner as additional insureds under the above insurance policies.

8.05   Owner will procure, at its expense, insurance against fire, flood and riot covering the entire Project including temporary structures and materials in place, to the full insurable value thereof and will maintain such insurance coverage during the course of construction and until completion of the Project.

8.05.1   Losses, if any, shall be payable to Owner and Contractor as their interests may appear. Owner and Contractor waive all rights against each other for damages caused by fire and other perils to the extent such losses are covered by insurance provided under this section except such rights as they may have to the proceeds of such insurance. Contractor shall require similar waivers from its subcontractors.

8.05.2   Owner shall not be required to provide insurance against loss, theft or mysterious disappearance of Contractor's materials, tools and equipment; and Contractor agrees to hold Owner harmless for any such loss, theft or mysterious disappearance.

SECTION 9. TAXES AND PERMITS

9.01   Contractor shall pay all sales, retailers' occupational, services use, excise, Old Age Benefit and Unemployment Compensation taxes, as well as any other taxes or duties upon the material and labor furnished under this Agreement, as required by any municipal, state or federal authority, department or agency having jurisdiction over the Project. All such taxes and levies are included in the compensation to be paid Contractor hereunder. All records maintained by Contractor pertaining to such taxes and levies shall be made available on reasonable notice to Owner, its designated employees, representatives or agents, and to representatives of any of the taxing bodies, at all reasonable times.

9.02   Contractor shall procure at its expense certificates of every kind (except the certificate of occupancy) which any municipal department or other agency may require or issue with respect to its work on the Project and deliver such certificates to Owner immediately upon completion of said work.

9.03   Contractor shall, at its own cost and expense, apply for and obtain all permits required in connection with the work covered by this Agreement, except the plan check and general building permit and sewer connection fee which shall be the responsibility of Owner. In the event Owner or Managing Contractor procures such permits for or on behalf of Contractor, it shall reimburse Owner or Managing Contractor for the costs thereof.

SECTION 10. CHANGES

10.01   Owner, at any time before completion of Contractor's work hereunder, may order additions, omissions or alterations to or in Contractor's work, materials, equipment or services. Such changes shall only be made pursuant to the prior express written order or authorization of Owner or Managing Contractor (hereinafter sometimes called "Change Order") and no payment shall be made for any changes made without such written order.

10.02   In the event additions, alterations or omissions are ordered and approved by Owner or Managing Contractor in writing during the course of construction and before completion of Contractor's work hereunder, the additions or reductions of fees and costs resulting therefrom shall be negotiated between Owner and Contractor prior to the making of such changes.

10.03   The costs and fees applicable to additions, alterations or omissions ordered by Owner or Managing Contractor shall be calculated and the Contractor agrees to perform such work on the basis of actual net cost of the material involved plus the cost of labor and all other costs at the following rates:

Labor rate per direct man hours $_____
Overhead per direct man hour $_____
Profit—Percentage of total cost _____%

SECTION 11. USE OF OWNER'S OR OTHER CONTRACTORS' EQUIPMENT

If Owner or other contractors have equipment, tools or appliances on the construction site which are not being used, Contractor may have use of such equipment, tools or appliances at no cost to Contractor, provided (i) prior permission in writing for such use has been granted by the owners of such equipment, tools or appliances to Contractor, and (ii) Contractor agrees to indemnify and hold Owner, Managing Contractor and the legal and rightful owners of such property harmless from all damages to or loss of the property and all losses, claims and damages arising from the use of such property or arising out of injury or death of persons or damage to property occasioned by the direct or indirect use of such property by Contractor.

SECTION 12. CONTRACTOR'S INDEMNITY

12.01   Contractor agrees to indemnify and hold Owner, Architect and Managing Contractor harmless (including reasonable attorney's fees) and, at the request of Owner, a Trustee and/or any lender or financial institution directly involved in the building or financing of the Project, from the following:

12.01.1   Any claims, damages, liabilities, actions, demands or losses for personal injury, death, or loss of or damage to property, arising out of or in any way connected with the performance of this Agreement by Contractor;

12.01.2   Any claims, demands, charges, costs, expenses or damages incurred or sustained by Owner, Architect or Managing Contractor or for which Owner, Architect or Managing Contractor may become liable by reason of anything to be supplied or used by Contractor under this Agreement being covered by a patent or license or any claim that any such item infringes upon a patent or license;

12.01.3   Any losses, judgments, expenses and liabilities resulting from any orders entered by any agency or governmental authority pursuant to any rule, statute, regulation or ordinance related to any claims for (i) death or injury to any person, including persons employed by Contractor or his subcontractors (if any) and its or their agents or employees and (ii) for any loss, damage or destruction of property.

12.02   Contractor agrees that all materials used by it on the Project will be purchased by it either for cash or on credit and that it will not execute any conditional bills of sale on account thereof so that such materials will be free from any claims, liens or encumbrances when placed in the Project.

12.02.1   Contractor shall protect and hold Owner, Architect and Managing Contractor harmless from all claims, damages, liens, losses and costs which Owner, Architect or Managing Contractor may suffer by reason of the filing of any notices, liens, security interests or encumbrances or from failure of Contractor to obtain cancellation and discharge thereof. If, at any time, there shall be evidence of any liens or claims for which Owner may become liable and which may in any event be chargeable to Contractor, Owner shall have the right to retain out of any payment due, or thereafter to become due Contractor, an amount sufficient to fully indemnify Owner against such lien or claim, unless Contractor shall cause such lien or claim to be covered by a bond by statutory procedure and removed as a cloud against the title of the Project or shall provide Owner with sufficient bonds or sureties, written by companies satisfactory to Owner, for one and one-quarter (1-1/4) times the amount of the claim.

12.02.2   In the event of a lien or claim after all payments are made, Contractor shall repay to Owner all sums which it may be compelled to pay in discharging such lien or claim, including legal fees and other costs. Contractor may, after bonding against any such claim at his own expense, undertake the legal defense against such claims in its own name or in the name of Owner without in any way reducing the effectiveness of the foregoing.

12.03   Should any party to this Agreement suffer injury or damage to person or property because of any act or omission of any other party to this Agreement, or any of their employees, representatives or agents, claims shall be made in writing to such other party within a reasonable time after the claiming party first has knowledge of such injury or damage.

SECTION 13. DEFAULT BY CONTRACTOR

13.01   The occurrence of any of the following events shall be a default by Contractor hereunder (hereinafter called "Events of Default"):

13.01.1   If Contractor shall default in the due observance and performance of any covenant or agreement contained herein and any such default shall continue unremedied for three (3) days after written notice of such default shall have been given by Owner or Managing Contractor to Contractor;

13.01.2   If Contractor shall (a) voluntarily terminate operations or consent to the appointment of a receiver, trustee or liquidator of Contractor for all or a substantial portion of its assets, (b) be adjudicated bankrupt or insolvent or file a voluntary petition in bankruptcy, or admit in writing its inability to pay its debts as they become due, (c) make a general assignment for the benefit of creditors, (d) file a petition or answer seeking reorganization or an arrangement with creditors or take advantage of any insolvency law, or (e) if action shall be taken by Contractor for the purpose of effecting any of the foregoing;

13.01.3   If any warrant, execution or other writ shall be issued or levied upon any property or assets of Contractor and shall continue unvacated and in effect for a period of thirty (30) days; or

13.01.4   If Contractor should, in the judgment of Owner or Managing Contractor, neglect to prosecute the work hereunder properly and with proper dispatch in accordance with the time schedule agreed upon between Contractor and Managing Contractor or Owner;

Then, if any such Event of Default continues for three (3) working days after written notice to the Contractor, Owner may, without prejudice to any other remedy it may have at law or in equity, (a) terminate this Agreement, suspend all payments otherwise due to Contractor hereunder, finish the work by such means as Owner may see fit, deducting from any balance due Contractor the cost of finishing the work and paying the excess, if any, to Contractor and in the event the cost of finishing Contractor's work exceeds the balance due Contractor, such excess shall be paid by Contractor to Owner within five (5) days of invoicing by Owner, or (b) at its option, Owner may remedy any Event of Default and deduct the cost thereof. The costs and expenses of completing the work of Contractor shall be computed and audited by Owner's designated representative, whose certification thereof shall be final and binding upon the parties. The audit shall be made in accordance with generally accepted accounting principles and Contractor shall pay all costs of such audit.

13.02   In the event Owner is required to perform or does perform or hires others to perform any work or services because of the default of the Contractor, the cost thereof, plus an additional ten percent (10%) to cover Owner's overhead expenses, shall be promptly paid by Contractor to Owner.

13.03   It is expressly agreed that Owner reserves the right to offset any and all claims made by Contractor for payment of its fees hereunder or the reimbursement of additional costs it incurred, with any claims that Owner might have against Contractor for failure to comply with any of the terms and conditions of this Agreement.

SECTION 14. TERMINATION OF AGREEMENT

14.01   Termination by Contractor.

14.01.1   Contractor may terminate this Agreement if:

(a)   the work on the entire Project shall be abandoned or indefinitely suspended by Owner or delayed for more than six (6) months through no fault of Owner, provided that such abandonment, suspension or delay was not caused by any act or failure to act by Contractor; or

(b)   Owner shall commit an act of bankruptcy, making an assignment for the benefit of creditors, be transferred to a receiver, be unable to pay its debts as they come due or sell all or a substantial portion of its assets, but only if any such act prevents Owner from completing its obligations under this Agreement.

14.01.2   Upon the occurrence of any of the events set forth in Section 14.01.1 above, Contractor shall give Owner five (5) days prior written notice of its intent to terminate this Agreement. Upon such termination by Contractor, Owner shall be liable to Contractor only for compensation earned and payable to the date of such termination and for any damages Contractor sustains resulting directly from such termination; provided, however, that Contractor shall not be entitled to prospective profits or fees for work unperformed or materials, labor or service not furnished.

14.02   Termination by Owner.

14.02.1   Owner may terminate this Agreement if:

(a)   Contractor shall commit an Event of Default as set forth in Section 13 above and Contractor shall not have remedied such default after five (5) days written notice by Owner; or

(b)   Owner shall elect to abandon the Project as set forth in Section 15 hereafter.

14.02.2   Owner shall give Contractor five (5) days prior written notice of its intent to terminate this Agreement. In the event of termination pursuant to sub-section (a) above, Owner shall be liable to Contractor only for compensation earned and payable to the date of such termination and for any damages Contractor sustains resulting directly from such termination; provided, however, that Contractor shall not be entitled to prospective profits or fees for work unperformed or materials, labor or services not furnished. If the termination is pursuant to sub-section (b) above, the provisions of Section 15 shall apply.

SECTION 15. ABANDONMENT BY OWNER

Notwithstanding anything to the contrary that may be contained herein, Owner shall have the absolute right to abandon or suspend construction of the Project and terminate Contractor's services hereunder upon five (5) days prior written notice. In such event, Owner shall be relieved, as of the effective date of said notice, of all further obligation to Contractor for compensation earned and payable to the date of such termination and for any damages Contractor sustains resulting directly from such termination; provided, however, that Contractor shall not be entitled to prospective profits or fees for work unperformed or materials, labor or services not furnished.

SECTION 16. MISCELLANEOUS PROVISIONS

16.01   Upon the request of Owner or Managing Contractor, Contractor will provide an authorized representative on the construction site who will be available during normal business hours until the completion of the entire Project. Such representative of Contractor shall be fully authorized by Contractor to act and commit for and on behalf of Contractor.

16.02   The term "Managing Contractor" as used in this Agreement shall refer to _____ and nothing contained herein shall be construed to create any contractual relationship between Managing Contractor and Contractor. Owner and not Managing Contractor shall be liable for any payments due Contractor hereunder unless Managing Contractor assumes such obligation in writing.

16.03   The Drawings and Specifications prepared by _____, A.I.A. together with the General Conditions, Instruction to Bidders, addenda to Specifications and other modifications thereto are hereby incorporated by reference and made a part of this Agreement.

16.04   All negotiations, proposals and agreement prior to the date hereof are merged into this Agreement and superseded hereby, there being no agreement or understanding other than as written or specified herein, unless otherwise provided herein.

16.05   Contractor shall, upon request, furnish to Owner a certificate or other satisfactory evidence of the authority of any officer or agent to execute this Agreement for or on behalf of Contractor.

16.06   The Owner, Managing Contractor and Architect of the Project shall not be responsible for any charges or back charges between Contractor and any other contractors on this Project in connection with the performance of this Agreement. Contractor shall be solely responsible for the collection of such charges.

16.07   This Agreement shall be of no force or effect unless and until Performance and Payment Bonds (as set forth in Section 7) satisfactory in form and substance to Owner have been delivered to Owner or Managing Contractor.

16.08   Should any part, term or provision of this Agreement be found by the courts to be illegal or in conflict with any law of the state where made, the validity of the remaining portions or provisions shall not be affected thereby.

16.09   All sections, paragraphs and descriptive headings are for convenience only and shall not affect the construction or interpretation of any of the terms hereof. All section numbers used herein refer to sections of this Agreement unless otherwise stated.

16.10   Upon the request by Managing Contractor, Contractor shall supply to Owner or Managing Contractor within fifteen (15) days of such request, the names and address of all contracts and/or purchase orders placed with materialmen or suppliers for materials to be used in the performance of the Agreement and to permit Owner or Managing Contractor to review the procurement dates of such material for compliance with performance schedules herein.

16.11   This Agreement shall not be assignable to Contractor without prior written approval of Owner. Owner shall have the right to assign this Agreement to Managing Contractor by notifying Contractor of such assignment, but no such assignment shall alter any obligation of Owner hereunder. This Agreement shall be binding upon and inure to the benefit of the heirs, executors, administrators, successors and assigns of the parties hereto.

16.12   This Agreement shall be governed by the laws of the State of _____ applicable to contracts made or performed therein.

16.13   All notices hereunder shall be made and submitted in writing, addressed as set forth below, or to such other address as the respective party may designate in writing. Such notices shall be served, charges prepaid, in the mails or by delivering them, charges prepaid, to a telegraph office. The date of mailing or the date of delivery of such notice to the telegraph office shall constitute the date of service. Payments due hereunder may also be made by mail, addressed as aforesaid.

If to Owner:

With a copy to:

If to Managing Contractor:

If to Architect:

If to Contractor:

CONTRACTOR

In WITNESS WHEREOF, the parties have executed this Agreement the day and year first above written.

By _____

Title: _____

MANAGING CONTRACTOR
OWNER'S REPRESENTATIVE
By _____
Title: _____

APPROVED AND AGREED TO:
OWNER
By _____
Title: _____

THE CONVENTIONAL CONSTRUCTION MANAGEMENT APPROACH

The conventional construction management approach generally encompasses the manager performing certain construction activities (as for example general conditions work). Specialty trades, however, normally subcontractors to a general contractor under the traditional construction delivery approach, are here put in place through multiple prime contracts with the owner. The efforts of these specialty and multiple businesses are coordinated by a general contractor acting as a construction manager.

Conventional Cost-Plus-a-Fee Construction Management Agreement

The following special form suggests an often-used variation of this conventional approach. Here an architect/engineering firm designs the project with the assistance and review of the construction manager as to both design and planning. The construction manager inspects the work throughout the construction project without assuming the responsibilities of the architect/engineer. Questions of interpretation are referred to the architect/engineer. The construction manager requests and receives proposals from the trade contractors and awards trade contracts by and for the account of the owner—after approval by the owner. The construction manager also performs certain portions of the work for the owner directly as authorized. The construction manager monitors the work and the costs, handles change orders, and pays the trade contractors.

The construction manager agrees that if requested by the owner to establish a guaranteed maximum price he will submit one. Assuming the guaranteed maximum price is acceptable to the owner, the owner agrees to assign the trade contracts to the construction manager in order to allow the construction manager to control the performance of the work.

What in reality this procedure accomplishes is the performance by the construction manager of the described duties for a fee plus the reimbursement of all costs (as defined with specificity) plus a kicker if the construction manager is required to tender a guaranteed maximum price.

AGREEMENT

AGREEMENT made this _____ day of _____ in the year Nineteen Hundred and Eighty-Two, between ABC COMPANY, INC. of _____, the Owner, and XYZ CONSTRUCTION COMPANY, INC. of _____, the Construction Manager.

For services in connection with the following described Project:

UVW Hotel _____, Wilmington, Delaware. Consisting of 500 room hotel, food and beverage facilities; approximately 500,000 square feet of gross building area including the construction of surface parking on _____, general landscaping on a lot bordered by _____ and appurtenant utility tie-ins adjacent to the site. See legal description attached as Exhibit A.

The Architect/Engineer for the Project is _____ .
The Owner and Construction Manager agree as set forth below:

ARTICLE I
THE CONSTRUCTION TEAM AND EXTENT OF AGREEMENT

The Construction Manager accepts the relationship of trust and confidence established between him and Owner by this Agreement. He covenants with the Owner to furnish his best skill and judgment and to cooperate with the Owner and all parties representing, or under contract with, the Owner (including the Architect/Engineer) in furthering the interests of the Owner. He agrees to furnish efficient business administration and superintendence and use his best efforts to complete the Project in the best and soundest way and in the most expeditious and economical manner consistent with the interest of the Owner and the terms hereof.

1.1   The Construction Manager, the Owner and the Architect/Engineer [called the "Construction Team"] shall work from the preconstruction phase through construction completion. The Construction Manager shall provide evaluations, support, administration and coordination to the Construction Team on all matters relating to construction.

1.2   This Agreement represents the entire agreement between the Owner and the Construction Manager and supersedes all prior negotiations, representations or agreements. When Drawings and Specifications are complete, they shall be identified by amendment to this Agreement. This Agreement shall be supplemented, but not be superseded, by any provisions of the documents for construction and may be amended only by written instrument signed by both the Owner and the Construction Manager.

1.3   The Project is the total construction and services (of which the construction and services to be performed under this Agreement form a part) and includes the Work. The Work includes all labor necessary to produce the construction required by the Drawings and Specifications and all materials and equipment incorporated or to be incorporated in such construction and therefore, by way of illustration, includes that part of the physical construction that the Construction Manager is to perform with his own forces as well as that part of the construction that a particular Trade contractor is to perform. Trade contractors include materialmen, contractors, subcontractors and all other persons furnishing or supplying equipment, labor and/or materials relative to the Project (other than the Construction Manager) and Trade contracts are agreements therewith, including purchase orders. The term day shall mean calendar day unless otherwise specifically designated.

ARTICLE II
CONSTRUCTION MANAGER'S SERVICES

The Construction Manager will perform the following services under this Agreement in each of the two phases described below:

2.1   *Pre-Construction (Design) Phase:*

2.1.1   Consultation During Project Development: Schedule and attend regular meetings with the Owner and the Architect/Engineer during the development of conceptual and preliminary design to advise on site use and improvements, selection of materials, building systems and equipment. Provide recommendations on construction feasibility, availability of materials and labor, time requirements for installation and construction and factors related to cost including costs of alternative designs or materials, preliminary budgets and possible economies.

2.1.2   Scheduling: Develop a Project Time Schedule that coordinates and integrates the Architect/Engineer's design efforts with construction schedules. Regularly update the Project Time Schedule incorporating a detailed schedule for the construction operations of the Project, including realistic activity sequences and durations, allocation of labor and materials, processing of shop drawings and samples and delivery of products requiring long lead-time procurement. Include the Owner's occupancy requirements showing portions of the Project having occupancy priority.

2.1.3   Project Construction Budget: Prepare a Project budget as soon as major Project requirements have been identified, and update periodically for the Owner's approval. Prepare an estimate based on a quantity survey of Drawings and Specifications at the end of the design development phase for approval by the Owner as the Project Construction Budget. Regularly update and refine this estimate for Owner's approval as the development of the Drawings and Specifications proceeds, and advise the Owner and the Architect/Engineer if it appears that the Project Construction Budget will not be met and make recommendations for corrective action.

2.1.4   Coordination of Construction Documents: Review the Drawings and Specifications as they are being prepared, recommending alternative solutions whenever design details affect construction feasibility or schedules without, however, assuming any of the Architect/Engineer's responsibilities for design.

2.1.5   Construction Planning:

(a)   Identify, recommend for purchase and expedite the procurement of long-lead items to ensure their delivery by the required dates.

(b)   Make recommendations to the Owner and the Architect/Engineer regarding the division of Work in the Drawings and Specifications to facilitate the bidding and awarding of Trade contracts, allowing for phased construction taking into consideration such factors as time of performance, availability of labor, overlapping trade jurisdictions and provisions for temporary facilities.

(c)   Review the Drawings and Specifications with the Architect/Engineer and the Owner to eliminate areas of conflict and overlapping in the Work to be performed by the various Trade contractors. Prepare for review and approval by the Owner, prequalification criteria for bidders and bid packages.

(d)   Develop Trade contractor interest in the Project and as working Drawings and Specifications are completed, take competitive bids on the Work of the various Trade contractors. After analysis of the bids, and review and approval thereof by the Owner, award Trade contracts as directed by, and for the account of, the Owner.

2.1.6   Notwithstanding anything to the contrary contained in this Agreement, no bid will be solicited or Trade contract awarded without the review, approval and direction of the Owner noted and required in paragraphs 2.1.5(c) and 2.1.5(d).

2.1.7   Compliance With Applicable Law; Equal Employment Opportunity—Determine applicable requirements for equal employment opportunity programs for inclusion in Project bidding documents. Comply with requirements for equal employment opportunity and comply with the provisions of Exhibit D, attached hereto and incorporated herein by reference.

2.2   *Construction Phase:*

2.2.1   Project Control:

(a)   Monitor the Work and coordinate the Work with the activities and responsibilities of the Owner, Architect/Engineer and Construction Manager to complete the Project in accordance with the Owner's objectives and requirements of cost, time and quality.

(b)   Maintain a competent full-time staff at the Project site to coordinate and provide supervision and general direction of the Work and progress of the Trade Contractors.

(c)   Establish, subject to the Owner's approval, on-site organization and lines of authority in order to carry out the overall plans of the Construction Team.

(d)   Establish, subject to the Owner's approval, procedures for coordination among the Owner, Architect/Engineer, Trade contractors and Construction Manager with respect to all aspects of the Project and implement such procedures.

(e)   Schedule and conduct progress meetings on a regular basis at which Trade contractors, Owner, Architect/Engineer and Construction Manager can discuss jointly such matters as procedures, progress, problems and scheduling.

(f)   Provide regular monitoring of the schedule as construction progresses. Identify potential variances between scheduled and probable completion dates. Review schedule for Work not started or incomplete and recommend to the Owner and Trade contractors adjustments in the schedule to meet the probable completion date. Provide summary reports of each monitoring and document all changes in schedule.

(g)   Determine the adequacy of the Trade contractors' personnel, performance and equipment and the availability of materials and supplies to meet the schedule. Recommend courses of action when requirements of a Trade contract are not being met.

2.2.2   Physical Construction: The Construction Manager shall perform construction management services with his own direct labor forces. However, it is anticipated that physical construction shall be performed under Trade contracts. Nonetheless, the Construction Manager may perform physical construction services in lieu of a Trade contractor but only when the Construction Manager has bid to perform such services, such performance is, per bid, of a competitive advantage to the Owner and the Owner has approved the same pursuant to paragraphs 2.1.6 and 4.1.

2.2.3   Cost Control: Develop and monitor an effective system of Project cost control. Revise and refine the initially approved Project Construction Budget, incorporate Owner-approved changes as they occur and develop cash flow reports and forecasts as needed. Identify variances between actual and budgeted or estimated costs and advise Owner and Architect/Engineer whenever projected cost exceeds budgets or estimates.

(a)   Maintain cost accounting records on authorized Work performed under unit costs, actual costs for labor and material or other bases requiring accounting records. Afford the Owner, and all parties so authorized thereby, access to these records and preserve them for a period of three (3) years after final payment. Deliver to the Owner copies of all such records requested thereby.

2.2.4   Change Orders: Develop and implement a system for the preparation, review and processing of Change Orders. Recommend necessary or desirable changes to the Owner and the Architect/Engineer, evaluate requests for changes, submit recommendations to the Owner and the Architect/Engineer and assist in negotiating Change Orders. Subject to Paragraph 9.3.1, all Change Orders must be approved by the Owner in writing.

2.2.5   Payments to Trade Contractors: Develop and implement a procedure for the review, processing and payment of applications by Trade contractors for progress and final payments.

2.2.6   Permits and Fees: Assist the Owner and Architect/Engineer in obtaining all building permits and special permits for permanent improvements, excluding permits for inspection or temporary facilities required to be obtained directly by the various Trade contractors. Assist in obtaining approvals from all the authorities having jurisdiction. Obtain, with the Owner's assistance, all certificates of occupancy or similar approvals.

2.2.7   Owner's Consultants: If requested by the Owner, assist the Owner in selecting and retaining professional services of a surveyor, testing laboratories and special consultants, and coordinate these services. Recommend to the Owner the retention of such parties when the Construction Manager believes the same to be necessary to the Project.

2.2.8   Inspection: Inspect the Work for defects and deficiencies in the Work without assuming any of the Architect/Engineer's responsibilities for inspection.

(a)   Review the safety programs of each of the Trade contractors and make appropriate recommendations. In making such recommendations and carrying out such reviews, the Construction Manager shall exercise diligence and reasonable care but shall not be required to make exhaustive or continuous inspections to check safety precautions and programs in connection with the Project or authorized to take remedial action of a non-emergency nature unless directed by the Owner. The performance of such services by the Construction Manager shall not relieve the Construction Manager or the Trade contractors of their responsibilities for the safety of persons and property or for compliance with all federal, state and local statutes, rules, regulations and orders applicable to their performance and conduct of the Project.

2.2.9   Document Interpretation: Refer all questions for interpretation of the documents prepared by the Architect/Engineer to the Architect/Engineer and the Owner's representative.

2.2.10   Shop Drawings and Samples: In collaboration with the Architect/Engineer and Owner's representative, establish and implement procedures for expediting the processing and approval of shop drawings and samples. Recommend to the Owner the preparation of shop drawings and samples when the Construction Manager believes the same to be necessary to the Project.

2.2.11   Reports and Project Site Documents:

(a)   Record the progress of the Project. Submit written progress reports to the Owner and Architect/ Engineer including information on the Trade contractors' Work, and the percentage of completion. Keep a daily log available to the Owner and the Architect/Engineer.

(b)   Maintain the following and all records related thereto at the Project site, on a current basis: Trade contracts, Specifications, Drawings, samples, purchases, materials, equipment, maintenance and operating manuals and instructions, and other construction related documents, including all revisions. Obtain data from Trade contractors and maintain a current set of record Drawings, Specifications and operating manuals. At the completion of the Project, deliver all of the foregoing to the Owner. As built Drawings and Specifications shall be prepared by the Architect/Engineer. The Construction Manager shall deliver to the Architect/Engineer, and cause the Trade contractors to deliver to the Architect/Engineer, all information necessary for the Architect/Engineer's preparation of said as built Drawings and Specifications.

2.2.12   Substantial Completion: Subject to written approval by the Owner, determine Substantial Completion of the Project or designated portions thereof and prepare for the Owner's review and approval a list of incomplete or unsatisfactory items and a schedule for their completion.

2.2.13   Start-Up: With the Owner's maintenance personnel, direct the check-out of utilities, operations systems and equipment for readiness and assist in their initial start-up and testing by the Trade contractors.

2.2.14   Final Completion: Subject to written approval by the Owner, determine final completion of the Project and provide written notice to the Owner and Architect/Engineer that the Project is ready for final inspection. Secure and submit to the Owner all required guarantees, affidavits, certificates of occupancy or similar approvals, releases, bonds and waivers and a detailed statement listing the Project cost, suitable for the Owner's computation of deduction and depreciation on a component parts method. Turn over to the Owner all keys, manuals, record drawings and maintenance stocks.

2.2.15   Warranty: If Work is performed by the Construction Manager's own forces or by Trade contractors under contract with the Construction Manager, the Construction Manager shall and does warrant that all materials and equipment included in such Work will be new, unless otherwise specified, and that such Work will be of good quality, free from improper workmanship and defective materials and in conformance with the Drawings and Specifications and the requirements of applicable law, rules and regulations of governmental bodies. With respect to such Work, the Construction Manager further agrees to correct all Work deficient and/or defective in material and/or workmanship to the date which is of one year from the Date of Substantial Completion of all, but not less than all, of the Project or for such longer periods of time as may be set forth with respect to specific warranties contained in the trade sections of the Specifications. The Construction Manager shall also collect and deliver to the Owner any specific written warranties given by others. The Construction Manager, on behalf of the Owner and at no expense to the Owner, shall cause Trade contractors to make a like warranty to the Owner and cause to be corrected any deficient Work, and/or defective material, of Trade contractors for a like period of one year from said Date of Substantial Completion or such longer periods as may be set forth in the Specifications.

2.3   *Additional Services:*

2.3.1   At the request of the Owner, the Construction Manager will provide the following additional services upon written agreement between the Owner and Construction Manager defining the extent of such additional services and the amount and manner in which the Construction Manager will be compensated for such additional services.

(a)   Services related to investigation, appraisals or valuations of existing conditions, facilities or equipment, or verifying the accuracy of existing drawings or other Owner-furnished information.

(b)   Services related to Owner-furnished equipment, furniture and furnishings which are not a part of this Work.

(c)   Services for tenant or rental spaces constituting leasehold improvements not a part of the Work.

(d)   Obtaining or training maintenance personnel or negotiating maintenance service contracts.

(e)   Construction management services with respect to the three-level parking garage containing approximately _____ parking spaces.

ARTICLE III
OWNER'S RESPONSIBILITIES

3.1   The Owner shall provide full information regarding his requirements for the Project.

3.2   The Owner shall designate a representative ("the Owner's representative") who shall be fully acquainted with the Project and has authority to approve Project Construction Budgets, Changes in the Project, render decisions promptly and furnish information expeditiously, all subject to such limitations as may be imposed by the Owner. The Owner initially designates (_____) as his representative; provided that such designation may be changed at any time by the Owner for any reason.

3.3   The Owner shall retain an Architect/Engineer for design and to prepare construction documents for the Work.

3.4   The Owner shall furnish for the site of the Project all necessary surveys describing the physical characteristics, soil reports and subsurface investigations, legal limitations, utility locations and a legal description.

3.5   The Owner shall secure and pay for necessary easements and governmental approvals, assessments and charges required for the construction and the use or occupancy of the completed structure, excepting certificates of occupancy or similar approvals which shall be the Construction Manager's responsibility to procure and the cost of which shall be included in the Cost of the Project.

3.6   The Owner shall furnish such legal service as may be necessary for providing the items set forth in paragraph 3.5 which are his responsibility, and such auditing services as he may require.

3.7   The Construction Manager will be furnished without charge all copies of Drawings and Specifications reasonably necessary for the execution of the Work. All copies shall be identified as having been approved by the Owner and note the date of such approval.

3.8   The services, information, surveys and reports required of the Owner by the above paragraphs shall be furnished with reasonable promptness at the Owner's expense, and the Construction Manager shall be entitled to rely upon the accuracy and completeness thereof, provided that the Construction Manager shall promptly notify the Owner upon the Construction Manager's discovery of any deficiencies therein.

3.9   If either the Owner or the Construction Manager becomes aware of any fault or defect in the Project or nonconformance with the Drawings and Specifications, he shall give prompt written notice thereof to the other.

3.10   Upon request, the Owner shall furnish reasonable evidence satisfactory to the Construction Manager that sufficient funds are available and committed for the entire cost of the Project. If such evidence is not presented within a reasonable time, the Construction Manager may stop his working on the Project upon fifteen (15) days' written notice to the Owner.

3.11   In the interest of singular responsibility and control, the Owner shall communicate with the Trade contractors only through the Construction Manager.

ARTICLE IV
TRADE CONTRACTS

4.1   All portions of the Work shall be performed under Trade contracts, except those portions of the Work that the Owner authorizes the Construction Manager in writing to perform with his own forces per paragraph 2.2.2. The Construction Manager shall request and receive proposals from Trade contractors and Trade contracts shall be awarded by the Construction Manager for the account of the Owner only after the proposals are reviewed and approved by the Owner.

4.2   If the Owner refuses to accept a Trade contractor recommended by the Construction Manager, the Construction Manager shall recommend a substitute acceptable to the Owner and the Cost of the Project or the Guaranteed Maximum Price, if established, shall be increased or decreased by the difference in cost occasioned by such substitution and an appropriate Change Order shall be issued.

4.3   Unless otherwise directed by the Owner in writing, each Trade contract will be between the Trade contractor and the Construction Manager for the account of the Owner. The form of the Trade contracts including the General and Supplementary Conditions shall be satisfactory to the Construction Manager and the Owner.

4.4 The Construction Manager shall be responsible to the Owner for the acts and omissions of the Construction Manager, his agents and employees, Trade contractors performing Work under contract or agreement with the Construction Manager, and such Trade contractors' agents and employees, and shall indemnify the Owner and its agents and employees from any claim, loss, expense (including reasonable attorneys' fees) or liability which they may individually or collectively sustain or occur as a result of any such act or omission.

ARTICLE V

SCHEDULE

5.1 The services to be provided under this Contract shall be in general accordance with the following schedule:

Pre-Construction (Design) Phase: Start activities— __/__/82

Design/Contract Phase: Start initial jobsite construction activities— __/__/82.

Full construction Document: Complete— __/__/82

Construction (Entire Project): Substantially complete— __/__/82

5.2 If a Guaranteed Maximum Price is established pursuant to Article VI, the Construction Manager and the Owner shall at such time also reevaluate, if necessary, and shall establish the Construction Completion Date, as revised if necessary (which Date shall be the Date of Substantial Completion of the Project) with reference to the theretofore estimated dates of completion of all Work items, progress of the Drawings, the Work performed to date and the balance of the Work yet to be performed. Once established, the Construction Manager agrees to satisfy the Construction Completion Date.

5.3 The Date of Substantial Completion of the Project or a designated portion thereof is the date when construction is sufficiently complete in accordance with the Drawings and Specifications so the Owner can occupy and utilize the Project or designated portion thereof for the use for which it is intended and when all required certificates of occupancy or similar approvals with respect to the Project or designated portion thereof are received by the Owner.

5.4 If the Construction Manager is delayed at any time in the progress of the Project by any act or neglect of the Owner or the Architect/Engineer or by any employee of either, or by any Trade contractor then employed by the Owner, or by changes ordered in the Project, or by labor disputes, fire, unusual delay in transportation, adverse weather conditions not reasonably anticipated, unavoidable casualties or any causes beyond the Construction Manager's reasonable control (excepting those of an economic or financial nature relative to the Construction Manager or persons under contract therewith) or by delay authorized by the Owner pending arbitration, the Construction Completion Date shall be extended by Change Order for an equitable length of time and, unless such delay is attributable to labor disputes, fire, unusual delay in transportation, weather conditions or unavoidable casualties, the Construction Manager's fee shall be equitably adjusted to compensate him for the additional personnel and support costs actually incurred due to the extended Project term.

ARTICLE VI

GUARANTEED MAXIMUM PRICE

6.1 If requested by the Owner (provided the design, Drawings and Specifications are sufficiently complete to perform accurate quantity surveys and obtain firm labor, material and equipment Trade contracts), the Construction Manager will establish, based upon such Trade contracts but subject to the Owner's approval, a Guaranteed Maximum Price, guaranteeing the maximum price to the Owner for the Cost of the Project and the Construction Manager's Fee. Such Guaranteed Maximum Price will be subject to modification for Changes in the Project as provided in Article IX and for additional costs arising from delays caused by the Owner or the Architect/Engineer as provided in paragraph 5.4.

6.2   At the time when the Construction Manager and the Owner establish a Guaranteed Maximum Price, all trade contracts issued by the Owner in his name or for or on the Owner's account or otherwise prior to the establishment of the Guaranteed Maximum Price will be promptly assigned to and assumed by the Construction Manager and will contain the necessary provisions to allow the Construction Manager to control the performance of the Work. As stated in paragraph 4.3, it is presently anticipated that Trade contracts will be issued by the Construction Manager for the account of the Owner.

6.3   The Guaranteed Maximum Price will only include those taxes in the Cost of the Project which are legally enacted at the time the Guaranteed Maximum Price is established.

ARTICLE VII
CONSTRUCTION MANAGER'S FEE

7.1   In consideration of the performance of this Agreement, the Owner agrees to pay the Construction Manager in current funds as compensation for his services a Construction Manager's Fee as set forth in subparagraphs 7.1.1 and 7.1.2.

7.1.1   For performance of the Pre-Construction (Design) Phase services, fees for personnel stationed at the home office shall be paid monthly per the following man hour rates, with the first payment due concurrent with the first payment due for services performed during the Construction Phase, but in no event later than _____, 1982:

Project Manager
Chief Estimator
Senior Estimator/Engineer
Estimator/Engineer

If required during the Pre-Construction (Design) Phase:
Project Superintendent
Field Engineer
Office Manager

Outside consultants, if required and approved by the Owner, shall be paid at direct cost.

However, in no event shall the Pre-Construction (Design) Phase services fee exceed $_____.

All transportation and travel expenses for personnel stationed at the home office performing services associated with the Pre-Construction (Design) Phase are incorporated into the Construction Phase fee in 7.1.2 below.

7.1.2   For work or services performed during the Construction Phase, a fee of _____ shall be paid proportionately to the ratio the monthly payment for the Cost of the Project bears to the Guaranteed Maximum Price if established (otherwise to a reasonable estimate thereof). Any balance of this fee shall be paid at the time of final payment.

7.2   Adjustments in Fee shall be made as follows:

7.2.1   For Changes in the Project as provided in Article IX, the Construction Manager's Fee shall be adjusted as follows: for additional change orders increasing the Cost of the Project or the Guaranteed Maximum Price, if established—direct job overhead plus 2.75% thereof.

7.2.2   For delays in the Project not the responsibility of the Construction Manager, there will be an equitable adjustment in accordance with paragraph 5.4 in the fee to compensate the Construction Manager for his actual increased expenses.

7.2.3   The Construction Manager shall be paid an additional fee in the same proportion as set forth in 7.2.1 if the Construction Manager is placed in charge of the reconstruction of any insured or uninsured loss not caused by the Construction Manager's act or omission or by the act or omission of any of his employees or agents, or parties under contract therewith or employees or agents thereof.

7.2.4   The Construction Manager's Fee shall be increased:

(a)   by _____ if a Guaranteed Maximum and Construction Completion Date are established by the parties; and

(b)   by _____ if the Construction Manager performs the services referred to in paragraph 2.3.1(e) [regarding the parking garage].

7.3   Included in the Construction Manager's Pre-Construction (Design) and Construction Services Fees are the following:

7.3.1   Management, engineering and estimating salaries of personnel stationed at the home office during the Pre-Construction (Design) Phase.

7.3.2   Payroll taxes, insurance, fringe benefits and home office facilities support for personnel stationed at the home office during the Pre-Construction (Design) Phase.

7.3.3   Transportation and travel expenses of personnel stationed at the home office during the Pre-Construction (Design) Phase.

7.3.4   Management, engineering, estimating and administrative salaries of personnel stationed at the jobsite during the Construction Phase.

7.3.5   Field engineering supervision is included in the Construction Manager's Fee. Field engineering work claimed by union crafts will be incorporated into Trade contracts or hired and charged as a Cost of the Project.

7.3.6   Payroll taxes, insurance and fringe benefits for its personnel stationed at the jobsite during the Construction Phase.

7.3.7   Transportation (including staff vehicles), travel and subsistence expenses of personnel stationed at the jobsite.

7.3.8   Construction Manager's indirect overhead and profit.

7.3.9   General operating expenses of the Construction Manager's principal and branch offices other than the field office.

7.3.10   Any part of the Construction Manager's capital expenses, including interest on the Construction Manager's capital employed for the Project.

7.3.11   Overhead or general expenses of any kind, except as may be expressly included in Article VIII.

7.3.12   Costs in excess of the Guaranteed Maximum Price.

ARTICLE VIII
COST OF THE PROJECT

8.1   The term Cost of the Project shall mean costs necessarily incurred in the Project during either the Pre-Construction (Design) or Construction Phase, and paid by the Construction Manager. Such costs shall be the items listed below in this Article.

8.1.1   Subject to and in accordance with the terms hereof, the Owner agrees to pay the Construction Manager for the Cost of the Project as defined in Article VIII (such payment to be in addition to the Construction Manager's Fee stipulated in Article VII); and, upon its receipt of a payment, the Construction Manager agrees, as the disbursing party for the Owner's account, to in turn satisfy its obligations under paragraph 11.4.

8.2   *Cost Items:*

8.2.1   Wages paid for labor in the direct employ of the Construction Manager in the performance of his Work under applicable collective bargaining agreements, or under a salary or wage schedule agreed upon by the Owner and Construction Manager, including such welfare or other benefits, if any, as may be payable with respect thereto, exclusive of the salaries for management, estimating, engineering and administrative personnel included in the Construction Manager's Fees per 7.3 above.

8.2.2   Cost of all employee benefits and taxes for such items as unemployment compensation and social security, insofar as such cost is based on wages, salaries, or other remuneration paid to employees of the Construction Manager and included in the Cost of the Project under subparagraph 8.2.1.

8.2.3   Cost of all materials, supplies and equipment incorporated in the Project, including costs of transportation and storage thereof.

8.2.4   Payments made by the Construction Manager to Trade contractors for their Work performed pursuant to Trade contracts approved by the Owner pursuant to this Agreement.

8.2.5   Cost, including transportation and maintenance, of all materials, supplies, equipment, temporary facilities and hand tools not owned by the workmen but owned by the Construction Manager, which are employed and consumed by the construction Manager in the performance of the Work, and cost less salvage value on such items used but not consumed. The Owner shall have the right, but not the obligation, to retain ownership of any such used but not consumed item. If the Owner does not elect to retain ownership, the Construction Manager shall sell such item and the salvage value shall be retained by the Owner as a credit. For the purpose of this Agreement, small tools shall be defined as hand tools not owned by workmen costing less than $300.00 and shall be charged at a flat rate of 3.50% of the Construction Manager's direct manual labor cost.

8.2.6   Rental charges of all necessary machinery and equipment, exclusive of hand tools, used at the site of the Project, whether rented from the Construction Manager or others, including installation, repairs and replacements, dismantling, removal, costs of lubrication, transportation and delivery costs thereof, at rental charges consistent with those prevailing in the area.

8.2.7   Cost of the premiums for all insurance and bonds which the Construction Manager is required to procure by this Agreement or by the Owner or which, subject to the Owner's approval, are deemed necessary by the Construction Manager. At the Owner's request, the Construction Manager shall promptly purchase bonds covering the Construction Manager's faithful performance of this Agreement and full payment of all obligations arising under the Contract Documents. Any bonds shall be written by a company, and be in form and substance acceptable to the Owner and the Owner's lender(s). The cost of premiums for insurance and/or bonds of Trade contractors shall not be included in the Cost of the Project as separate items but shall be included in any awarded Trade contract.

8.2.8   Sales use, gross receipts or similar taxes related to the Project imposed by any governmental authority, and for which the Construction Manager is liable.

8.2.9   Permit fees, licenses and tests and, unless caused by the negligence of the Construction Manager or of persons under contract therewith, royalties, damages for infringement of patents and costs of defending suits therefor and lost deposits. If royalties or losses and damages, including costs of defense, are incurred which arise from a particular design, process or the product of a particular manufacturer or manufacturers specified by the Owner or Architect/Engineer, and the Construction Manager has no reason to believe there will be infringement of patent rights, such royalties, losses and damages shall, as between the Construction Manager and the Owner, be paid by the Owner and not considered as within the Guaranteed Maximum Price.

8.2.10   Losses, expenses or damages to the extent not compensated by insurance or otherwise (including settlement made with the written approval of the Owner).

8.2.11   The cost of corrective work subject, however, to the Guaranteed Maximum Price.

8.2.12   Minor expenses such as telegrams, long-distance telephone calls, telephone service at the site, expressage and similar petty cash items in connection with the Project.

8.2.13   Cost of removal of all debris.

8.2.14   Cost incurred due to an emergency affecting the safety of persons and property.

8.2.15   Cost of data processing services required in the performance of the services outlined in Article II.

8.2.16   Reasonable legal costs properly resulting from prosecution of the Project for the Owner.

8.2.17   All costs directly incurred in the performance of the Project and not included in the Construction Manager's Fees as set forth in Article VII; it being understood, notwithstanding anything to the contrary herein contained, that the cost items in this paragraph 8.2 (and accordingly the Cost of the Project) do not include any labor, materials or other items necessary or incidental to the Construction Manager's performance of those services compensated by the Fees described in Article VII.

8.3   *Lump Sum Performance:* In the event that the Construction Manager shall perform physical construction services in lieu of a Trade contractor, the amount payable to the Construction Manager therefor (i.e., the amount included in the Cost of the Project therefor) shall not exceed the Construction Manager's bid price therefor (see paragraphs 2.2.2 and 4.1).

ARTICLE IX
CHANGES IN THE PROJECT

NOT PERTINENT.

ARTICLE X
DISCOUNTS

All discounts for prompt payment shall accrue to the Owner. The Construction Manager shall promptly advise the Owner of the availability of prompt payment discounts so that the Owner shall have the opportunity (but not the obligation) to obtain same. All trade discounts, rebates, and refunds, and all costs from sale of surplus or defective materials and equipment, shall be credited to the Cost of the Project.

ARTICLE XI
PAYMENTS TO THE CONSTRUCTION MANAGER

11.1   On or about the end of a month in which Work is done, the Construction Manager shall submit to the Owner a request for payment setting forth in detail all moneys paid out, costs accumulated or costs incurred on account of the

Cost of the Project during the previous month and the amount of the Construction Manager's Fee as provided in 7.1.1 and 7.1.2 above. Such request for payment shall separately itemize all taxes (e.g., sales tax) included therein. Copies of all bills and from Trade contractors (including material and equipment suppliers) shall be submitted with the Construction Manager's monthly request for payment for the Owner's review and approval. In addition, in connection with each request for payment after the first, the Construction Manager shall deliver to the Owner affidavits, lien waivers and such other documentation as may be requested by the Owner or his title insurer or lender(s), from the Construction Manager and all Trade contractors and other persons furnishing labor and materials covered by the request for payment for the preceding monthly period.

11.2   Within fifteen (15) days after receipt of each monthly statement, the Owner shall pay directly to the Construction Manager the appropriate amount for which request for payment is made therein, less 10% of the appropriate amount to be retained by the Owner until final payment; provided, however, that no retainage shall be made from the Construction Manager's Fee or for monthly non-personnel overhead items (e.g., trailer rental) of the Construction Manager. However, when the Project is 50% complete, if the performance of the Construction Manager and Trade contractors is satisfactory, a 5% retainage of subsequent draws shall thereafter be withheld by the Owner. The Owner, in the case of any Trade contractor, may modify the aforesaid retainages (10% and 5%) or period of the retainage (final payment), provided that any such modification must be evidenced by a written instrument, executed in all cases by the Owner.

11.3   Final payment constituting the unpaid balance of the Cost of the Project and the Construction Manager's Fee shall be due and payable when the entire Project is delivered to the Owner, ready for occupancy, or when the Owner occupies the entire Project, whichever event first occurs, provided in either event that the Project be then substantially complete, all required certificates of occupancy or similar approvals have been issued and received by the Owner and this Agreement has been substantially performed. If there should remain minor items to be completed, the Construction Manager and Owner shall list such items and the Construction Manager shall deliver, in writing, his unconditional promise to complete said items within a reasonable time thereafter. The Owner may retain a sum equal to 150% of his estimated cost of completing any unfinished items, provided that said unfinished items are listed separately and the estimated cost of completing any unfinished items likewise listed separately. Thereafter, Owner shall pay to Construction Manager, monthly, the amount retained for incomplete items as each of said items is completed, subject to the delivery of affidavits, lien waivers and other documentation referred to in paragraph 11.1.

11.4   The Construction Manager shall pay all the amounts to (a) Trade contractors, on behalf of the Owner, and (b) all persons with whom the Construction Manager has an agreement or contract, upon receipt of payment from the Owner, the application for which includes amounts due such Trade contractors and/or such persons. Within ten (10) days of written demand from the Owner, the Construction Manager shall (without cost to the Owner and not as a Cost of the Project) bond (or otherwise cause to be cancelled of record) any lien which is filed against the subject property, or any interest therein, and which has reference to any Trade contractor or party with whom the Construction Manager has an agreement for labor, equipment and/or materials. Before issuance of final payment, the Construction Manager shall submit satisfactory evidence that all payrolls, materials bills and other indebtedness connected with the Project have been paid or otherwise satisfied and furnish affidavits and lien waivers all in a form satisfactory to the Owner and to Owner's title insurer and lender(s).

11.5   Notwithstanding anything to the contrary contained in this Agreement, the Owner's making of any payment to the Construction Manager or otherwise shall not operate as a waiver of any of the Owner's rights.

ARTICLE XII

INSURANCE, INDEMNITY AND WAIVER OF SUBROGATION

NOT PERTINENT.

ARTICLE XIII

TERMINATION OF THE AGREEMENT AND OWNER'S RIGHT TO PERFORM CONSTRUCTION MANAGER'S OBLIGATIONS

13.1   *Owner's Right to Perform Construction Manager's Obligations and Termination by the Owner for Cause:*

13.1.1   If the Construction Manager fails to perform any of his obligations under this Agreement including any obligation to perform Work with his own forces, the Owner may, after seven (7) days' written notice during which period the Construction Manager fails to perform such obligation, make good such deficiencies. The Cost of the Project (and, if established, Guaranteed Maximum Price) shall be reduced by the cost (direct or otherwise) to the Owner of making good such deficiencies and any deficiency shall be promptly paid by the Construction Manager to the Owner.

13.1.2   If the Construction Manager is adjudged a bankrupt, or if he makes a general assignment for the benefit of his creditors, or if a receiver, debtor in possession or trustee is appointed on account of his insolvency, or if he persistently or repeatedly refuses or fails, except in cases for which extension of time is provided, to supply enough properly skilled management, engineering and administrative personnel, workmen or proper materials, or if he fails to make prompt payment to Trade contractors or for materials or labor, or disregards laws, ordinances, rules, regulations or orders of any public authority having jurisdiction, or otherwise is guilty of a substantial violation of a provision of the Agreement, then the Owner may, without prejudice to any right or remedy and after giving the Construction Manager and his surety, if any, seven (7) days' written notice, during which period Construction Manager fails to cure the violation, terminate the employment of the Construction Manager and take possession of the site and of all materials, equipment, tools, construction equipment and machinery thereon owned by the Construction Manager and may finish the Project by whatever method he may deem expedient. In such case, the Construction Manager shall not be entitled to receive any further payment until the Project is complete nor shall he be relieved from his obligations under this Agreement, including without limitation his obligations under Articles V and VI.

13.2   *Termination by Owner Without Cause:*

13.2.1   The Owner has the right to terminate this Agreement without cause. If the Owner terminates this Agreement without cause (i.e., other than pursuant to subparagraph 13.1.2) he shall only be obliged to: reimburse the Construction Manager for any unpaid Cost of the Project due him under Article VIII, plus (since the Construction Manager's Fee is stated as a fixed sum) such an amount as will increase the payment on account of his fee to a sum which bears the same ratio to the said fixed fee sum as the Cost of the Project at the time of termination bears to the Guaranteed Maximum Price if then established (otherwise to a reasonable estimate thereof), and pay to the Construction Manager fair compensation, either by purchase or rental at the election of the Owner, for any equipment retained and not included in the Cost of the Project. In case of such termination of the Agreement the Owner shall further assume and become liable for contractual obligations and commitments and unsettled contractual claims that the Construction Manager has previously undertaken or incurred in good faith with the Owner's approval as herein required in connection with said Project. The Construction Manager shall, as a condition of receiving the payments referred to in this paragraph 13.2, execute and deliver all such papers and take all such steps, including the legal assignment of his contractual rights, as the Owner may require for the purpose of fully vesting in him the rights and benefits of the Construction Manager under such obligations or commitments.

13.3   *Termination by the Construction Manager:*

13.3.1   If the Project is stopped for a period of thirty (30) days under an order of any court or other public authority having jurisdiction or as a result of an act of government, such as a declaration of a national emergency making materials unavailable, through no act or fault of the Construction Manager, or if the Project should be stopped for a period of thirty (30) days by the Construction Manager for the Owner's failure to make proper payment thereon, then, provided that neither party has demanded or does not demand arbitration pursuant to Article XV, the Construction Manager may, upon seven (7) days' written notice to the Owner and the Architect/Engineer (at the end of which period such condition still exists), terminate this Agreement and recover from the Owner payment for all Work executed by the Construction Manager to date, the Construction Manager's Fee earned to date, and for any proven loss sustained upon any materials, equipment, tools, construction equipment and machinery, including reasonable profit and damages.

ARTICLE XIV
ASSIGNMENT AND GOVERNMENT LAW

14.1   Neither the Owner nor the Construction Manager shall assign his interest in this Agreement without the written consent (which consent shall not be unreasonably withheld) of the other except as to the assignment of proceeds and/or except as to an assignment by the Owner to his lender(s) pursuant to the terms and conditions of the loan documents.

14.2   This Agreement shall be governed by the law of the place where the Project is located.

14.3   The Construction Manager warrants and represents that it is and will remain a corporation in good standing under the laws of _____, is and will remain qualified to do business in _____, is and will remain in good standing under the laws of _____, and is and will remain empowered to fulfill all of the obligations to be performed by it hereunder. The Construction Manager shall promptly provide a corporate resolution authorizing this transaction and the execution of all documents required in connection therewith by the officer(s) executing same.

ARTICLE XV
ARBITRATION

NOT PERTINENT.

ARTICLE XVI
CONTRACT DOCUMENTS

16.1   The contract documents are defined as and shall consist of:

16.1.1   This Agreement when executed by the parties.

16.1.2   Exhibit A attached hereto, being a legal description of the Owner's real estate.

16.1.3   Exhibit B, the construction documents (Drawings and Specifications) as annexed hereto per 1.2 above.

16.1.4   Exhibit C attached hereto, being the Amendment to the Owner–Construction Manager Agreement affixing the Guaranteed Maximum Price and Construction Completion Date.

16.1.5   This Agreement and the Exhibits hereto shall be interpreted so as to avoid conflicts between the same. However, in the event of a conflict between this Agreement and an Exhibit hereto, the former shall govern.

ARTICLE XVII
AUDIT

The Owner shall have the unlimited right to audit, at his sole expense, the books and records of the Construction Manager, pertaining to the Project.

ARTICLE XVIII
NOTICES

Written notices shall be deemed to have been served if deposited in the United States mails and addressed to (i) the Owner at _____;
and (ii) the Construction Manager at _____ .

This Agreement executed effective and as of the day and year first above written.

Date of Execution: _____   OWNER:

ABC COMPANY, INC.

By: _____

CONSTRUCTION MANAGER:

XYZ CONSTRUCTION COMPANY, INC.

By: _____

EXHIBIT C
AMENDMENT TO OWNER-CONSTRUCTION MANAGER AGREEMENT

Pursuant to Articles, V, VI and XVI of the original Agreement, dated effective and as of _____, 1982 between ABC COMPANY, INC. (the Owner) and XYZ CONSTRUCTION COMPANY, INC. (the Construction Manager), for: UVW Hotel (the Project), the Owner desires to fix a Guaranteed Maximum Price and a Construction Completion Date for the Project and the Construction Manager agrees that the design, plans and specifications are sufficiently complete for such purpose. Therefore, the Owner and Construction Manager agree as set forth below:

ARTICLE I
GUARANTEED MAXIMUM PRICE

The Construction Manager's Guaranteed Maximum Price for the Project, including the Cost of the Project as defined in Article VIII and the Construction Manager's Fee as defined in Article VII is _____ Dollars ($_____).
This price is for the performance of the Project in accordance with the original

Agreement, including the documents listed and attached to such Agreement and marked Exhibit B, as the same may have been amended in writing.

At the completion of the Project, if the Cost of the Project plus the Construction Manager's Fee, as adjusted by approved Change Orders, is more than the Guaranteed Maximum Price, the difference, representing the overrun shall be borne entirely by the Construction Manager.

At the completion of the Project, if the Cost of the Project plus the Construction Manager's Fee, as adjusted by approved Change Orders, is less than the Guaranteed Maximum Price, the difference, representing the savings shall be distributed 75% to the Owner, 25% to the Construction Manager; provided, however, that in no event shall the Construction Manager's portion of any such savings exceed 1% of the Cost of the Project.

ARTICLE II
TIME SCHEDULE

The Construction Completion Date established by this Amendment is ——————————, 19———.

OWNER:

ABC COMPANY, INC.

Date of Execution: ——————   By: ———————————————

CONSTRUCTION MANAGER:

XYZ CONSTRUCTION COMPANY, INC.

By: ———————————————

THE TWO-PHASED AGREEMENT

There are in reality two distinct phases in the construction management relationship. Phase one has to do with planning and design—developing a general estimate, planning, scheduling, developing a control numerical model (analysis of each item of construction expense), preparing detailed estimates, monitoring design development and the costing of the alternative system, and determining work divisions. The second phase is the construction phase—developing a construction schedule; supervising the work; inspection; coordinating; and supervising job safety, equal employment opportunity, and other statutory compliance. In addition there is the task of handling progress payments, processing change orders, and conducting job meetings.

A frequently used method of construction delivery in today's marketplace, and one preferred by many owners, is a combination of the more traditional approach with the more contemporary construction management approach. It is a method whereby the construction manager during phase one works on a fee basis as a professional consultant to the owner and works with the design professionals during the design and planning stages as a link between the architect/engineer and the owner.

Then, when the design, drawings, and specifications are complete to a point where the construction manager can furnish an absolute, guaranteed maximum price (and completion time) to the owner, the construction manager and owner enter phase two. The construction manager becomes a general contractor and enters into a general contract with the owner. It then contracts directly with its supplier of labor and materials.

This two-phased approach gives the owner all the advantages during the design and planning stages of a professional manager as well as the security during construction of a guaranteed completion cost and date. Moreover, it allows the owner to look to one entity for warranty guarantees. But "there ain't no free lunch"—the construction manager will certainly add a contingency to its guaranteed maximum price and date to protect itself.

Owner Services Agreement for a Small Private Project

An example of this two-phased agreement for a smaller project follows.

AGREEMENT

THIS IS AN AGREEMENT made _____, 1982 between ABC MANU-FACTURING COMPANY, INC. (the "Owner"), and XYZ CONSTRUCTION COMPANY, INC. (the "Construction Manager").

ARTICLE 1.
THE PROJECT

1.1   The Project forming the subject of this Agreement is the construction of a warehouse at _____ Newark, New Jersey from Drawings, Specifications and Contract Documents prepared by DEF AND ASSOCIATES (the "Architect"). (The Contract Documents, as the phrase is used in this Agreement, include the proposal of Construction Manager as approved by Owner.)

1.2   The Construction Manager will provide services in connection with the Project in two phases as follows:

1.2.1   Phase 1 will consist of the Construction Manager assisting the Owner and the Architect, in whatever manner necessary as determined by the Owner and Architect, in the completion of the "Schematic Design Phase," Design Development Phase," and "Construction Documents Phase," as set forth in paragraphs 1.2, 1.3 and 1.4 respectively of the Agreement between the Owner and Architect dated _____. The preceding sentence is not to be construed as shifting the ultimate responsibility for the design of the Project from the Architect to the Construction Manager.

1.2.2   Phase II will consist of the Construction of the Project.

ARTICLE 2.
PHASE I

2.1   The Construction Manager will provide sufficient qualified personnel to work with the Architect and Owner, at such times and places as designated by the Owner or Architect, during Phase I as set forth in paragraph 1.2.1 of this Agreement or Owner. This will include, without limitation, advice on such matters as:

2.1.1   Practicality of the Architect's design and choice of systems and materials with respect to site conditions, installation, cost, labor availability, labor jurisdiction and prevailing codes.

2.1.2   Review of design, determination and production with respect to construction timing, scheduling and possible phasing of demolition, pre-bid purchasing and bidding of construction subcontracts.

2.2   The Construction Manager will prepare construction cost estimates according to the following schedule:

2.2.1   One estimate of probable construction cost based on the Architect's first sketches of the schematic design as set forth in paragraph 1.2 of the Agreement between the Owner and the Architect will be delivered to the Architect and Owner within 14 days after receipt by Construction Manager of the first sketches of the schematic design.

2.2.2   One estimate of probable construction cost, highlighting the changes from the immediately preceding estimate, based upon the Architect's final schematic design, as approved by the Owner will be delivered to the Owner and the Architect within 14 days after receipt by the Construction Manager of the Owner approved schematic design drawings.

2.2.3   One estimate of probable construction cost, highlighting the changes from the immediately preceding estimate, based upon the Preliminary Drawings and Specifications prepared during the design development phase, will be delivered to the Owner and the Architect within 14 days after receipt by the Construction Manager of the Preliminary Drawings and Specifications.

2.2.4   The final construction cost (the "Guaranteed Maximum Cost") highlighting the changes from the immediately preceding estimate will be delivered to the Owner and the Architect prior to the completion by the Architect of the Construction Documents.

2.3   At the request of the Owner or Architect, the Construction Manager will revise the cost estimates, so that such estimates will be consonant with current information and circumstances.

2.4   The Construction Manager will perform in such a manner to allow the Owner to occupy the Project no later than _____ , 1985.

ARTICLE 3.
PHASE II

3.1   Upon acceptance by the Owner of the Construction Manager's Guaranteed Maximum Cost, the Construction Manager will construct the Project in accordance with (i) the Plans and Specifications prepared by the Architect upon the advice of the Construction Manager, (ii) the Standard Form of Agreement between Owner and Contractor, AIA document A111, attached as Appendix A (the "Construction Agreement") and (iii) the General Conditions of the Contract for Construction which forms a part of the Construction Agreement and is attached as Appendix "B."

ARTICLE 4.
PROJECT REQUIREMENTS

4.1   The Owner will provide full information regarding his requirements for the Project and will not unreasonably withhold or delay his determination of an acceptable cost of the Project or his approval of designs and specifications, and of resultant estimates which do not exceed said acceptable cost.

ARTICLE 5.
FEE

5.1   The Owner will pay the Construction Manager for the performance of services defined in Phase I in Article 2 of this Agreement, a Fee of $10,000 which will, without limitation, include profit, overhead and all other costs and which will be paid in five equal installments on or about the first day of each month following the date of this Agreement.

ARTICLE 6.
TERMINATION

6.1   Prior to the commencement of Phase II of this Agreement, performance of this Agreement may be terminated by the Owner upon seven days' written notice if the Contractor, in the opinion of the Owner, fails to perform in accordance with its terms.

6.2   The Owner, regardless of any provision of this Agreement or of Appendix A, may elect not to accept the Guaranteed Maximum Cost presented by the Construction Manager in accordance with the provisions of subparagraph 2.2.4 of this Agreement in which event, this Agreement will be terminated upon payment by the Owner to the Construction Manager of an amount equal to the difference between the amount previously paid by the Owner to the Construction Manager and $10,000, which payment will be final settlement of all rights which the Construction Manager may have in respect of its relationship with the Owner in connection with the Project.

ARTICLE 7.
ARBITRATION

7.1   The provisions of paragraph 7.10 of AIA Document A201 will apply to this Agreement as if herein written in full.

THIS AGREEMENT, executed this day and year first written above:

Owner:	Construction Manager:
ABC MANUFACTURING	XYZ CONSTRUCTION
COMPANY, INC.	COMPANY, INC.
By:_____	By:_____
President	President
Attest:_____	Attest:_____
Secretary	Secretary
SEAL	SEAL

APPENDIX A—Use AIA Document A111, *Standard Form of Agreement between Owner and Contractor.*

APPENDIX B—Use AIA Document A201, *General Conditions of the Contract for Construction.*

CHAPTER SIX

FORMS FOR EFFECTIVE CONTRACT MANAGEMENT TO CONTROL THE PROJECT AND MAKE IT WORK

The forms in this chapter have been selected to illustrate how a variety of construction contractors and managers have dealt with the problems of compiling their management information, providing for detailed and over-all project cost and management control, and accumulating the necessary accounting data. They are intended to be illustrative but, for obvious reasons, they cannot be exhaustive nor are they designed to be automatically put into place or use. Nevertheless, the information on the forms is basic and must, in some way, be generated and systematically recorded in most contract operations. Where illustrative schedules are presented, they have been simplified to convey the idea without resorting to excessive detail.

FORMS FOR OVERALL MANAGEMENT CONTROL

Charting the Organization

Because no two construction projects or businesses are alike, each one must develop its own personalized organization chart. To be efficient, as it has to be in today's competitive market, the contractor must be so organized that the lines of responsibility and authority are clearly drawn. The following are several illustrative organization charts for contractors of varying sizes.

Very small organization. The very small contractor works directly on the job. He keeps the time, buys the materials, and does all the jobsite receiv-

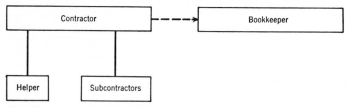

Organization Chart for Very Small Organization

ing. He usually writes checks for payroll and purchases, and turns these basic data over to a public accountant or contract bookkeeper. His staffing problems are virtually nonexistent unless he is trying to find a helper who is capable of developing into a foreman or superintendent. Subcontractors deal directly with the owner/contractor. (The terms "owner/contractor" and "contractor" are used more or less interchangeably, because in many instances the subject matter could apply equally to either and because, as to certain subcontractors, the owner and the contractor are sometimes the same entity.)

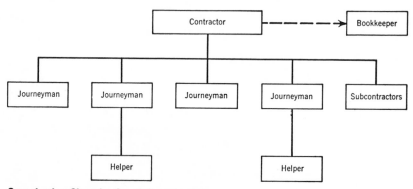

Organization Chart for Small Organization

Small organization. The small contractor supervises several journeymen, and usually does his own buying and timekeeping. The journeymen receive materials and supplies delivered at the jobsite and turn the signed delivery receipts over to the owner/contractor when he visits the job. The owner/contractor makes the payrolls, checks the delivery data, writes both the payroll checks and the vendor checks, and turns the basic records of the transactions over to the bookkeeper. Subcontractors deal directly with the owner/contractor.

Small to medium-sized organization. The small to medium-sized owner/contractor works as foreman or superintendent on one large job while he has

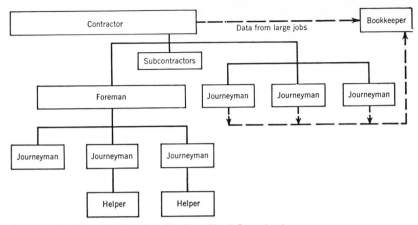

Organization Chart for Small to Medium-Sized Organization

other smaller jobs in progress under foremen or journeymen. The flow of accounting data tends to be about the same, but in operations of this size there is usually a company bookkeeper who assembles the data, keeps the books, and writes checks for the contractor's signature in the company office. Usually, too, in this type of operation, the foreman on the large job keeps the time and receives the materials, and turns the time sheets and signed delivery slips over to the contractor for checking. Subcontractors, for the most part, still deal directly with the contractor, though on the big job some of the dealings will be with the foreman.

Medium-sized organization. In the somewhat larger operation the contractor supervises several foremen who keep the time and receive the materials

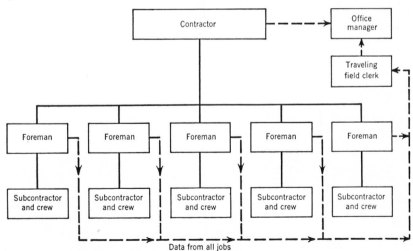

Organization Chart for Medium-Sized Organization

and supplies delivered to the jobsite. Accounting data are picked up and summarized by a field clerk and delivered to the office manager for final processing in the general office. It is at about this point that staffing the jobs begins to present special problems. To keep an operation of this size going, the contractor cannot depend so completely on his own personal observations to follow job progress, but must make greater use of reports. In addition to their technical skill, the foremen must be chosen on the basis of cost consciousness, their ability to manage a crew, and their ability to make independent decisions. When an operation reaches this size, the contractor will often have his foremen picked before bidding a job. In the same way, it is quite common for foremen to have a crew that follows them from job to job. In operations of this size, it is usual for subcontractors to make their original deals with the contractor, but supervision of their work is left to the contractor's foremen.

Many larger construction companies, whose jobs normally cost $500,000 or less, operate on substantially the same pattern, except that they may substitute a superintendent or a construction or area manager in the place of the contractor shown in the diagram.

As the jobs grow larger—say from $1 million to $5 million—the pattern tends to remain much the same, except that the larger jobs are usually in the charge of a construction manager who may have several foremen and subcontractors under him.

Large organization.   As the individual jobs pass the $5 million mark, there is an increasing tendency for each job to have its own organization, headed by a project manager or superintendent. With centralized accounting, a job clerk and a combination timekeeper–pickup truck driver will usually be enough accounting personnel. With a decentralized system, the job clerk

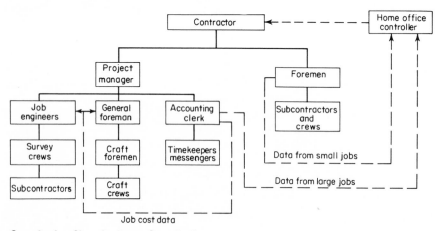

Organization Chart for Large Organization

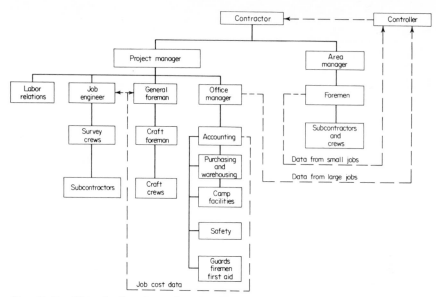

Organization Chart for Very Large Organization

would be replaced by a job accountant, and if there are as many as 100 men on the payroll, there may be a separate pickup truck driver and a combination timekeeper-paymaster, who would probably then spend at least part of his time operating a small warehouse. If the job is cost-plus-fixed-fee, in addition to an office manager, it may be necessary to have full-time payroll, accounts payable, and warehouse clerks. This pattern tends to apply up to about $5 million, depending on whether the job is centralized or decentralized.

Projecting Job Financial Requirements

Every time a new job is undertaken, it is necessary to provide financing to operate it, and in order to have money on hand when it is needed, it is necessary that cash requirements be forecast as soon as possible. This is a simple matter of spreading the estimated costs to the various weeks in which they are expected to occur, and it can usually be done by the estimator as soon as the bid is complete. When the anticipated cost figures are combined with the schedule of expected progress payments, it is possible to estimate how much cash will be needed to complete the job.

In order to minimize the amount of money the contractor has invested in the job, it is common practice to unbalance the bid or the payment schedule so that the contractor's estimated profit is received in the first few progress payments. In the schedules that follow, a form for forecasting cash requirements is illustrated, together with examples of balanced and unbalanced bids.

Ecks Construction Co.
Cash Forecast for the Six Months Ending
June 30, 19—

Expected Available Cash

	Jan.	Feb.	Mar.	Apr.	May	June	Total
Cash balance Dec. 31	$30,000						
Monthly balances brought forward		$22,500	$19,500	$ 19,500	$ 22,500	$22,500	$ 30,000
Expected receipts:							
Job A—retention		15,000					15,000
B—fee	3,000	3,000	3,000				9,000
C—progress payments			28,500	36,000	45,000	19,500	129,000
D—progress payments	18,000	9,000					27,000
D—retention					7,500		7,500
E—progress payments		6,000	15,000	30,000	30,000	12,000	93,000
F—fee	1,500	1,500	1,500	1,500	1,500	1,500	9,000
Insurance claim				13,500			13,500
Warehouse scrap sales	1,500						1,500
Total cash before bank loans	$54,000	$57,000	$67,500	$100,500	$106,500	$55,500	$334,500
Bank loans	15,000	30,000	22,500				67,500
TOTAL EXPECTED CASH	$69,000	$87,000	$90,000	$100,500	$106,500	$55,500	$402,000

Expected Cash Requirements

	Jan.	Feb.	Mar.	Apr.	May	June	Total
Job B—None (revolving fund)							
C—Costs		$12,000	$36,000	$12,000	$ 9,000	$15,000	$ 84,000
D—Costs	$12,000	6,000					18,000
E—Costs	18,000	15,000	21,000	18,000	18,000		90,000
F—None (revolving fund)							
G—Costs					12,000	9,000	21,000
Equipment payments	6,000	6,000	6,000	6,000			24,000
Taxes	3,000			4,500		4,500	12,000
Insurance		21,000					21,000
Overhead	7,500	7,500	7,500	7,500	7,500	7,500	45,000
SUBTOTALS	$46,500	$67,500	$70,500	$48,000	$46,500	$36,000	$315,000
Loan repayments				30,000	37,500		67,500
TOTAL CASH REQUIRED	$46,500	$67,500	$70,500	$78,000	$84,000	$36,000	$382,500
Cash balance carried forward	$22,500	$19,500	$19,500	$22,500	$22,500	$19,500	$ 19,500
TOTAL LOANS DUE TO BANK	$15,000	$45,000	$67,500	$37,500	—	—	—

Cash Forecast for Six Months

Week	Direct labor	Incorporated materials	Equipment rentals	Subcontracts	Supplies, etc.	Distributed overhead	Estimated progress billings	Cumulative estimated cash investment in job
1	$ 500					$ 750		$ 1,250
2	750					1,250		3,250
3	1,000					1,000		5,250
4	1,250					800		7,300
5	2,000	$ 2,000	$ 1,500	$15,000	$1,000	800	($ 29,300)	300
6	2,000					700		3,000
7	3,000	14,000	4,500		1,250	700		26,450
8	3,000					700		30,150
9	3,000	4,000	5,000	20,000	1,500	700	(63,400)	950
10	3,500					600		5,050
11	4,000	16,000	5,000		1,750	600		32,400
12	4,000					600		37,000
13	2,000	10,000	4,000	20,000	1,500	600	(83,400)	(8,300)
14	1,500					500		(6,300)
15	1,000		2,000		500	500		(2,300)
16	500					400		(1,400)
17				17,000		300		15,900
18							(21,900)	(6,000)
19								
20								
21				8,000			(22,000)	(20,000)
	$33,000	$46,000	$22,000	$80,000	$7,500	$11,500	($220,000)	-0-

NOTE: This schedule is oversimplified for purposes of illustration.

Forecast of Cash Requirements on Cost Basis

Week	Direct labor	Incorporated materials	Equipment rentals	Subcontracts	Supplies, etc.	Distributed overhead	Estimated progress billings	Cumulative estimated cash investment in job
1	$ 500					$ 750		$ 1,250
2	750					1,250		3,250
3	1,000					1,000		5,250
4	1,250					800		7,300
5	2,000	$ 2,000	$ 1,500	$15,000	$1,000	800	($ 30,800)	(1,200)
6	2,000					700		1,500
7	3,000	14,000	4,500		1,250	700		24,950
8	3,000					700		28,650
9	3,000	4,000	5,000	20,000	1,500	700	(66,600)	(3,750)
10	3,500					600		350
11	4,000	16,000	5,000		1,750	600		27,700
12	4,000					600		32,300
13	2,000	10,000	4,000	20,000	1,500	600	(87,600)	(17,200)
14	1,500					500		(15,200)
15	1,000		2,000		500	500		(11,200)
16	500					400		(10,300)
17				17,000		300		7,000
18							(15,000)	(8,000)
19								
20								
21				8,000				
	$33,000	$46,000	$22,000	$80,000	$7,500	$11,500	($200,000)	($20,000)
24	On acceptance by owner, retention release						($ 20,000)	
	TOTAL CONTRACT PRICE, CASH RECEIPTS, AND PROFIT						($220,000)	($20,000)

NOTE: The same schedule can be completed for each job, then the calendar weeks added to arrive at total company cash flow.

Forecast of Cash Requirements—Unbalanced Payment Schedule

Work item	Estimated cost	Estimated profit	Bid price
1	$ 28,000	$ 2,800	$ 30,800
2	22,000	2,200	24,200
3	12,000	1,200	13,200
4	23,000	2,300	25,300
5	10,000	1,000	11,000
6	20,000	2,000	22,000
7	15,000	1,500	16,500
8	25,000	2,500	27,500
9	21,000	2,100	23,100
10	24,000	2,400	26,400
	$200,000	$20,000	$220,000

Payment Schedule—Balanced Bid

Work item	Estimated total	Costs to date	Percentage completion
1	$28,000	$21,000	75
2	22,000	5,500	25
3	12,000	1,200	10
4	23,000	1,150	5
5	10,000	750	7.5
	$95,000	$29,600	

Percentage Completion of Work Items

Work item	Unbalanced payment schedule	Percentage completion by items	Progress billings
1	$ 34,500	75	$25,875
2	27,100	25	6,775
3	14,800	10	1,480
4	28,600	5	1,430
5	10,000	7.5	750
6	20,000	—	—
7	15,000	—	—
8	25,000	—	—
9	21,000	—	—
10	24,000	—	—
TOTAL	$220,000		$36,310
Less	10% Retention		3,631
			$32,679

NOTE: Figures in *italics* reflect estimated cost given in Schedule 4-4 plus entire profits for job, prorated among these four items.

Computation of Progress Billings on Unbalanced Basis

Job number and name	Contract price	Projected total cost	Margin	Cost to date	Percent complete	Earned revenue	Billed to date	Dr. revenue	Cr. revenue
	(A)	(B)	(C)	(D)	(E)	(F)	(G)	(H)	(I)
1908 Remlap Office Ctr.	208,000	193,000	15,000	45,000	.23	47,840	80,000	32,160	-0-
1102 Coombs Towers	400,000	360,000	40,000	180,000	.50	200,000	250,000	50,000	-0-
1103 Redner Apartments	150,000	140,000	10,000	126,000	.90	135,000	130,000	-0-	5,000
1104 Bruzek Shopping Ctr.	600,000	540,000	60,000	135,000	.25	150,000	160,000	10,000	-0-
1105 Bodie Towers	1,000,000	900,000	100,000	-0-	-0-	-0-	-0-	-0-	-0-
TOTALS*	$2,358,000	$2,133,000	225,000	$486,000	N/A	$532,840	$620,000	$92,160	$5,000
	(J)			(K)		(L)			

Formulas

$E = D \div B$

$F = E \times A$

H or $I = F - G$

Contract backlog $= J - L$

Profit in backlog $= (J - L) - (B - D)$

Journal entry

Dr. revenue (net) $87,160

Dr. earned revenue in excess of billed
   revenue (asset) 5,000

Cr. billings in excess of earned revenue
   (liability) $92,160

*If a section were added showing completed jobs, the totals would tie in to the related figures on the financial statements. This type of schedule is commonly used for presentation to banks and bonding companies.

Worksheet for Adjusting Jobs From Billed to Earned Revenue

Long-Range Planning

Just as planning is necessary for individual jobs, it is also necessary for the company as a whole. One of the more satisfactory ways of making long-term plans is to prepare alternative plans and then compare them. Following is a simple form for making such a comparison.

	Last year	Expected results next year under			Year after next under plan selected
		Plan 1	Plan 2	Plan 3	
Gross billings	$	$	$	$	$
Job costs					
Job profits	$				
Home office overhead	$				
Expected profit before taxes	$				
Federal and state taxes	$				
Expected profit after taxes	$				
Expected net increase in retained earnings	$				

A Simple and Effective Means of Comparing Projected Results of Alternative Plans

Internal Auditing

While internal auditing is not normally a function that resides within the construction manager's company, it is well for the construction manager to have a good understanding of the internal audit function and the types of things that internal auditors look for. It is not uncommon for the construction manager and the contractors to be subject to an audit, which quite frequently is performed by the internal audit group of the owner. This is particularly true when the terms of the contract provide for cost-plus payment, which is common in fast-track jobs and in contractual arrangements that provide for a construction manager.

The following pages illustrate a typical audit program; it is presented here primarily to provide the reader with an understanding of the internal audit function.

---

1. Determine audit objectives which generally fall into the following three areas:
   a. Financial or compliance
   b. Operating procedures and efficiencies
   c. Program results
2. Obtain an overview of the project, department, or other unit to be audited.
   a. How does it fit into the overall company organization?
   b. What are the backgrounds and personalities of the key management?
   c. What are the objectives of the project, department, or other unit being audited?
   d. What are the contractual terms—financial and operational?
   e. What are the known problems?
   f. Where are the "soft spots" or what should we look out for?
3. Select the staff for the assignment. Match individual abilities of the members of the internal audit group with the audit objectives and information obtained in step 2.
4. By inquiry and reference to policy manuals and similar sources, obtain a detailed understanding of the systems and procedures in effect at or in the project, department, or other unit to be audited.
5. By use of flowcharts and written narratives, record the data obtained in step 4.
6. Select representative transactions for each function or system cycle recorded in step 5, and, by taking the transaction through the system, determine that the underlying facts are correctly understood.
7. Identify key financial or operating controls (depending on audit objectives). Financial controls are those that help ensure the safeguarding of company assets and the accurate accumulation of data for financial reporting. Operating controls are those that show whether company assets are efficiently used or employed and whether data are accurately gathered and reported for management decision making.
8. Test the operation of the key controls (identified in step 2). Where the volume of data is great, use statistical sampling of documents to test the control.
9. Prepare a report on any control weaknesses noted during previous steps. For a maximum impact with management, state the results in terms of dollars: that is, show what it might cost the company if a control is lacking or not operating.
10. Based on steps 1 through 9, determine extent of additional testing needed to validate the results of the examination and to achieve overall assignment objectives.
11. Prepare final report.
12. The internal audit department manager should review audit procedures and findings to ensure adequate documentation to support findings.

Systems Approach—Audit Program for Internal Audit

Job Audit Report

Job:
Location:
Dates Visited:

	Yes	No	Not appli- cable	See com- ment
A. *Cash on hand*				
1. Did you count it?				
2. Did you reconcile it with the books?				
3. Did you consider all possible sources?				
4. Do you feel certain that all cash from miscellaneous sources is being recorded?				
5. What did you do to satisfy yourself that it is (or is not) being recorded?				
6. Do we cash personal checks from cash on hand?				
B. *Cash in bank*				
1. Did you reconcile all bank accounts?				
2. Did you trace out the reconciling items in last month's reconciliations?				
3. Are any items from last month still open?				
4. If so, have you satisfied yourself that they are proper?				
5. Did you test-check the signatures on the checks?				
6. In reconciling the payroll account, did you compare the signatures and endorsements on termination checks with those on the two preceding paychecks?				
7. Did you make a test-check of the endorsements on the checks?				
8. Any exceptions not satisfactorily explained?				
9. Did you make a test comparison of the checks against the supporting payrolls and vouchers? What steps were followed in the test?				
10. Did you compare the deposits as shown on the statements with the books, the detail of the duplicate deposit slips, and the underlying estimates, invoices, cash reports, and so on?				
11. Are receipts from extra work being properly accounted for?				

Checklist for Internal Auditor

	Yes	No	Not appli- cable	See com- ment
12. Did all duplicate deposit slips bear the bank's stamp and receipt?				
13. Did any duplicate deposit slips show signs of alteration?				
14. Was there any substantial lag between receipt and deposit of funds?				
15. Did any part of your examination give indication of signature irregularities, such as checks going through on only one signature?				
16. Did you account for all check numbers?				
17. Is the name of the payee visible to the signer when checks are signed?				
18. Did you notice any weakness in the control of cash that should be strengthened?				
C. *Purchasing and vouchering*				
1. Do you have a list of persons authorized to sign requisitions?				
2. Were any requisitions not signed by an authorized signer?				
3. Were requisitions coded?				
4. Did you test the coding for reasonableness?				
5. Were any purchase orders not covered by requisitions?				
6. If there were any, was the omission satisfactorily explained?				
7. Are purchase orders priced and coded?				
8. If not, was omission satisfactorily explained?				
9. Do you have a list of individuals authorized to sign purchase orders?				
10. Did you test-check the purchase orders for authorized signatures?				
11. Any exceptions?				
12. Are purchase orders issued at the time of purchase?				
13. If not, are confirming orders sent out before receipt of the invoice?				
14. Did you investigate the procedure for handling back orders?				
15. Is it adequate to prevent duplicate shipments?				
16. Are purchase orders checked, before				

Checklist for Internal Auditor (*continued*)

	Yes	No	Not appli-cable	See com-ment
being issued, by someone not under the purchasing agent's control?				
17. Is numerical control maintained on purchase orders?				
18. Are blanket purchase orders used?				
19. If so, are there proper safeguards to prevent misuse?				
20. Are receiving memoranda prepared at the time goods are received?				
21. Are they compared with the purchase orders by someone not under the control of the individuals controlling the warehouse or the purchasing?				
22. Are vendors' invoices received by the accounting department before anyone else?				
23. Are vendors' invoices compared with the purchase order and receiving memoranda?				
24. Did you test-check vouchers to see that the system is being followed?				
25. Did your test-check of vouchers include a test of pricing, extensions, and discounts claimed?				
26. Did you see evidence of adequate follow-up on vendor's invoices bearing a discount to assure the taking of all discounts?				
27. Is there any preaudit of vouchers and checks before checks are released?				
28. If so, is it effective?				
29. Does the project manager or superintendent see all vouchers?				
30. Does he actually review them or just sign them?				
31. Do you have a list of authorized check signers?				
32. Did you see any evidence of presigning of checks?				
33. Are documents supporting paid vouchers stamped to prevent reuse?				
D. *Receiving and warehousing*				
1. Is responsibility for receipt of materials and supplies fixed?				
2. Is an adequate system in effect to check incoming materials and supplies?				

	Yes	No	Not appli-cable	See com-ment
3. Are adequate receiving memoranda prepared?				
4. Is there any procedure for checking on the person who receives materials and supplies?				
5. Is there any attempt to keep a perpetual warehouse inventory?				
6. If so, is the account kept in terms of units rather than money?				
7. If not, is there a satisfactory control on warehouse stocks by other means? (If other means are used, describe in comments.)				
8. If no warehouse account is maintained, is there an adequate system for controlling charges for purchased materials and supplies to work items?				
9. Is the system being followed? What did you do to check this point?				
10. On the basis of your review of the procedures, personnel, and records, do you believe that all incoming materials and supplies are being properly checked and accounted for?				
11. Do you believe that materials and supplies being taken from the warehouse are being accounted for correctly?				

E. *Timekeeping and payroll*

	Yes	No	Not appli-cable	See com-ment
1. Have you watched the operation of the system for reporting field time—brass, cards, foreman's report, or some combination of the three? Describe briefly the system in use and its operation.				
2. Did you check at least one payroll?				
3. Did you supervise the distribution of at least one week's paychecks?				
4. Were there any complaints from the men on the accuracy of their checks for the last two pay periods?				
5. If so, did you secure satisfactory explanations of the claimed differences?				
6. Are the personnel records in satisfactory condition?				
7. Did you make test-checks on the hiring and termination procedures?				

Checklist for Internal Auditor (*continued*)

	Yes	No	Not appli-cable	See com-ment
8. If the job uses field timekeepers, did you check the terminations on their reports? If not, explain how attendance of employees is checked by the job.				
9. Is such a check effectively maintained?				
10. Did you check the rates of pay against the union agreement?				
11. If such a check is made, does it include variable wage rates?				
12. When a man is transferred from one crew to another, does the second foreman turn in his time for the entire day?				
13. Is there any possibility that both foremen might report his time?				
14. Is there an adequate control on identification badges?				
15. When time comes in from the field, is it subjected to any check before being entered in the records?				
16. Would there be any way of detecting duplications or omissions?				
17. If the job is subject to the Davis-Bacon Act, have you checked the hourly rates with the specifications?				
18. Did you note any possible violations of the Copeland antikickback law?				
19. Did you check the specifications for requirements regarding overtime?				
20. If there are any, are they being observed?				
21. Did you trace any checks to the employee's earnings record?				
22. Did you secure satisfactory explanations for all split checks (when the employee was paid with two or more checks for the same period)?				
23. If facsimile signatures are used, is a satisfactory control kept on the machines and plates?				
24. Is there any postaudit on payrolls?				
25. Is it made by someone outside the payroll department?				
26. Does it include a check of all documents supporting earnings of all employees paid off during the week and test-checks of comparable documents of other employees?				

	Yes	No	Not appli-cable	See com-ment
27. Have you looked into the procedure for handling garnishments and attachments?				
28. Have you any suggestions for strengthening internal control of payroll and timekeeping?				

F. *Cost accounting*

	Yes	No	Not appli-cable	See com-ment
1. Has the original estimate been revised to reflect changes in the manner of doing the job?				
2. Are labor costs reported daily?				
3. Are equipment costs reported daily?				
4. Did you look into the possibilities that costs or quantities of materials, or both, might exceed estimates?				
5. Is the cost of extra work segregated?				
6. Is extra work incurred only on written order signed by or for the client?				
7. If the job is using ready-mixed concrete, dry mix, or similar materials, are comparisons being made between quantities shown on vendors' billings and quantities actually used?				
8. Are labor, materials, and equipment charges coded by the engineering department and checked by the accounting department?				
9. If so, is the check effective?				
10. If any attempt were made to divert costs from one work item to another, would present procedures disclose and correct the practice?				
11. Have you test-checked the cost codings and traced them to the cost ledger?				
12. Is the cost ledger up to date?				
13. How often are cost ledger postings made?				
14. If unit costs as of last night were required on any particular work item, could they be provided some time today?				
15. If not, how long would it take and why?				
16. Can purchase commitments be determined at the close of any particular day?				
17. Can the cost per hour to operate any particular piece of equipment or any particular type of equipment be determined quickly and accurately?				

Checklist for Internal Auditor (*continued*)

	Yes	No	Not appli-cable	See com-ment
18. If the job is a joint venture, did you check the rentals paid to each joint venturer against equipment time records and the rental schedule in the joint venture agreement?				
19. Can repair and maintenance costs on all equipment be determined quickly from the records?				
20. Can repair and maintenance costs be determined by individual units and by types of equipment?				
21. Did you check equipment time records against rental schedules?				
22. If so, was there any evidence of idle equipment that should be terminated?				
23. If there is a mess hall or a camp, did you check the costs per meal and the costs per man-day?				
24. Did you check the computations of percentages of completion?				
25. If so, what method was used? Do you consider it sound in view of the nature of the job?				
26. Did you inspect the work visually and compare what you saw with the percentages of completion shown for the various work items?				
27. Did you encounter any confusion about which work item should be charged with certain costs?				
28. How were such items being handled?				
29. Would it be feasible to record man-hours and equipment hours as well as dollar costs?				
30. If you were the project manager, would you feel safe in making decisions based on our cost figures? If not, why?				
31. If you were an estimator, would you feel safe in relying on our cost figures in bidding a comparable job? If not, why?				
32. Have you any suggestions for improving our job cost accounting?				
G. General accounting				
1. Did you make test-checks of the accuracy of coding vouchers?				
2. Did you trace the entries to the books of original entry?				

	Yes	No	Not appli- cable	See com- ment
3. Did you test the footings of the books of original entry?				
4. Did you test-check postings and footings in the general ledger?				
5. Did you check the figures in the last financial statements with the general ledger?				
6. Did they agree?				
7. Did you check all general journal entries?				
8. If so, were they proper?				
9. Were they properly authorized?				
10. Were explanations complete and accurate?				
11. Did you analyze all deferred income and deferred cost accounts?				
12. If so, were the deferred items properly supported?				
13. Does the job maintain an equipment ledger?				
14. Does the job maintain a subcontract ledger?				
15. Did you review subcontract retentions?				
16. Were any released in advance without consent of the subcontractor's bonding company?				
17. Are any being held that should be released?				
18. Did you reconcile all subsidiary ledgers with the related general ledger control accounts?				
19. If any inventory accounts are carried, did you see the actual inventory to see if it looked reasonable in relation to the book amount?				
20. Did you review receivables and payables for questionable items?				
21. Do the income accounts and the related receivables reflect all items of sundry income and extra work?				
22. Are you satisfied that the receivables are in order?				
23. What did you do to satisfy yourself on the point?				
24. Did you review the correspondence file for indications of unusual or questionable items?				

Checklist for Internal Auditor (*continued*)

	Yes	No	Not appli-cable	See com-ment
H. Office procedure				
1. Are files orderly and complete?				
2. Is filing kept up to date?				
3. Do employees keep desk files?				
4. Are papers put away and desks cleared at night?				
5. Are blank checks controlled?				
6. Is the office work being properly scheduled?				
7. Are insurance reports made promptly at the end of the month?				
8. Are financial statements prepared promptly?				
9. Is the work so scheduled as to allow ample time for the preparation of tax returns required to be filed and tax data required by the general office?				

Comments

Question
No.

Checklist for Internal Auditor (*concluded*)

PROGRESS BILLINGS AND CHANGE AND WORK ORDER FORMS

Contractors seem to have an inherent dislike for paperwork, and this weakness most frequently shows itself in the failure to submit progress billings in a logical format and to submit them in a timely fashion. Even more troublesome for the contractor is keeping track of change orders and extra work orders, submitting and getting approvals of them on time, and rendering billings for these orders in a timely fashion. Smart contractors make money with change orders—those who have dealt with the government learned this long ago. It is not uncommon for construction contractors dealing in the private sector to put off discussing the money side of change orders until the job is finished. When the job is finished, memories are dim and many a dollar of profit has been lost.

Per Article XI of Contract
Dated March 4, 19___

Contract price	$75,000	
One-third due when rafters are in place		$25,000
Less 10 percent retention		2,500
Due on this billing		$22,500

APPROVED:

_____ Architect

Simple Progress Billing

Work item No. Description	Total amount	Percent complete	Amount earned	Previously billed	Earned this period
1 Site preparation	$ 1,000	100	$1,000	$1,000	—
2 Excavation	2,500	100	2,500	1,500	$1,000
3 Forms	2,400	75	1,800	400	1,400
4 Reinforcing	1,500	60	900	100	800
5 Concrete	13,300	10	1,330	—	1,330
6 Roof	9,100	—	—	—	—
7 Plumbing	3,100	5	155	—	155
8 Electrical	4,300	5	215	—	215
9 Cement finishing	1,000	20	200	—	200
10 Sash and doors	3,900	—	—	—	—
11 Painting	2,500	—	—	—	—
12 Cleanup	500	—	—	—	—
	$45,100		$8,100	$3,000	$5,100
Less: 10 percent retention			810	300	510
TOTALS			$7,290	$2,700	$4,590

APPROVED:

_____ Architect

Progress Billing on Percentage-of-Completion Basis

CHANGE ORDER NO. _____

Date . . . .

To . . . .

. . . .

. . . .

Sirs:

Please make the following changes in the contract between

. . . .

(owner's name)

and yourselves dated . . . . .and
more particularly described as . . . .

. . . .

Description	Contract Increase (Decrease)
	$
Total this change order	
Previous contract amount	
Revised contract amount	$

Signed: . . . . .Title . . . .
We hereby accept the foregoing changes and ratify all parts of the subject
contract as amended thereby.

Signed: . . . . .Title . . . .

Change Order. This form is prepared in quadruplicate on 8½ × 11 paper.
The original is accepted and returned to the owner by the contractor. The
first copy is retained by the contractor. The second copy is sent to the job
office. The third copy is sent to the architect.

ECKS & WYE
General Contractors

DISPOSITION OF PENDING CHANGE ORDER

Contract No.........................Date.................
Job description...
....
Please refer to our "Notice of Pending Change Order No..........,"
dated...................., issued in connection with the above-
described job.

The proposed change has been:

☐ Abandoned
☐ Ordered in the form originally described
☐ Ordered in an amended form, see attached.

Will you please proceed as follows:

☐ Continue under present plans and specifications
☐ Proceed at once with change. Formal change order is being
    prepared
☐ Do not proceed with change pending issuance of formal change
    order.

Signed...

Disposition of Pending Change Order. Once a change order has been abandoned or definitely ordered, a notice of its disposition should be given to subcontractors.

ECKS & WYE   No. 123
General Contractors

NOTICE OF PENDING CHANGE ORDER

Contract No.....................Date........................
Job description ....
....
We have been notified by the owner's representative on the above-
described job that the following change(s) are to be ordered:

Location of change(s)	Nature of change(s)

To the extent reasonably possible, you are requested to avoid performing work that will be affected by such change(s).

Signed.........................

Notice of Pending Change Order. As soon as notice of a pending change order is received, a change estimate number should be assigned and a change estimate file should be set up. In this file are placed all working papers and other data supporting the change estimate. Before negotiation of the change order is completed, the estimate may be revised one or more times. When the change order affecting the contract price is issued, the job estimate should be adjusted accordingly.

CHANGE ESTIMATE

Form 9 - 10M 5-53 CPCO

ESTIMATED BY_____ DATE_____ PROJECT_____ SHEET_____ OF_____

CHECKED BY_____ DATE_____ LOCATION_____ DESIGNED BY_____

Account No.	Description	Quantity	Unit	Labor	Material & Expense	Equip. & Trans.	Sub C	Labor	Material and Expense	Equipment and Transporting	Sub Cont.	Total

Change Order Estimate. Change order estimates are often required when the cost of the change is an important factor in deciding whether the change is to be made.

WO Date....................   WO No.....................

WORK ORDER

Job No............Description.............................

....

Location..

Description of work — Give full details

Charge this work order for all labor, materials, equipment rentals, and other direct costs incurred in performing the work described below:

Work ordered by.............................Title..........

Work to be done for...

Accepted by.............................for................

Work Order. This form is prepared in quadruplicate on 8½ × 11 paper. The original goes to the person or firm that is to do the work, the first copy to the person ordering the work, the second copy to the job office, and the third copy to the architect. This same form may be used to cover extra work requested by the prime contractor from the subcontractor or vice versa, as well as to compile costs on a pending change order.

CHANGE ESTIMATE FOLDER

Change   Job
estimate No............name...........Contract No...........
Description
of changes..
....

Estimate   Requested
made by...............Date.............. by...............
Date of   Date   Amount
request................quoted............quoted...........
Date change   Amount of
order issued............change order.....................
Date job   Date purchasing
advised...............department advised..................

Record of subcontract and purchase order changes made necessary by this change:

Vendor or subcontractor	Amount		P. O.		Subcontract		Work order
	Add	Deduct	No.	Date	No.	Date	

Instructions:

When a change estimate is requested, all supporting data will be compiled in this folder until the estimate is submitted. At that time the file will be sent to the purchasing department. When written instructions to proceed are received the necessary changes in the purchase orders and subcontracts will be made and the file sent to accounting.

Change Estimate Folder. This 8½ × 11 tagboard is used to accumulate all the documents supporting a change estimate. If the job is large enough to require a job engineer, the estimate is usually prepared on the job. Otherwise, it is prepared in the contractor's office. The use of such a folder will generally not be justified on very small jobs, but in this event some record must be kept to ensure that subcontractors are notified of any changes in their work.

FORM 6T 3300 2-66										
				CHANGE ORDER RECORD						
P. O. NO.	P. O. DATE	C. O. NO.	VENDOR	REASON	AMOUNT OF C. O.	DATE C. O.	DATE MAILED	DATE EXEC'D	REMARKS	

Job Change Order Control Sheet

DAILY EXTRA WORK REPORT

CONTRACT NO _____ LOCATION _____   EXTRA WORK ORDER NO _____

DESCRIPTION OF WORK _____ DATE _____

NUMBER OF MEN	LABOR: OCCUPATION OR ITEM	HOURS	RATE	AMOUNT
	TOTAL LABOR			

NUMBER	EQUIPMENT: DESCRIPTION	HOURS	RATE	AMOUNT
	TOTAL EQUIPMENT			

NUMBER	MATERIAL: DESCRIPTION	UNIT	PRICE	AMOUNT
	TOTAL MATERIAL			
	TOTAL			

ACCEPTED:

_____ BY _____
FOR OWNER

Daily Extra Work Report. This form provides data for billing clients and others for work not specified in the main contract.

FORMS FOR PURCHASING AND INVENTORY CONTROL

Purchasing Forms

Theoretically the purchasing process should start with a requisition on which a person authorized to order the purchase of materials, supplies, or equipment notifies the purchasing department that a purchase is required. As a practical matter, only the largest contractors use requisitions.

The purchase order, however, is another matter. The purchase order is the contract between the contractor and suppliers and should contain all the terms and conditions the contractor wants to impose on purchases. Following is a form of purchase order in common use.

Purchase Order. Purchase orders can, and probably should, contain the terms of purchase to take full advantage of the Uniform Commercial Code. Where back ordering can create problems, a provision prohibiting back ordering would be desirable. Terms may be placed on the reverse of the form and tied in by a statement above the signature reading: "The terms and conditions on the reverse hereof are an integral part of this order and are incorporated herein by this reference.

FORM 5F 2M SETS—2-56—HPCO

VENDOR

FIELD PURCHASE ORDER

Date_____ 19____

To_____

Address_____

Please Deliver The Following Order To:

Address_____

Terms:_____

QUANTITY	DESCRIPTION	PRICE

F.O.B. Shipping Point ☐ — F.O.B. Destination ☐   (Check One)

Issue Three Copies Of Invoice At Time Of Purchase Or Mail To Above Address.

Our Order Number Must Show On Your Invoice.
Prices On This Order Are Not Subject To Increase.

We Will Only Accept Charges For Merchandise Ordered By Person Whose Signature Is Authorized In Writing By The General Office.

Charge_____

Job No._____

By_____

Job Name_____

Nº 3502

Field Purchase Order. Copies are made in quadruplicate for vendor, jobsite office, central purchasing, and accounting. This form is for use only for small pickup purchases at the jobsite.

Form 107A

LUMBER, MILLWORK, AND OTHER CARPENTRY MATERIALS (7A)
JOB NO.   COST ESTIMATE

DESCRIPTION	AMOUNT	UNIT PRICE	TOTAL ESTIMATED COST
A. BUILDING LAYOUT			
TOTAL			
B. FORMS AND SCREEDS			
TOTAL			
C. WOOD FLOORS AND FRAMING			
TOTAL			
D. EXTERIOR WALLS			
TOTAL			
E. ROOF FRAMING AND SHEATHING			
TOTAL			
F. EXTERIOR TRIM			
TOTAL			
SUB-TOTAL -- ESTIMATED COST OF LUMBER, MILLWORK, AND OTHER CARPENTRY MATERIALS (TRANSFER TO FORM 107B)			

Materials List for Lumber. This form is often used as a takeoff sheet. If the material takeoff has been prepared on working paper, this kind of sheet is used to summarize the materials required so that purchase orders may be written as needed. If a materials budget is used, one sheet of this type may be used for each kind of lumber included in the budget.

```
┌─────────────────────────────────────────────────────────────┐
│                              EQUIPMENT LOCATION CARD          │
│                                                               │
│   DATE    TRANSFER NO.    FROM        TO         CONDITION     │
│                                                               │
│                                                               │
│                                                               │
│                                                               │
│                                                               │
│                                                               │
│   DATE PURCHASED_____FROM_____           │
│                                                               │
│   COST_____NEW ☐   USED ☐                       │
│                                                               │
│   MONTHLY DEPRECIATION RATE_____MONTHLY RENTAL RATE│
│                                                               │
│   DESCRIPTION                          SERIAL NO.  EQUIP'T. NO.│
└─────────────────────────────────────────────────────────────┘
```

Equipment Location Card. This card is used to control power or pneumatic hand tools and similar small equipment. One or two drawers in a cabinet usually suffice for the cards listing equipment used on any one job, and the cards on any additional items in the warehouse are kept in drawers reserved for warehouse stock. Thus it is possible, at a moment's notice, to tell what items are immediately available at the end of each job.

```
OVER, SHORT AND DAMAGE REPORT
                                          No. . . . . . . .
                            Req. No. . . . . . . . . . . . . . . .
Vendor. . . . . . . . . . . . . . . . . . . . . . .   P.O. No. . . . . . . . . . . . . . . .
Address. . . . . . . . . . . . . . . . . . . . . .   Date received . . . . . . . . . . . . . . . . . . . .
Shipper. . . . . . . . . . . . . . . . . . . . . .   Delivery ticket No. . . . . . . . . . . . . . .
Car No. . . . . . . . . . . . . . .   Collect ☐
Name of carrier. . . . . . . . . . . . . . . . .   Prepaid ☐   Charges (if collect). . . .
```

Quantity		Unit	Description of Material			
Received	Charged		Items	Units over	Units short	Nature of damage

Report prepared by. . . . . .Checked and received by. . . . .

. . . .
Signature

Typical "Over, Short, and Damage" Report

No. 2572

Hayward, Calif., _____ 19 _____

Sold to _____

Address _____

_____ Truck No. _____

Job Location _____

QUANTITY	DESCRIPTION	PRICE	AMOUNT

Gross

Tare

Net

It is agreed that the above described materials and labor are to be paid from funds, received from the above described job, but not in any event later than 30 days from date of delivery. Close Construction Co., Inc. will not be responsible for damage due to delivery inside of curb line.

TRUCK TIME

Leave:

Return:

Stand by: _____ Hrs. _____ Min.

Received

By _____

8% interest will be charged after 30 days.

Reasonable Attorney Fees to be allowed in event of suit to collect.

℗r.

OFFICE COPY

Combination Delivery Ticket and Invoice. This snap-out form published by Moore Business Forms, Inc., has single-use carbon paper. The original is the office copy, the first copy is the invoice, and the second and third are delivery memoranda.

REQUISITION ON PURCHASING DEPARTMENT

REQ. NO. _____

DATE ORDERED _____

ORDER NO _____

TO
PURCHASING AGENT:

PLEASE ORDER FROM _____

DATE WANTED _____

SHIP TO _____ VIA _____

QUANTITY	DESCRIPTION	FOR WHAT PURPOSE

APPROVED	ORDERED
BY	BY

PRINTED IN U.S.A.   REG. U.S. PAT OFF.   STD. REQUISITION ON PURCHASING DEPT.   FORM C536

Purchase Requisition

```
                THIS IS YOUR INVOICE
                NO OTHER WILL BE SENT




Cust.
Order No.  565              Date  July 24        19

M      INVESTO CORP
Address    1025 N San Fernando Rd
           Burbank, Calif.

SOLD BY  CASH  C.O.D.  CHARGE  ON ACCT.  MDSE. RETD.  PAID OUT

QUANTITY           DESCRIPTION              PRICE    AMOUNT
    4      8116  AMSTAN CLOSETS       27.35   109 40
    2      385    ✓      LAVS.        36.50    73 60
    1      7602   ✓    Laundry Try    25.65    25 65
   10     Closet Bolts CompleR         .11      1 10
    8     ½"CP Angle Stops             .81      6 48
    4     ½" galu Nips                 .07        28
    8     1½"CP Sure Grip flanges      .07        56
   12'    1½" CP Tubing Casing         .76        26
   6"     2"    ✓      ✓      ✓        .81      4 86
    2     1¾x1¼ CP Traps             2.53      5 06
   8'     7/6"  C.P Tubing             .15      1 80
   16     7/16" CONE WASHERS           .03        48
   12     Putty                        .08        96
    2     1½" galu Nips               .19        38
   22     14x2" Wood Screws           .02        48
                              TAX              330 75
              1954            TOTAL              8 07
                                               238 82
     All claims and returned goods MUST be accompanied by this bill.

No. 11516      Rec'd by  ppt  199.91 ✓
                         FORM 6718  SUNSET BUSINESS FORMS  OAKLAND  LOS ANGELES
```

Materials Billing Form. This is the second (accounting department) copy of a form used by a small local plumbing contractor. When materials are taken from stock, they are recorded on a hand-operated billing machine on the stockroom counter. Four copies are made. The first is the customer's copy of the invoice. The second is sent to the bookkeeper and serves as the office copy of the customer's billing. The third copy is the cost copy. The fourth copy remains in the machine as a chronological record.

Inventory Control Forms

Article _____

When balance on hand is _____ verify, count and notify office.

RECEIVED	DATE	WITHDRAWN	BALANCE

CHARLES E. HADLEY CO., PAPERMAKERS, LOS ANGELES, OAKLAND, PORTLAND   REG. U.S. PAT. OFF.
PRINTED IN U.S.A.   STANDARD BIN TAG   FORM C548

A. "In and Out" Bin Card

LOCATION _____   STOCK NO. _____

ARTICLE

Verify count and notify office when balance on hand is _____

Send Office WARNING NOTICE when balance on hand is _____

DATE 19	REQUISITION NO.	BALANCE
	BALANCE FORWARD	

CHARLES E. HADLEY CO., PRINTFINDERS, LOS ANGELES, SAN FRANCISCO, SAN DIEGO, CHICAGO   REG. U.S. PAT. OFF.
PRINTED IN U.S.A.   STANDARD BIN TAG   FORM C51R

B. Inventory Bin Card. Additions or deductions can be made in the balance column, or the adjusted balance can merely be carried forward.

MATERIAL CREDIT

CREDIT TO _____

LOCATION _____

DATE _____ NO. _____

QUAN.	ARTICLE	UNIT COST	AMOUNT

RECEIVED BY _____   STOCK CLERK

CHARLES E. HARLEY CO. PATHFINDERS (AMERICA) ... REG. U. S. PAT. OFF.   STANDARD MATERIAL CREDIT FORM G511

D. Material Credit

MATERIAL ORDER

CHARGE TO _____

LOCATION _____

DATE _____ 19____

QUAN.	UNIT	ARTICLE	UNIT COST	AMOUNT

RECEIVED BY _____

CHARLES E. HARLEY CO. PATHFINDERS (AMERICA) ... REG. U. S. PAT. OFF.   STANDARD MATERIAL ORDER FORM C573
PRINTED IN U.S.A. —

C. Material Order. This form is frequently used to authorize deliveries from, and returns to, the stockroom or warehouse. It may be made in as many copies as the accounting needs of the business dictate.

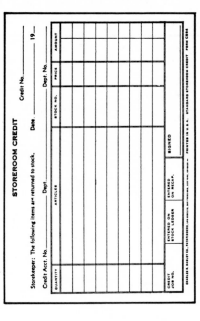

F. Storeroom Credit

E. Storeroom Requisition. This form is similar in use to the material order and material credit forms.

G.   Perpetual Inventory Card. This card is used by one of the largest construction companies in the United States. It is reported to be very accurate and a great timesaver if volume is heavy. Computer systems, of course, use a different form.

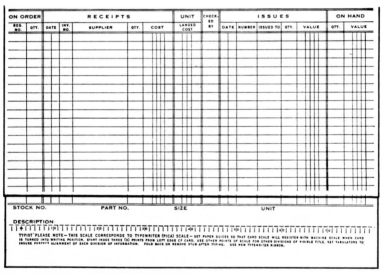

| | ARTICLE | | | | LOCATION | | | UNIT | | DATE | MAX. | MIN. |

(Inventory Card form with columns)

ORDERED			RECEIVED			SOLD													
DATE	ORD. NO.	QUAN.	DATE	ORD. NO.	QUAN.	DATE	ORD. NO.	QUAN.	BAL.	DATE	ORD. NO.	QUAN.	BAL.	DATE	ORD. NO.	QUAN.	BAL.

H. Inventory Card

(Stock Card form)

ON ORDER		RECEIPTS					UNIT	CHECK-ED	ISSUES					ON HAND	
REG. NO.	QTY.	DATE	INV. NO.	SUPPLIER	QTY.	COST	LANDED COST	BY	DATE	NUMBER	ISSUED TO	QTY.	VALUE	QTY.	VALUE

STOCK NO.   PART NO.   SIZE   UNIT

DESCRIPTION:

TYPIST PLEASE NOTE—THIS SCALE CORRESPONDS TO TYPEWRITER (PICA) SCALE—SET PAPER GUIDES SO THAT CARD SCALE WILL REGISTER WITH MACHINE SCALE WHEN CARD IS TURNED INTO WRITING POSITION. START INDEX THREE (3) POINTS FROM LEFT EDGE OF CARD. USE OTHER POINTS OF SCALE FOR OTHER DIVISIONS OF VISIBLE TITLE. SET TABULATORS TO INSURE PERFECT ALIGNMENT OF EACH DIVISION OF INFORMATION. FOLD BACK OR REMOVE STUB AFTER TYPING. USE NEW TYPEWRITER RIBBON.

I. Stock Card for Visible-Index Use

Dated................................   **TRANSFER OR SALES REPORT**   No................

From..(Contract No.................................)
CONSIGNOR

To..(Contract No.................................)
CONSIGNEE

Address of Consignee...Via................

UNITS	DESCRIPTION	ACCT'G.	CREDIT ACCOUNT	DEBIT ACCOUNT	UNIT PRICE	AMOUNT

Approved:..   Correct:...
SUPT.   MATERIAL CLERK

Received above described material,....................., 19......,   By...

FORM 11   1M SETS 12-33—CP   TITLE

Transfer or Sales Report. This is a combination form for use in reporting and recording either transfers of materials, supplies, or equipment from one job to another, or sales of such items. Five differently colored copies are prepared.

FORMS FOR CONTROLLING SUBCONTRACTOR COSTS

It might appear to many inexperienced contractors that once the subcontract is let, there are few, if any, problems in controlling and accounting for these costs. To a large degree this statement is true. On very small jobs the subcontractor is engaged using very simple standardized subcontract forms, performs the work, and renders a bill at the end of the job. If subcontracting is on a cost-plus basis, the subcontractor submits periodic billings for costs and fees, and again the cost is accounted for in the period when the billing takes place. If jobs cover a longer period and involve progress billings, retained percentages, change orders, and special work orders, or if there are second- and third-tier subcontractors, jobsite security regulations, or labor problems, control procedures must be broadened to meet the broader needs. The larger and more numerous the subcontracts become, the greater are the problems of controlling them.

The forms that appear on the following pages are those typically used by large, successful contractors who manage and control subcontractors.

SUB-CONTRACTOR'S PARTIAL PAYMENT ESTIMATE NO._____

For Period Ending_____

Sub-Contractor:

Description of Work:

	Total Amount	% Completed to Date	Amount to Date
Original Sub-Contract Amount:	$		
Change Orders:			
Total Amount of Contract:	_____	_____	_____
Less Retained Percentage (%)			_____
Less Previous Payments			_____
Net Amount to Date:			_____
Less:			
Payment Now Due			_____

Subcontractor's Payment Estimate. This form is designed to be either typed or handwritten.

FORM 74—5M—5-56 CPCO

Date _____

SUB-CONTRACT PARTIAL PAYMENT ESTIMATE NO. _____

JOB NAME _____

JOB NO. _____

FOR PERIOD FROM _____ 19 ___   TO _____ 19 ____

SUB-CONTRACTOR_____   SUB-CONTRACT NO. _____

DESCRIPTION OF WORK _____   ACCOUNT NO. _____

ORIGINAL CONTRACT PRICE	CHANGE ORDERS TO DATE	NET CONTRACT PRICE	% COMPLETED TO DATE	VALUE OF WORK COMPLETED TO DATE
$ _____	$ _____	$ _____	_____ %	$ _____

CHANGE ORDER RECORD

C. O. NO.	DATE	INCREASE	DECREASE	
_____	_____	$ _____	$ _____	
_____	_____	$ _____	$ _____	
_____	_____	$ _____	$ _____	
_____	_____	$ _____	$ _____	
_____	_____	$ _____	$ _____	$ _____

PARTIAL PAYMENT RECORD

	PREVIOUS	THIS PERIOD	TO DATE
TOTAL AMOUNT ESTIMATED	$ _____	$ _____	$ _____
RETAINED AMOUNT (%)	$ _____	$ _____	$ _____
NET PAYMENT DUE	$ _____	$ _____	$ _____

ESTIMATE PREPARED BY_____ APPROVED BY_____

APPROVED FOR PAYMENT BY_____

Summary of Subcontractor's Partial Payment Estimate. This form is designed for separate computation of percentage of completion on change orders. If this procedure is not followed, changes can be lumped into one amount or the work items can be adjusted to take the changes into account.

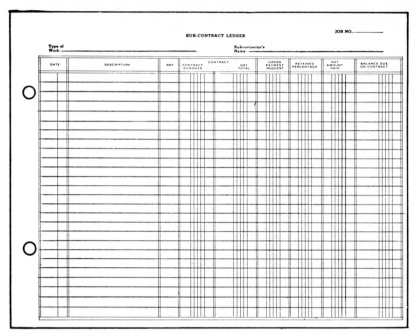

Subcontractor's Ledger. A special variety of accounts payable ledger, this form is particularly useful when there are numerous subcontracts, when these constitute an important part of job costs, and when the amounts owed are important in the contractor's balance sheet.

Subcontractor's Field Work Order. This form was designed to make it easy to reduce to writing agreements between the prime contractor and subcontractor and thus prevent the disputes that arise from oral instructions and unrecorded "deals."

FORMS FOR CONTROL OF LABOR AND SUPPORT COSTS

Before an employee is hired, someone has to decide that his services are in fact needed. Just how much paperwork is required to control workforce levels and to start the new employee working depends on such things as the size of the contractor's oganization, the nature of the construction project, the nature of a particular job that the employee is to do, the requirements of the tax and labor laws, and the extent to which employment is subject to labor union rules and union contracts.

In a small organization there would be few formalities and the contractor would probably do the hiring. In larger organizations the superintendent would issue a requisition which would be initialed by the project manager and sent to the personnel department to authorize hiring the number and classification of workers needed.

Labor cost is the most difficult in terms of cost management since it is the most variable. Support costs such as camp and mess hall facilities are also difficult for the average contractor and construction manager, particularly if they have had little experience in performing jobs in out-of-the-way places where these support facilities are necessary.

The forms on the following pages are typical of those used by successful contractors to provide for adequate management and control over basic hiring, labor cost control, and support facilities.

Hiring and Payroll Forms

Employment application forms are not frequently used by construction companies except for office and high-level engineering personnel. However, the risks are great if, by reason of asking the wrong questions on an application form, the company becomes involved in an affirmative action suit. The rules vary by states, and any employer would do well to check state regulations before drafting (or using) an employment application form. Following is a form that would be acceptable in most states.

Application for Employment

GENERAL INFORMATION

Print name in full _____ Soc. Sec. No. _____

Address _____ Telephone _____

Date of birth _____ Are you a U. S. citizen _____

Name and address of parent or guardian (if applicant is a minor)

The company does not permit employment of relatives or spouses in the same department.

Marital Status (check) Single _____ Married _____ Widowed _____ Divorced _____ Separated _____

Number of children _____ Number dependent upon you for support _____

Do you rent, own your own home, live with your parents, or board? _____

Position applied for _____ Monthly salary expected _____

Are you employed at present? _____ How soon can you report if engaged _____

EDUCATION

Name and location of school	Dates From – to	Did you Graduate	Nature of course taken, or degree	Studying now?

What studies did you like best?

EXPERIENCE

Give below, *starting with present or most recent position, complete record of your business experience.*

Firm	Period	Describe your duties and position	Name immediate superior and his title	Why did you leave?
Name	From			
Address	To			
Business	Salary			
Name	From			
Address	To			
Business	Salary			
Name	From			
Address	To			
Business	Salary			

Application Blank—Face

Why do you desire to change your position?_____

Were you ever discharged from any position?_____Explain _____

REFERENCES

GIVE THREE REFERENCES OTHER THAN RELATIVES OR PAST EMPLOYERS

Name	Business or profession	Position	Address	Telephone	How long known
1					
2					
3					

MISCELLANEOUS

Will you be engaged in any other income producing activity, if employed? _____

Explain _____

Have you ever been bonded?_____Has bond ever been refused you? Explain_____

The answers to the foregoing questions are true and correct and I have not
knowingly withheld any fact or circumstance which would, if disclosed,
affect my application unfavorably.

Applicant's signature _____

Date of application _____

Application Blank—Reverse. A number of states have fair employment laws which prohibit questions on an application which disclose sex, marital status, dependents, pregnancy, birth control, names or addresses of relatives of adult applicants, or whether the applicant lives with parents.

Assignment Slip for New Employees
(GIVE THIS TO YOUR FOREMAN)

EMPLOYEE'S   P. R. No. ....

Name.. Ident. No..................

Occupation.................................... Rate.............. per..............

Report to Foreman..
                                    (name)
                                                                            P.M.
At Div.................................... To Start Work.................. A.M.

WILL HAVE MEALS AND LODGING IN CAMP   YES
                                                                           NO

Assignment Slip for New Employees. This form is used to provide verification to the foreman that hiring procedures are complete and that the individual may begin work. It also indicates work assignment and approval for meals and lodging.

PAYROLL CHANGE NOTICE

Please check —

☐ Enter on payroll
☐ Change rate
☐ Transfer to.............Entered (Time Dept.)....Date........
☐ Pay off and remove from payroll

Name................................S.S.A/C No...............

Dept..................Shift................Clock No..............

Date effective............................Hour............. A.M.
                                                              P.M.

Old rate.............Per......New rate.............Per.......

Remarks:...
....

Record   ☐ Discharged   ☐ Left   ☐ Laid off   Would you   Yes ☐
                                                    re-employ   No ☐

	Excellent	Good	Fair	Poor
Ability				
Conduct				
Attendance				
Production				

Approved..........................   ....
                        Supt.   *Foreman*

Payroll Change Notice. This form is used to hire a new employee, to change an assignment or rate after employment, or as a termination.

Job No........ Ecks & Wye Construction Co.										
TIME REPORT FOR THE WEEK ENDING										
Badge No.	Name	S O	Sat	Sun	Mon	Tue	Wed	Thu	Fri	Total
		S								
		O								
		S								
		O								
		S								
		O								
		S								
		O								
		S								
		O								
		S								
		O								
		S								
		O								
		S								
		O								
		S								
		O								
		S								
		O								
Signed		S								
.... Foreman		O								

Weekly Time Report. This form is designed for a payroll week ending Friday evening.

Daily Time Report. This form is prepared by the foreman and is then completed in the payroll department.

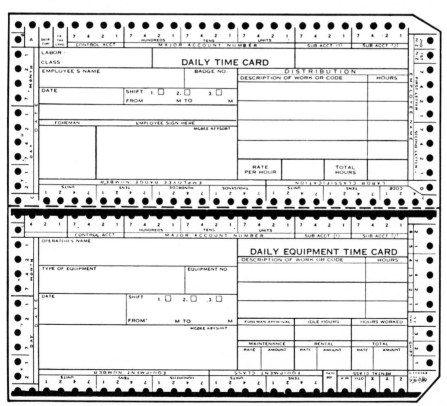

Keysort Labor and Equipment Time Card. This card is used in large operations where mechanical sorting is required. The daily time card may be used independently.

Always required

Form W-4—employee's withholding exemption certificate. Note that the forms show name, home address, and social security number. Form W-4 is a U. S. Treasury Department form, but the data serve most state and local payroll tax purposes as well.

May be required

Union "clearance slip" may or may not be required, depending on the union agreement. If it is required, the unions will usually have their own forms.

State withholding tax certificate may or may not be required, depending on the state. California requires such forms.

Optional

Separate employee data card
Combined employee data card and employee earnings record
Application blank, used almost exclusively for executive and office employees
Equipment issue receipts are used when the contractor provides such items as hard hats, safety shoes, raincoats, boots, blankets, mattresses, pillows, sheets, tools of the trade, and so on. Such a receipt in the man's file may be used to support a deduction upon his separation.
Camp and mess hall clearance slip serves as a receipt when blankets, mattresses, bedding, and so on are turned in on termination, and when any mess hall charges are paid. If a camp and mess hall are maintained, this slip would have to be presented before the final check is issued.
Assignment slip for new employees
Payroll change notice

Checklist of Hiring Forms

```
┌─────────────────────────────────────────────────────────────┐
│                  ECKS & WYE CO.          No.........          │
│                  General Contractors                          │
│  Manpower Requisition                    Date..............   │
├─────────────────────────────────────────────────────────────┤
│  To: Personnel Dept.                                          │
│  Please provide the following personnel for................   │
│                                          (Name of Job)        │
│  Number                                              Date     │
│  Needed              Classification                  Needed   │
│                                                               │
│                                                               │
│                                                               │
│                                                               │
│                                                               │
│                                                               │
│              Signed.......................................    │
│               Title.......................................    │
│  OK to hire.......................................Project Mgr. │
└─────────────────────────────────────────────────────────────┘
```

Manpower Requisition. The original is sent from the project manager to the personnel department. The duplicate remains in the superintendent's book.

Employee's Withholding Allowance Certificate

The explanatory material below will help you determine your correct number of withholding allowances, and will indicate whether you should complete the new Form W-4 at the bottom of this page.

How Many Withholding Allowances May You Claim?

Please use the schedule below to determine the number of allowances you may claim for tax withholding purposes. In determining the number, keep in mind these points: If you are single and hold more than one job, you may not claim the same allowances with more than one employer at the same time; If you are married and both you and your wife or husband are employed, you may not claim the same allowances with your employers at the same time. A nonresident alien other than a resident of Canada, Mexico or Puerto Rico may claim only one personal allowance.

Figure Your Total Withholding Allowances Below

(a) Allowance for yourself—enter 1 . . . . _____

(b) Allowance for your wife (husband)—enter 1 . . . . _____

(c) Allowance for your age—if 65 or over—enter 1 . . . . _____

(d) Allowance for your wife's (husband's) age—if 65 or over—enter 1 . . . . _____

(e) Allowance for blindness (yourself)—enter 1 . . . . _____

(f) Allowance for blindness (wife or husband)—enter 1 . . . . _____

(g) Allowance(s) for dependent(s)—you are entitled to claim an allowance for each dependent you will be able
    to claim on your Federal income tax return. Do not include yourself or your wife (husband)* . . . . _____

(h) Special withholding allowance—if you have only one job, and do not have a wife or husband who works—
    enter 1 . . . . _____

(i) Total—add lines (a) through (h) above . . . . _____
    If you do not plan to itemize deductions on your income tax return, enter the number shown on line (i) on
    line 1, Form W-4 below. Skip lines (j) and (k).

(j) Allowance(s) for itemized deductions—If you do plan to itemize deductions on your income tax return, enter
    the number from line 5 of worksheet on back . . . . _____

(k) Total—add lines (i) and (j) above. Enter here and on line 1, Form W-4 below . . . . _____

*If you are in doubt as to whom you may claim as a dependent, see the instructions which came with your last Federal income tax return
or call your local Internal Revenue Service office.

See Table and Worksheet on Back if You Plan to Itemize Your Deductions

Completing New Form W-4

If you find that you are entitled to one or more allowances in addition to those which you are now claiming, please increase your number of allowances by completing the form below and filing with your employer. If the number of allowances you previously claimed decreases, you must file a new Form W-4 within 10 days. (Should you expect to owe more tax than will be withheld, you may use the same form to increase your withholding by claiming fewer or "0" allowances on line 1 or by asking for additional withholding on line 2 or both.)

▼ Give the bottom part of this form to your employer; keep the upper part for your records and information ▼

- -

Form **W-4** (Rev. Aug. 1972) Department of the Treasury Internal Revenue Service	**Employee's Withholding Allowance Certificate** (This certificate is for income tax withholding purposes only; it will remain in effect until you change it.)
Type or print your full name	Your social security number
Home address (Number and street or rural route)	Marital status ☐ Single ☐ Married
City or town, State and ZIP code	(If married but legally separated, or wife (husband) is a nonresident alien, check the single block.)

1 Total number of allowances you are claiming . . . . _____

2 Additional amount, if any, you want deducted from each pay (if your employer agrees) . . . . $_____

I certify that to the best of my knowledge and belief, the number of withholding allowances claimed on this certificate does not exceed the number to which I am entitled.

Signature ▶ _____ Date ▶ _____ 19 _____

Employee's Withholding Allowance Certificate. This form is supplied by the Internal Revenue Service. Care should be taken to ensure that new employees are signed up properly.

Most states which have personal income tax also have withholding. Below are examples of the California forms. However, care should be taken to see that new employees are signed up on the most recent forms issued by the state in which the employment occurs.

California Residence Tax Form. This form and the next one are withholding tax forms used by the state of California. Other states, and even some cities, have personal income taxes, which may be collected by withholding.

California Nonresidence Tax Form

NAME														EMPLOYEE NO.			

ADDRESS _____ PHONE _____ SOCIAL SECURITY NO. _____

DATE			JOB CLASSIFICATION	DEPARTMENT	WORK IS			FULL TIME HRS.	RATE	PER	EQUIV-ALENT RATE PER HR.	RATE OK'D BY	TAX STATUS		SEPARATION				
MO.	DAY	YR.			REG.	TEMP.	PART TIME						DATE	NO. EXEMPT.	DATE			REASON	O. K. TO RE-EMPLOY?
															MO.	DAY	YR.		

MALE ☐ FEMALE ☐   MARRIED ☐ SINGLE ☐

YEAR AND QTR.	WORKED		EARNINGS					DEDUCTIONS						TAXABLE EARNINGS REPORTED			
	WKS.	DAYS OR HOURS	REGULAR			TOTAL	F.I.C.A.	STATE UNEMP. DIS.	INC. TAX					UNEMPLOY. DIS. INS.		FEDERAL OLD-AGE
														MONEY	OTHER	
19 (1)																	
(2)																	
(3)																	
(4)																	
TOTALS																	
19 (1)																	
(2)																	
(3)																	
(4)																	
TOTALS																	
19 (1)																	
(2)																	
(3)																	
(4)																	
TOTALS																	
19 (1)																	
(2)																	
(3)																	
(4)																	
TOTALS																	

DATE OF BIRTH _____   IN CASE OF ACCIDENT NOTIFY _____
PLACE OF BIRTH _____   ADDRESS _____
AGE AT TIME RECORD PREPARED _____   PHONE _____   RELATIONSHIP _____

BUSINESS SYSTEMS INCORPORATED L. A. - S. F.   STANDARD EARNINGS RECORD FOLDER FORM M801

Combination Employee's Data Card and Earnings Record. This form is used as a file folder for copies of the payroll checks as well as a data card and earnings record. In computerized systems the data on this record are ordinarily compiled on the computer.

EMPLOYEE DATA CARD

Date employed			Job classification	Rate	Per	Rate OK'd by	Tax status			Separation		OK to rehire
Mo.	Day	Year					Date	No. exemp.		Date	Reason	

Married ☐ Single ☐ Separated ☐
Male ☐ Female ☐
Place of birth. . . . .

Work is: Permanent ☐ Temporary ☐
Date of birth. . . . .
U. S. Citizen: Yes ☐ No ☐

In case of accident notify. . . . .

Withholding
exemptions. . . . .

Address. . . . . Phone. . . . .

Name. . . . . Badge
No. . . . .
Social
Sec. No. . . . .

Separate Employee's Data Card

Time Week Ending _____ 19___

NAMES	S	M	T	W	T	F	S	Total Time	Rate	AMOUNT	
										$	Cts.

A Sheet from a Foreman's Time Book

DAILY TIME CARD

Date _____ From _____ To _____

NAME _____

Classification _____ Rate _____

Distribution	Hours

CORRECT   This is to Certify that my Classification, Rate & Time shown above are correct.

Foreman _____   Employee Sign Here.

Daily Time Card. This simple card is suitable for almost any kind of job except one that is so large that mechanical sorting would be required because of the number of employees involved.

PAYROLL LEDGER & EMPLOYEES RECORD OF EARNINGS

JOB NO _____ JOB NAME _____

DATE HIRED _____ TERMINATED _____ WITHHOLDING EXEMPTIONS CLAIMED _____ SINGLE ☐ MARRIED ☐ OCCUPATION _____

NAME _____ ADDRESS _____ S S NO _____

W'E	DAILY HOURS	TOTAL HOURS	RATE	AMOUNT EARNED	DEDUCTIONS				
					F O A B	INC. TAX	S.U.I.		
	YEAR TO DATE	BRO. FWD.		$					
									NET CHECK
									NO.
		GROSS EARNINGS		$					$
									NET CHECK
									NO.
		GROSS EARNINGS		$					$
									NET CHECK
									NO.
		GROSS EARNINGS		$					$
									NET CHECK
									NO.
		GROSS EARNINGS		$					$
		TO DATE		$					
									NET CHECK
									NO
		GROSS EARNINGS		$					$
									NET CHECK
									NO.
		GROSS EARNINGS		$					$
									NET CHECK
									NO.
		GROSS EARNINGS		$					$
									NET CHECK
									NO.
		GROSS EARNINGS		$					$
		TO DATE		$					
									NET CHECK
									NO.
		GROSS EARNINGS		$					$
									NET CHECK
									NO.
		GROSS EARNINGS		$					$
									NET CHECK
									NO.
		GROSS EARNINGS		$					$
									NET CHECK
									NO.
		GROSS EARNINGS		$					$
									NET CHECK
									NO.
		GROSS EARNINGS		$					$
		QTR TOTAL		$					
	YEAR TO DATE CARRIED FWD			$					

Combination Payroll Ledger and Employee's Earnings Record. In computerized systems this form is usually generated in connection with preparation of the paychecks and the employees' earnings statement.

ECKS & WYE CO.
Time Control Record

Date....................

01			51			101			151			201			251		
02			52			102			152			202			252		
03			53			103			153			203			253		
04			54			104			154			204			254		
05			55			105			155			205			255		
06			56			106			156			206			256		
07			57			107			157			207			257		
08			58			108			158			208			258		
09			59			109			159			209			259		
10			60			110			160			210			260		
11			61			111			161			211			261		
12			62			112			162			212			262		
13			63			113			163			213			263		
14			64			114			164			214			264		
15			65			115			165			215			265		
33			83			133			183			233			283		
34			84			134			184			234			284		
35			85			135			185			235			285		
36			86			136			186			236			286		
37			87			137			187			237			287		
38			88			138			188			238			288		
39			89			139			189			239			289		
40			90			140			190			240			290		
41			91			141			191			241			291		
42			92			142			192			242			292		
43			93			143			193			243			293		
44			94			144			194			244			294		
45			95			145			195			245			295		
46			96			146			196			246			296		
47			97			147			197			247			297		
48			98			148			198			248			298		
49			99			149			199			249			299		
50			100			150			200			250			300		

Time Control Record. This form is used by field timekeepers checking badge numbers on the job.

DAILY LABOR COST REPORT

Job:.............................Date:.....................

Work Item	Unit	YESTERDAY				JOB TO DATE			
		Units Compl.	Man Hrs.	Total Cost	Unit Cost	Units Compl.	Man Hrs.	Total Cost	Unit Cost

Time checked & priced by......	Pricing checked, extensions made by...............	Cost coded, unit priced by..................

Daily Labor Cost Report. When labor costs are computerized, this form is normally not needed. However, the basic data must be compiled manually and placed in the computer.

Form 2—1M Sets—3-56—CPCO

PAYROLL VOUCHER

Payroll Week Ending_____   Payroll Number_____

Project Name_____   Project Number_____

GENERAL LEDGER

DEBITS			CREDITS		
A/c#	Items	Amount	A/c#	Items	Amount

BASIS FOR ACCRUALS

A/c#___ ___ % of $_____   A/c#___ ___ % of $_____   A/c#___ ___ % of $_____

Detail or Cost Ledger Distribution--Debits (Credits)

Entered in Vo. Register_____ Posted to Cost Ledger_____

Payroll Voucher. To avoid unnecessary writing or typing on regular vouchers, many firms use specially printed vouchers to record the check that is issued to reimburse the payroll bank account.

FORM 4B IOM 4-57 CPCO

O-OVERTIME   R-REGULAR TIME   PAYROLL   CONTRACT_____
                                    LOCATION OF JOB_____
MADE BY_____ APPROVED BY_____   PAYROLL FROM_____TO_____ SHEET____

(Blank payroll grid form with column headers: BADGE NOS., EMPLOYEES NAMES – SOC. SEC. NOS., OCCUPATIONS, DAY OF WK., HOURS, TOTAL, RATES, GROSS EARNINGS (AT EACH RATE, TOTAL), DEDUCTIONS (S.U.I., F.O.A.B., INCOME TAX), NET PAYMENTS, CHECK NOS. Each employee row marked O and R.)

Payroll Form. This is one of the many forms on which payrolls are recorded. This form is more than usually complete, since it shows the hours worked each day. This kind of payroll is usually required on United States government jobs because of the need to comply with the various laws governing payment to workers on government work. Computerized payrolls usually generate this form as a part of the regular recording process.

TERMINATION SLIP

CONTRACT NUMBER	STATE	HOUR	DATE

| SOCIAL SECURITY NO. | TERMINATION DATE | PAYROLL NUMBER | IDENTIFICATION NUMBER |

| LAST NAME | FIRST NAME | INITIAL | OCCUPATION | RATE | PER |

REASONS FOR SEPARATION

- [ ] 1 TRANSFERRED
- [ ] 2 LAID OFF
- [ ] 3 QUIT FOR OWN REASONS
- [ ] 4 SICK OR INJURED
- [ ] 5 DIED
- [ ] 6 REMOVED BY OWNER

TIME TODAY _____ HOURS

7 DISCHARGED REASON

Termination Slip. In states where the unemployment insurance rate may be affected by contesting employee claims, a termination slip can be helpful. Many firms use slips which provide space for signatures by the supervisor and the employee.

IMPORTANT

Enclosed is a Withholding Tax Statement (Form W-2), showing your earnings on the contract indicated. If your record of earnings is different from the enclosed Form W-2, please fill out this card and return.

Name_____

Social Security Number_____

Correct Earnings_____W-2 Earnings_____

Termination Date_____

Location of Work _____

Information Card. If this notice is enclosed with the employee's copy of the Form W-2, payroll errors and frauds may be discovered.

Support Facility Forms

Bread rack
Butcher block
Can opener, hotel model with spare knives and gear
Coffee maker, urn
Coffee pots, 3 gal
Dishwasher with necessary tables
Doughnut cutters
Egg slicer
Fans, blower type for ranges
Fans, electric for kitchen and mess hall
Flour sifter
Food storage containers
Freezer, deep freeze
Freezer, walk-in type
French fry cutter
French fryer
Fryer, electric
Garbage cans, 10, 20, and 36 gal
Grater, 4-sided, $9 \times 4 \times 4$ in
Griddle, automatic

Hand truck
Hot food tables
Ice cream freezer
Ice machine with storage bin and spare parts
Ice pick
Juice extractor
Knives, butcher, 12 and 14 in
Knives, French forged bolster, 10, 12, and 14 in
Lamb splitter, 12 in
Meat and bone saw
Meat slicer
Mixing spoons
Mixers, 5 and 30 qt
Muffin pan, 12 cup, $14 \times 10 \times 1\frac{1}{2}$ in
Oven, bake
Pans, baking
Pans, long-handled
Pans, loop-handled
Pastry tube

Pie server
Pot racks
Potato peeler
Pots, brazier, heavy aluminum, with 2 loop handles
Pots, sauce, with cover, various sizes
Proof box, 16-gauge galvanized, on casters
Ranges, fuel to suit available supply
Refrigerators, walk-in
Refrigerators, electric or gas
Refrigerator truck
Roast beef slicer
Scale, kitchen
Scale, bakers'

Scoops, sugar and flour
Sinks
Spatulas, bakers' 10 and 12 in
Steak cuber
Steak hammer
Steel, butchers'
Stove, electric for coffee maker
Tables, cooks' work
Tables, bakers'
Tenderizer, meat, with cleaning and sharpening tools
Toasters, electric
Tongs, stainless steel utility
Trays, serving, 14×18 in plastic
Waste baskets

Checklist of Kitchen and Mess Hall Equipment

Aprons, cooks' and waiters'
Ash trays
Bowls, 5 and 6 in
Bowls, grapefruit
Bowls, gravy
Bowls, serving
Brooms
Brushes, bottle
Brushes, cleaning and scrub
Caps, cooks' and waiters'
Cruets, oil and vinegar
Cups
Dishtowels
Dustpans
Foil, aluminum, 18 and 24 in
Griddle bricks
Ice cream dishes
Ice cream scoops
Ice cream spade
Mats, door
Mats, hot
Mops, heads and handles
Mustard pots with covers
Napkins, paper, with dispensers

Platters, china, 15½ × 10, 13¾ × 8, 12 × 6, and 11 × 5½ in
Plates, china, 10 and 8 in
Pot lifters
Plumbers' helpers, rubber suction cups
Salt and pepper shakers
Saucers
Scouring pads
Silverware, knives, forks, spoons, soup spoons, serving spoons, steak knives, etc.
Serving dishes
Soap, bar
Soap, powder
Soup plates
Sponges
Spoons and ladles, cooking
Sugar dispensers
Syrup dispensers
Trays, celery
Uniforms, cooks' and waiters'
Vegetable bowls
Water glasses
Wooden bowls

Checklist of Kitchen and Mess Hall Supplies

Beds, single, steel
Bedsprings, single, steel
Chairs
Coffee tables (for recreation hall)
Lamps, floor and table
Mattresses, innerspring
Motion picture equipment (recreation hall)

Side tables
Upholstered chairs and sofas (recreation hall)
Wardrobes
Washing machine and laundry equipment
Writing desks

Checklist of Camp Equipment

Bathmats
Bathtowels and hand towels
Bedpads, single
Bedspreads, single
Blankets, single, woolen and cotton
Bleach (Clorox or equivalent)
Brooms
Clothesline and clothespins
Disinfectant (Lysol or equivalent)
Dust mops and supplies
Dust pans
Floor brushes
Insect spray
Lye

Mattress covers
Mops
Padlocks and extra keys
Pails
Pillows and pillowcases
Rat poison
Saniflush
Sheets, contour and plain
Soap, hand and bath
Soap, liquid
Soap, laundry, bar and powder
Toilet paper
Washcloths

Checklist of Camp Supplies

Stock Card   CAMP, MESS HALL AND COMMISSARY   Foodstuffs

Date	ON ORDER			RECEIPTS			ISSUES			BALANCE		
	Req. No.	P. O. No.	Quantity	P. O. No.	Quantity	Invoice cost	Issue ticket	Quantity	Charge-out price	On hand	Physical count	Date counted

Item	Unit	Stock No.	Loca-tion	Reorder at

Inventory Stock Card. This card is designed for use in a visible-index perpetual inventory record. Note that no effort is made to extend a value for the items on hand after each transaction. Note, too, that "charge-out price" can be so set that a margin for loss and spoilage can be included.

INVENTORY OF CAMP SUPPLIES AND EQUIPMENT

Date taken.............By................................Mgr...................

Sheet No.
....
....

Items	Storeroom location	Total to account for	Location				Explanation
			Storeroom	Bunkhouse	Laundry and repair	Short	
Bedding Blankets, single, cotton	Sec. 1	100	19	40	40	1	1 burned by cigarette
" double "	"	50	10	20	20		
" single, wool	"	40	18	20	2		
Mattress covers	"	40	18	20	2		

Camp Supplies and Equipment Inventory

ECKS AND WYE
CONTRACT 51   MOUNTAIN CITY
CAMP AND MESS TRANSACTIONS

Name...Date...............

Badge No...............Barracks...........................

Cash sale	Charge sale	Sub-contractor charge	Paid on account	Return sale	Memo

Quantity	Description	Price	Amount

No. S- 5024

Charge Ticket. This slip is used in the camp manager's counter billing machine.

Ecks & Wye
Construction Company

MEAL TICKET NO. 1234

Sold to................................

Not Transferable

Meal Ticket. This meal ticket is good for twenty-one meals (one week).
After it is punched out, the ticket itself is turned in for the last meal. It
is wallet-sized and printed on durable card stock.

No. 1234

Good for **ONE MEAL** at the
Ecks & Wye
Construction Company
Mess Hall

Sold to ....
Not transferable

Binding Margin

Meal Book. This coupon is sold in books of twenty. The cover is
turned in for the last meal in the week.

```
┌─────────────────────────────────────────────────────┐
│                    ECKS & WYE                         │
│          Contract No. 51   Mountain City              │
│        CAMP MANAGER'S DAILY REPORT    Date.......     │
├────────────────────────┬──────────────────────────────┤
│      Men in Camp       │        Meals Served          │
│  Last night    ......  │  Breakfast         ......    │
│  Arrivals      ......  │  Lunches (mess hall) ......  │
│  *Departures   ......  │  Lunches (put up)  ......    │
│  Tonight       ......  │  Dinners           ......    │
│  * Clearance slips attached │  Total meals     ......  │
└────────────────────────┴──────────────────────────────┘
```

CASH RECEIPTS AND DISBURSEMENTS

Change fund on hand this morning $........

Cash received (receipts attached) ....

Total to be accounted for $........

Less:

Cash disbursed — supporting documents attached $..........

Tomorrow's change fund ....

Cash (over) or short ....

Cash accompanying this report $........

Net sales on account — sales slips attached $........

Purchases on account — delivery memoranda attached $........

Checks written:

Payee	Purpose	Amount
....		.........$........

Camp Manager's Daily Report

N⁰ ◯ 160

NON-RESIDENT
M E A L   T I C K E T
Good For 1 Meal

Nonresident Meal Ticket

N°   1100

LODGING CARD

(No Refund if Lost or Stolen)

NAME ....

PAYROLL — Week Ending ....

Badge No. ....

Lodging Card

MEAL TICKETS ISSUED

Name			Badge No.		
Date	Tic. No.	Charge	Date	Tic. No.	Charge

Summary of Meal Tickets Issued

BEDDING RECEIPT

Badge No. . . . .

Wool blankets	. . . .	Cotton blankets	. . . .
Pillows	. . . .	Pillow slips	. . . .
Sheets	. . . .	Mattress	. . . .

I received the above items on: . . . .

*Payroll deductions for meals and
lodging are hereby authorized.*   . . . .

(Signed)

Equipment Issue Receipt—Face

Name   Check-in date   Badge No.

Home address

                            Contractor

Occupation

Barracks No.   Room No.   Bed No.

                Check-out date

Equipment Issue Receipt—Reverse

Ecks & .Wye Construction Co.
CAMP AND MESS HALL CLEARANCE

Date. . . . . .Hour. . . . .

To the Paymaster:

This is to certify that. . . . .
. . . . .Badge No. . . . .
has turned in all items charged to him except:. . . . .
. . . .
. . . .
He is to be charged/credited as follows:. . . . .
. . . . : . . . .
. . . .
. . . .

                                                             Camp

Signed:. . . . . Mgr.

Camp and Mess Hall Clearance Slip

FORMS FOR CONTROL OF EQUIPMENT

Control of equipment starts when the equipment is purchased, but the problems peculiar to the construction environment begin when an item is sent to a job or is put into service at the head office. On small jobs, equipment costs are often negligible, and there is little need for formal control over such costs. However, as the size of the job increases and equipment costs become substantial, equipment cost control becomes essential.

<table>
<tr><td colspan="2">

DAILY TIME CARD

DATE_____ FROM_____ M. TO_____ M.
NAME (PRINT)

_____ BADGE NO._____

</td><td colspan="3">

DAILY EQUIPMENT TIME CARD

DATE_____
NAME (PRINT)

SHIFT FROM_____ M. TO_____ M.

</td></tr>
<tr><td>OCCUPATION</td><td>PER HOUR RATE</td><td>TOTAL HOURS</td><td colspan="2">TYPE OF EQUIPMENT</td><td>EQUIP. NO.</td></tr>
<tr><td>DISTRIBUTION</td><td>HOURS</td><td>AMOUNT</td><td>HOURS IDLE</td><td>HOURS REPAIR</td><td>HOURS WORKED</td></tr>
<tr><td></td><td></td><td></td><td colspan="3" align="center">TIME DISTRIBUTION</td></tr>
<tr><td></td><td></td><td></td><td colspan="3">1</td></tr>
<tr><td></td><td></td><td></td><td colspan="3">2</td></tr>
<tr><td>FOREMAN'S O.K.</td><td colspan="2">EMPLOYEE SIGN HERE</td><td>FOREMAN'S O.K.</td><td colspan="2">EMPLOYEE SIGN HERE</td></tr>
<tr><td colspan="3" align="center">FILL IN CARD COMPLETE</td><td colspan="3" align="center">FILL IN CARD COMPLETE</td></tr>
</table>

Daily Time Card. This form is used by equipment operators. It is a combination equipment time card and operator's time card.

THE PRIME COMPANY
Equipment Receiving Report

Job No. . . . . Location. . . . . Date. . . . . No. . . . .

Equipment No.		Motor or serial No.	Owner	Description	Condition — Specify any condition not satisfactory in detail	Insurance value
Job No.	Owner's No.					

Received by. . . . . Condition checked by. . . . .

Equipment Receiving Report. This form is prepared in quadruplicate, in different colors. The white copy is sent to the job file, the yellow to the general office, the pink to the equipment timekeeper, and the blue to the accounting department.

Form E75-250-4/55

EQUIPMENT LOG

Project _____   Job No. _____

Equipment Rented to Job By_____

RECEIVED	TERMINATED	DESCRIPTION OF EQUIPMENT	RATE	OWNER

Equipment Log

EQUIPMENT LIST

JOB NO. ____
JOB NAME ____
SHEET ____ OF ____
LOCATION ____ DATE ____ MADE BY ____

JOB EQUIP. NO.	TYPE AND DESCRIPTION OF EQUIPMENT	MOTOR NO.	SERIAL NO	MODEL OR SIZE	OWNER	INS. BY	VALUE	RENTAL RATE	RENTAL AGREE NO.	DATE STARTED	DATE TERMI-NATED	DATE SHIPPED OUT	SHIP TO	JOB RATE HOUR WEEK	ACCOUNT NO.

Equipment List. This form illustrates one of the methods used to keep a current control of equipment on hand.

PRIME CONTRACTOR

EQUIPMENT RECORD

PROJECT_____   EQUIPMENT NO._____

JOB OWNED EQUIPMENT DATA

	$			
ORIGINAL COST				
IMPROVEMENTS				
TOTAL COST				
DEPRECIATION TAKEN				
PRICE SOLD				

OWNERSHIP EXPENSE RATE

ANNUAL	I	SHIFTS II	III	
DEPRECIATION RATE	%	%	%	
MAJOR REPAIRS	%	%	%	
INS., INT. AND TAXES	%	%	%	
TOTAL ANNUAL RATE	%	%	%	
AVERAGE USE PER YEAR	MOS.	MOS.	MOS.	
MONTHLY OWNERSHIP RATE	%	%	%	
MONTHLY OWNERSHIP EXPENSE	$			

JOB OWNED EQUIPMENT CHARGE

MONTHLY RATE*		

*USE THIS RATE ON OPPOSITE SIDE TO DETERMINE JOB RATE PER HOUR.

VOUCHER / TIME RECORD

NO.	PERIOD FROM	PERIOD TO	SHIFT I WK	I	R	SHIFT II WK	I	R	SHIFT III WK	I	R

DESCRIPTION OF EQUIPMENT

P. O. NO.		DATE	
VENDOR			
ADDRESS			
CITY		STATE	
SERIAL NO.		ENG. NO.	
MODEL		YEAR	
MANF.			
WEIGHT			
REMARKS			

RENTED EQUIPMENT

PURCHASE ORDER RATE

SHIFT	MONTH	WEEK	DAY	HOUR
I				
II				
III				

MINIMUM RENTAL TIME	
MINIMUM RENTAL CHARGE	

JOB RATE PER HOUR

	SHIFT I	II	III
OWNED OR RENTED RATE			
REPAIRS AND MAINT. %			
GAS AND OIL GALS/DAY			
MISCEL.			
TOTAL JOB RATE PER HOUR			
ON AND OFF EXPENSE			
FREIGHT			
OPERATOR			
HELPERS			
INS. AND TAXES ON LABOR			
ADDED OPERATING CHARGE			

USE TO DETERMINE OPERATING CHARGE FOR SUB-CONTRACT RENTAL

JOB RATE BASED ON_____ HOURS PER DAY
_____ DAYS PER WEEK _____ WEEKS PER MONTH

RENTAL PAID	DEPRECIATION TAKEN	INTEREST AND TAXES	MAJOR REPAIRS	JOB REPAIRS

Equipment Record. Another method of controlling equipment is to keep a ring binder, with a sheet like this one for each piece of equipment.

RENTED EQUIPMENT MONTHLY TIME REPORT

Date _____ Los Angelas Calif.

MONTH OF _____

Address _____

Firm _____

Owner's Equip. No.	Project Equip. No.	Rental Agree No.	Description of equipment				1	2	3	4	5	6	7	8	9	10	11	12	13	14	15	16	17	18	19	20	21	22	23	24	25	26	27	28	29	30	Total hrs	Basic Rate		Total Rental Due
			Make	Model	Kind	Meter No. Serial No.																																Amount	Unit	

I certify that the above is a true and correct statement of the hours of the operation of the equipment above.

Signed _____ Date _____

Equipment Payroll. This form provides all the data needed when equipment is rented by the month. It is somewhat more complete than those used by many contractors.

DAILY EQUIPMENT CARD

Equipment No.	Type	

Job Distribution	Hours Worked
TOTAL HOURS WORKED	
Repair Time	
Idle Time	
State Reasons—	

Please Note on Back
Any Remarks re State and ....
Performance of Equipment   Operator

Equipment Time Card. This card does not have an operator's time card attached. A combination equipment time card and operator's time card is shown elsewhere.

Daily Equipment Report. An equipment payroll can be prepared from this time sheet.

	DATE	REMARKS	REF.	SHOP LABOR	PARTS & SUPPLIES	OUTSIDE REPAIRS	EQUIPMENT RENTAL	JOB OVERHEAD	TOTAL

EQUIPMENT COST LEDGER

Project No. _____

Description _____   Class of Equipment _____

EQUIPMENT NO. _____

Equipment Cost Ledger. This form provides a convenient way of summarizing information if costs are kept by individual pieces of equipment or by groups of similar equipment.

FORM 34 SM 5-48

PRIME CONTRACTOR

EQUIPMENT EARNINGS RECORD

Owner_____

Street_____

City_____State_____

Date Started . . _____

Date Terminated __ _____

P. O. No._____Equip. No._____

Description_____

Earnings From_____To_____.

P. O. Rate	Day	Week	Month
1st Shift			
2nd Shift			
3rd Shift			

DAILY TIME RECORD

DATE	1st SHIFT			2nd SHIFT			3rd SHIFT			DATE	1st SHIFT			2nd SHIFT			3rd SHIFT		
	WK	I	R	WK	I	R	WK	I	R		WK	I	R	WK	I	R	WK	I	R

Amount Earned:

Remarks and Repair Charges:

Equipment Earnings Record. Essentially a time report on which space is provided to compute total earnings and offset repair and maintenance costs, this form is best adapted to systems in which equipment costs and time are kept on a memo basis.

Equipment Repair Order. This form is manufactured principally for use in repair shops maintained by automobile dealers. However, it is an excellent form for recording the cost of construction equipment repairs and maintenance. It is manufactured by Reynolds & Reynolds Co.

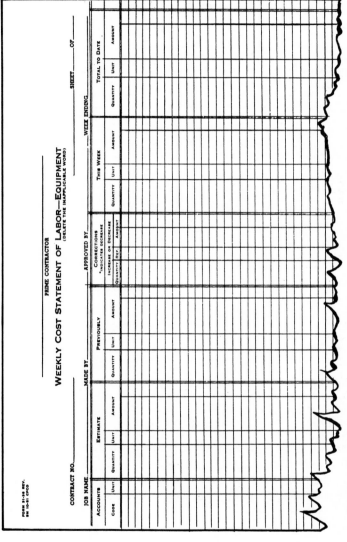

Weekly Cost Statement. The function of this weekly report is to compare the costs actually incurred with those provided in the estimate. It can be used for either equipment or labor costs.

GRADE FOREMAN REPORT

Shift:_____ to _____   Date:_____

Work Acct. No._____   Work Acct. No._____

Location:_____   Location_____

Type Equip.	Equip. No.	Total Loads	Size Load	Total Cu. Yds.	Type Equip.	Equip. No.	Total Loads	Size Load	Total Cu. Yds.

Total_____   Total_____

Remarks:_____

Foreman:_____

Form 1—Desert Trail—1M 5-52

Grade Foreman's Report. An example of the specialized forms that each company may devise for its own particular needs, this form is designed to provide job management with current information on an earthmoving job.

DAILY EQUIPMENT COST REPORT

Job:........................Date:......................

Work Item	Unit	YESTERDAY				JOB TO DATE			
		Units Compl.	Equip. Hrs.	Total Cost	Unit Cost	Units Compl.	Equip. Hrs.	Total Cost	Unit Cost

| Time checked and priced by .... | Pricing checked, extensions made by .... | Cost coded, units priced .... |

Index Card for Controlling Equipment Rental Billings. If volume is heavy, Kardex or similar cards may be useful.

```
┌─────────────────────────────────────────────────────────┐
│              Request for Equipment Termination             │
├─────────────────────────────────────────────────────────┤
│  To:...............................(Equipment timekeeper)  │
│  From.....................................................  │
│  Please terminate the following piece of equipment effective:  │
│  .........................................................  │
│  Equipment No.......................Serial No.............  │
│  Owner...................................................  │
│  Description.............................................  │
│  .........................................................  │
│  Condition...............................................  │
├─────────────────────────────────────────────────────────┤
│  Requested by............................................  │
│  Approved by.............................................  │
│  Date....................................                 │
└─────────────────────────────────────────────────────────┘
```

Request for Equipment Termination. When equipment is being rented, it must be kept busy or its cost will be lost to the job. Most rental contracts require notice of termination of rented equipment. This form is satisfactory for that purpose.

| | Time In Use | | Total | | Rental | Date |
Rented To	From	To	Time	Rate	Earned	Billed

EQUIPMENT RENTAL CONTROL

Item.......................... Equipment Number...........

Daily Equipment Cost Report. This form is useful on jobs for which equipment costs are large in relation to total cost.

VOUCHERING AND DISBURSEMENT CONTROL FORMS

W. O. No._____ Voucher No._____ Month_____

Name_____

Address_____

_____

TOTAL NET INVOICES _____

LESS CONTRA CHARGES _____

NET AMOUNT PAID _____

Paid_____ 19___ Check No._____

VOUCHER DISTRIBUTION

Invoice Amount

FORM 381-6014-6-88

Voucher Envelope. One of the functions of vouchers is the assembling in one place of the various papers and documents supporting a transaction. A convenient way of accomplishing this end is to use a manila envelope for a voucher.

FORM 2 5M SETS 8-51

ACCOUNTS PAYABLE VOUCHER

PROJECT NO._____	OFFICE OR PROJECT_____	VOUCHER NO_____	
MADE BY_____	VENDOR_____	DATE PAY'T DUE_____	
AUDITED BY_____	ADDRESS_____	DATE PAID_____	
APPROVED BY_____	CITY_____ STATE_____	CHECK NO._____	

VENDOR'S INVOICE		DESCRIPTION	P. O. NO.	AMOUNT INVOICED	DEDUCTIONS (ADDITIONS)		DISCOUNT		AMOUNT PAYABLE
DATE	NUMBER				EXPLANATION	AMOUNT	%	AMOUNT	
									1
									2
									3
									4
									5
									6
									7
									8
									9
									10
									11
									12
									13
									14
									15
									16
									17
									18
									19
									20
									21

DETAIL OR COST LEDGER DISTRIBUTION—DEBITS (CREDITS)   GEN. LEDGER DIST.

A/C NO.   AMOUNT   DEBITS

CREDITS

ENTERED IN VO. REGISTER_____ POSTED TO COST LEDGER_____

Accounts Payable Voucher. This is a simple voucher form suitable for a small contractor. Supporting documents are stapled to this form after it has been completed.

Accounts Payable Voucher Folder. This voucher jacket is a file folder with spaces for analyzing the transactions with a particular vendor for the month. Supporting papers are stapled or fastened inside.

DISBURSEMENT VOUCHER REGISTER

Project No. _____ Name _____ Page No. _____

Address _____ City _____ Month of _____ 19__

Date Voucher Made (1)	Creditor (2)	Paid by Check			Credits							Debits						
		Date (3)	Check Number (4)	Voucher Number (5)	Accounts Payable Vendors (6)	Wages Payable (7)	Social Sec. Taxes Payable (8)	ACS Payable Subcontractors		Miscellaneous		Construction Costs			Sub-contract Retentions (15)	(16)	Accounts Receivable (17)	Miscellaneous	
								Current Estimates (9)	Retentions (10)	Account Number	Amount (11)	Labor (12)	Materials & Expenses (13)	Sub-Contracts (14)				Account Number (18)	Amount (19)

Disbursement Voucher Register. This is one of the many types of columnar distribution voucher registers. It is made in duplicate to permit a copy to be sent to the general office in lieu of a job report.

PAY TO THE ORDER OF:

CHECK N°

DATE

$

DOLLARS

COPY
NOT NEGOTIABLE

DETACH STATEMENT BEFORE DEPOSITING
DO NOT CHANGE OR ALTER

WHEN DETACHED AND PAID THE ABOVE CHECK BECOMES A
RECEIPT IN FULL PAYMENT OF THE FOLLOWING ACCOUNT
NO OTHER RECEIPT NECESSARY

DATE	DESCRIPTION	AMOUNT	DISCOUNT %	AMOUNT	OTHER DEDUCTIONS FOR	AMOUNT	NET AMOUNT

DISTRIBUTION

ACCOUNT	DETAIL	✓	AMOUNT	ACCOUNT	DETAIL	✓	AMOUNT

COMPILED BY   AUDITED BY   APPROVED BY

Voucher Check. The analysis of the payment is on the face of the check, and the stub is used as a simple voucher. If a carbon copy is made, the copy can serve as a voucher and the stub as a remittance advice.

Form 56 Rev. 2M 12-51 - CPCO

WEEKLY REPORT

JOB NO._____ NAME_____ DATE_____
CONTRACT NO._____ LOCATION_____
WEEK ENDING_____

DAY	DATE	WEATHER							MEN EMPLOYED					
		TEMP.		Clear	Cloudy	Fog	Rain	Snow	PRIME CONTRACTOR			SUB-CONTRACTORS		
		Min.	Max.						On Payroll	Hired	Termin.	On Payroll	Hired	Termin.
Sunday														
Monday														
Tuesday														
Wednesday														
Thursday														
Friday														
Saturday														

GENERAL NOTES AND REMARKS (Re weekly trend on labor and equipment costs; detailed information as to what crews are doing; crafts in which there is a labor shortage; etc.)_____

(Continued on Reverse Side)

Weekly Report—Face. This report is designed for project managers or superintendents. It is a good idea to supplement it with a foreman's daily diary.

DELAYS (State steps. taken to stop these delays)_____

P R O G R E S S	WORK COMPLETED %	ELAPSED TIME %
PREVIOUS—WEEK ENDING		
THIS PERIOD—WEEK ENDING		
TOTAL TO DATE		

Original Contract   .   .   .   . $_____
Change Orders & Supplemental Agreements issued, signed and received $_____
Change Orders & Supplemental Agreements approved, in process of
   being issued and signed, but NOT received   .   .   . $_____
TOTAL CURRENT CONTRACT   .   .   .   . $_____
Change Estimates for work proceeding without Change Orders or
   Supplemental Agreements   .   .   .   .   .   . $_____
Change Estimates quoted, which will probably be approved
   (Omit all dead or abandoned Change Estimates)   .   .   . $_____
TOTAL CHANGE ESTIMATES PENDING   .   .   .   .   .   . $_____
TOTAL ANTICIPATED CONTRACT   .   .   .   . $_____

Dates on which Partial Payment Requests or Progress Estimates are to be made_____
Net Progress Payments paid to date of this report   .   .   .   .   . $_____
Retention on above Progress Payments   .   .   .   .   .   . $_____
Gross Partial Payment Requests or Progress Estimates submitted
   but not paid to date of this report:
No._____submitted_____195___for work completed to_____195___ $_____
No._____submitted_____195___for work completed to_____195___ $_____
No._____submitted_____195___for work completed to_____195___ $_____
Total gross Partial Payment Requests or Progress Estimates due but not paid   .   .   .   . $_____
TOTAL PROGRESS PAYMENT REQUESTS or PROGRESS ESTIMATES TO DATE OF THIS REPORT $_____

CASH PAID OUT	THIS WEEK	TO DATE	ESTIMATED FOR NEXT WEEK
PAYROLLS . . .	$_____	$_____	$_____
SUB-CONTRACTORS . .	$_____	$_____	$_____
VENDORS . . .	$_____	$_____	$_____
MISCELLANEOUS . .	$_____	$_____	$_____
TOTALS . . .	$_____	$_____	$_____

CASH IN BANK .   .   .   . $_____
COLLECTIONS ESTIMATED FOR NEXT WEEK
   PENDING PARTIAL PAYMENTS   .   .   .   . $_____
   MISCELLANEOUS ACCOUNTS RECEIVABLE   .   .   . $_____
   TOTAL ESTIMATED COLLECTIONS   .   .   .   . $_____
REMARKS (Concerning Changes, Progress Payments, Disbursements and Collections)__._____

_____
Project Manager
Superintendent

Weekly Report—Reverse.

BIBLIOGRAPHY

"Advantage in Hiring Architect ahead of CM," *Building Design & Construction*, vol. 19, July 1978, p. 35.

American Institute of Architects: Document B801, *Standard Form of Agreement between Owner and Construction Manager*, 2d ed., July 1980.

"Arnold says CM Leads to Design—Construct," *Constructor*, January 1972, p 24.

Associated General Contractors: Document 8, *Standard Form of Agreement between Owner and Construction Manager*, 2d ed., July 1980.

"Build Team Reviews Its Performance on Phoenix Job," *Building Design & Construction*, vol. 17, July 1976, p. 12–13, illus. Hartford Insurance Corporate Building; Guirey, Srnka, Arnold & Sprinkle.

"CM: The Only Way To Go Fast Track," *Architectural Record*, vol. 156, December 1974, p. 69; vol. 157, January 1975, p. 69.

"CM Approach Stresses Adaptability," *Building Design & Construction*, vol. 18, August 1977, p. 66–69. Kitchell Contractors.

"CM Clients Look for Quality," *Engineering News Record*, vol. 205, Oct. 9, 1980, pp. 82–83.

"CM Payoff; Industrial Facility Built under Budget in 8 Months," *Building Design & Construction*, vol. 19, January 1978, pp. 70–75, illus., plans. Clark Parts Distribution Center, Kalamazoo, Mich.

"CM Provides Sharper Focus on Hotels' Design Options," *Building Design & Construction*, vol. 17, December 1976, pp. 38–40, illus.

"CM Strengths, Problems Explored at Conference," *Building Design & Construction*, vol. 18, December 1977, pp. 16–18.

"The Case for Early CM Involvement," *Building Design & Construction*, vol. 19, December 1978, p. 39.

"The Changing Role of the General Contractor," *Building Design & Construction*, April 1971, p. 46.

"Construction Management, in a Miami Test, Saves $1.5 Million," *Architectural Record*, vol. 161, January 1977, pp. 75–78, illus., plans, table. Dade County High School; Caudill Rowlett Scott.

"Construction Management with a British Accent," *Progressive Architecture,* vol. 53, January 1972, pp. 30f.

"Construction Manager: Heir Apparent?" *Progressive Architecture,* vol. 55, June 1974, p. 26f.

Coombs, William E., and William J. Palmer: *Construction Accounting and Financial Management,* McGraw-Hill Book Company, New York, New York, 1977.

Cornell: "Construction Administration Changes Hands," *Construction Specifier,* January 1972, p. 39.

Cushman, Robert F., and William J. Palmer: *The Dow Jones Businessman's Guide to Construction,* Dow Jones Books, Princeton, N.J., 1981.

Davis, W., and White, L.: "How To Avoid Construction Headaches," *Harvard Business Review,* vol. 87, March–April 1973.

DeMars, Richard B.: "A Contractor Looks at Construction Management," *Architectural Record,* vol. 151, January 1972, pp. 55–56.

Engineering News Record, Apr. 17, 1980, pp. 78–80; July 31, 1980, p. 30.

Farrell, Paul B., Jr.: "Construction Manager: Menace to the Architect?" *Architectural and Engineering News,* vol. 10, April 1968, pp. 25–43, table.

Forwalter, J.: "Construction Management Techniques Serve Owner—Save Money, Time," *Food Processing,* vol. 40, June 1979, pp. 52–53.

Foxhall, W.: *Professional Construction Management and Project Administration,* Architectural Record, New York, 1976.

———: "Accelerated Project and/or Construction Management Means: Sorting out the Tasks, Putting Them on Track, Time/Cost/Quality Control, Client-Architect-Consultant-Contractor Unity," *Architectural Record,* vol. 148, October 1970, pp. 160–161, illus., plans.

Gallagher, Gerard B.: "The Case for Clarity in CM, Design/Build Contracts," *Building Design & Construction.* vol. 20, December 1979, p. 39.

———: "Legal Seminar Looks at Construction Management," *Building Design & Construction.* vol. 18, June 1977, pp. 12–13.

General Services Administration, Public Building Services: *Construction Contracting Systems, a Report,* Mar. 17, 1970.

———: *Construction Manager Contract,* rev. ed., Apr. 15, 1975.

———: *The GSA System for Construction Management",* October 1977.

———: Order No. ADM/P545039 B, Oct. 20, 1980.

"The Guaranteed Maximum Price: A Sampling of Views," *Architectural Record,* vol. 155, April 1974, pp. 65f.

Halpern, Richard C: "2-Faced CM [Should a CM Firm Offer Design or Construction Services]," *Building Design & Construction,* vol. 19, February 1979, pp. 70–73f., illus., pors.

Harack, Tom: "Contractor Finds In-House Preparation an Essential CM Prerequisite," *Building Design & Construction,* vol. 17, October 1976, pp. 28–30, illus.

Hart, B.: "Construction Management, 'CM for Short': The New Name for an Old Game," *The Forum,* vol. 8, 1972, p. 210.

Hastings, R. F.: "Proposal: A New and Comprehensive System for Design and Delivery of Buildings," *Architectural Record,* November 1968.

Heery, George T.: "Let's Define Construction Management," *Architectural Record,* vol. 155, March 1974, pp. 69, 71.

Heinly, David R.: "GAO Report on CM, Phased Construction Proves Inconclusive," *Building Design & Construction,* vol. 19, January 1978, p. 27.

"How To Increase Profits On Construction Projects [Construction Management]," *Management Review,* vol. 66, December 1977, p. 6.

"Hybrid Design/Build—CM Plan To Be Modified," *Building Design & Construction,* vol. 20, February 1979, pp. 1–2. Staten Island Hospital, N.Y.

Justin, J. Karl: "An Architect's Notes on Construction Management," *Architectural Record,* vol. 20, 155, January 1974, pp. 75, 77.

Koehler, R. E.: "Where the Office Becomes a Classroom for Continuing Education," *AIA Journal,* May 1971, p. 37.

Lammers: "Construction Manager: More Than a Hard Hat Job," *AIA Journal,* May 1971, p. 31.

Marshall: "Old Hand Turner Company Talks about CM," *Constructor,* February 1972, p. 27.

McLaughlin, Herbert, and Cynthia Ripley: "Partial Guide to Painless Construction Management Projects," *Architectural Record,* vol. 165, May 1979, pp. 65f.; June 1979, pp. 69f.

Maevis, A. C.: "Pros and Cons of Construction Management," *American Society of Civil Engineers Proceedings.* vol. 103, June 1977, pp. 169–177; December 1977, pp. 668–670; vol. 104, June 1978, pp. 256–257.

Meathe, Philip J.: "It's a Wide Open Field: Construction Management," *AIA Journal,* vol. 59, March 1973, pp. 41–43.

Nash, Ralph C.: "Innovations in Federal Construction Contracting," *George Washington Law Review,* vol. 45, 1977, p. 309.

Nelson, N. W.: "Renovation of the Field Museum of Natural History: Cost-Effective Modernization Achieved by Construction Management," *Technology & Conservation,* vol. 4, Summer 1979, pp. 12–15.

"New GSA System for Construction Management," *Architectural Record,* vol. 157, June 1975, pp. 69–70.

Note, "The Roles of Architect and Contractor in Construction Management," *University of Michigan Journal of Law Reform,* vol. 6, Winter 1973, p. 447.

Perkins, Bradford: "A/E-CM Relations: Approaching a Modus Vivendi?" *Architectural Record,* vol. 154, October 1973, pp. 67–68.

"PBS Chief Tells Plan, Scores CM," *Building Design & Construction,* vol. 20, November 1979, pp. 19–20, port.

"Professional Construction Management Services," *American Society of Civil Engineers Proceedings,* vol. 105, June 1979, pp. 139–156.

"A Rapidly-Maturing Concept; Geupel DeMars President Sees Quality of CM Practice Improving under Test of Competition," *Building Design & Construction,* vol. 18, August 1977, pp. 70–71.

Reed, Campbell: "CM in Federal Building Construction," *Constructor,* June 1979, pp. 33–35.

Reed, Campbell: "CM's Conflict of Interest Question," *Constructor,* June 1980.

Ricchin, John A.: "Regulation and the Construction Manager: A Call for Government Licensing," *Construction Specifier,* vol. 33, February 1980, pp. 56–64.

Rosenfeld, S. H.: "Contract Documents Further Define Professional Construction Management," *Architectural Record,* vol. 158, July 1975, pp. 51–52.

"Roundtable Discussion: Construction Management and GSA," *Constructor,* November 1971.

Scarano, Joe: "Construction Management Ensures Flexibility in Project Delivery," *Hospitals,* vol. 51, Nov. 1, 1977, pp. 99–104.

Siatt, Wayne: "Emphasis on CM Helps Fuller Revive Its Corporate Fortunes," *Building Design & Construction,* vol. 17, June 1976, pp. 36–38, illus.

———: "GSA Systems—A Diamond or a Deed?" *Building Design & Construction,* vol. 19, January 1981, pp. 36–41, illus., plans, diags., tables. Federal Building, Norfolk, Va. VVKR Partnership.

Sneed, William R.: "The Construction Manager's Liability," in *Construction Litigation,* Practicing Law Institute, 1981.

"Special Report: Construction Management: Putting Professionalism into Contracting," *Construction Methods and Equipment,* March 1972.

Stokes, McNeill: "Statement on Construction Management by American Subcontractors Association," American Subcontractors Association, Landover, Maryland, 1981.

Strafford, J. E.: "Construction Project Management," *Accountancy,* vol. 88, September 1977, pp. 56f.

"Staten Island Hospital Will Use CM Approach," *Building Design & Construction,* vol. 17, June 1979, p. 2. Rogers, Butler & Burgun.

Tatum, Rita: "Heery Associates Reveals Its Recipe for CM Files," *Building Design & Construction,* vol. 17, March 1976, pp. 34–36, illus., table.

University of Richmond Law Review, vol. 14, p. 791.

Wagner, W. F., Jr.: "Where, Oh Where Are the Management Skills?" *Architectural Record,* December 1971, p. 9.

Wright, Gordon: "Construction Management: How It Works," *Design & Construction,* vol. 19, August 1978, pp. 94–119.

Wulfsburg, H. James: "Liability on a Construction Project," University of Missouri (Kansas City) Law Center.

Yee, Roger: "Technics: Construction Management," *Progressive Architecture,* vol. 57, February 1976, pp. 90–95.

INDEX

About the Authors

ROBERT F. CUSHMAN, a partner in the national law firm of Pepper, Hamilton & Scheetz, is a recognized specialist and lecturer on all phases of construction and real estate law. He serves as legal counsel to numerous trade associations and construction, development, and bonding companies.

Mr. Cushman is the editor and coauthor of *The McGraw-Hill Construction Business Handbook, The Construction Industry Formbook, The Dow Jones Businessman's Guide to Construction, The Dow Jones–Irwin Business Insurance Handbook, Avoiding Liability in Architecture, Design and Construction,* published by John Wiley & Sons, Inc., and *Planning, Financing and Constructing Health Care Facilities* for Aspen Systems is presently editing *The Businessman's Guide to Commercial Real Estate Transactions* and *Doing Business in America* for Dow Jones–Irwin.

Mr. Cushman, who is a member of the bar of the Commonwealth of Pennsylvania and who is admitted to practice before the Supreme Court of the United States and the U.S. Court of Claims, has served as Executive Vice President and General Counsel to the Construction Industry Foundation and Regional Chairman of the Public Contract Law Section of the American Bar Association. He is a member of the International Association of Insurance Counsel.

WILLIAM J. PALMER is a partner in the multinational public accounting firm of Arthur Young & Company and is chairman of that firm's Construction Industry Group. He is a recognized specialist and lecturer on the accounting and financial aspects of the construction industry. His construction industry activities have involved audits, analysis, system studies, and special reviews of more than 100 major construction projects throughout the world. He is coauthor of *Construction Accounting and Financial Management,* McGraw-Hill Book Company, 1977; coeditor/author of *Businessman's Guide to Construction,* Dow Jones, 1980; and a contributing author of the *Construction Handbook,* McGraw-Hill, 1978. He was vice chairman of the American Institute of Certified Public Accountants' Committee which prepared the Audit and Accounting Guide for the construction industry. He is a frequent expert witness and consultant in major construction contract claims and litigation.

WILLIAM R. SNEED, III (Sam) joined the national law firm of Pepper, Hamilton & Scheetz after a judicial clerkship. Mr. Sneed graduated *cum laude* from Yale University and received his juris doctor degree from New York University School of Law, where he was Note and Comment Editor of the *Review of Law and Social Change*. Mr. Sneed, a recognized specialist in all phases of construction law, is a coauthor of *Avoiding Liability in Architecture, Design and Construction*, John Wiley & Sons, Inc., 1983; a coeditor and coauthor of *Construction Litigation*, Practising Law Institute, 1981; and a coauthor of *Construction Contracts 1982: Rights and Responsibilities of the General Contractor, Subcontractor and Material Supplier*, Practising Law Institute, 1982. In addition to these recent publications, Mr. Sneed lectures regularly on construction litigation, is a member of the American Bar Association's Section of Tort and Insurance Practice, the Fidelity & Surety Law Committee, the Forum Committee on the Construction Industry, and the Public Contract Law Section, is a member of the Pennsylvania Bar Association's Public Contract Law Committee, and specializes in construction and surety litigation for lenders, owners, architects, construction managers, general contractors, subcontractors, suppliers, and sureties.

ALAN B. STOVER, AIA, received his B.Arch from Cornell University and his law degree from Georgetown. Prior to joining AIA, his architectural experience included four years in multidisciplinary medium-sized firms with design responsibility for residential projects of all types and sizes; small- and medium-scale commercial, industrial, and institutional projects; and master planning and site development responsibility for various types and sizes of projects. Mr. Stover joined AIA in 1974 as Director of the Documents Division, where he served as editor of the *Handbook of Professional Practice*, authored the new AIA *Standardized Accounting for Architects*, and staffed the liability program. He became Deputy General Counsel in early 1978 and served as counsel for PSAE and the National Judicial Committee, among other responsibilities. In March 1980 he was appointed Acting General Counsel, and in late 1980, he became the Institute's General Counsel. Mr. Stover teaches as a guest lecturer in professional practice courses, is a contributing author to texts on construction law, and serves as vice-chairman of the ABA Committee on Professional, Officers' and Directors' Liability.